Performance and Modification of Wood and Wood-Based Materials

Performance and Modification of Wood and Wood-Based Materials

Editors

Tomasz Krystofiak
Pavlo Bekhta

MDPI • Basel • Beijing • Wuhan • Barcelona • Belgrade • Manchester • Tokyo • Cluj • Tianjin

Editors
Tomasz Krystofiak
Department of Wood Science and Thermal Technics
Poznan University of Life Sciences
Poznań
Poland

Pavlo Bekhta
Department of Wood-Based Composites, Cellulose and Paper
Ukrainian National Forestry University
Lviv
Ukraine

Editorial Office
MDPI
St. Alban-Anlage 66
4052 Basel, Switzerland

This is a reprint of articles from the Special Issue published online in the open access journal *Forests* (ISSN 1999-4907) (available at: www.mdpi.com/journal/forests/special_issues/performance_modification_wood).

For citation purposes, cite each article independently as indicated on the article page online and as indicated below:

LastName, A.A.; LastName, B.B.; LastName, C.C. Article Title. *Journal Name* **Year**, *Volume Number*, Page Range.

ISBN 978-3-0365-7667-1 (Hbk)
ISBN 978-3-0365-7666-4 (PDF)

Contents

Preface to "Performance and Modification of Wood and Wood-Based Materials" . . . . **vii**

Pavlo Bekhta and Tomasz Krystofiak
Performance and Modification of Wood and Wood-Based Materials
Reprinted from: *Forests* **2023**, *14*, 963, doi:10.3390/f14050963 . . . . **1**

Yulei Gao, Liyuan Zhao, Jinghui Jiang, Zhu Li and Jianxiong Lyu
Water Absorption Properties in Transverse Direction of Heat-Treated Chinese Fir Wood Determined Using TD-NMR
Reprinted from: *Forests* **2021**, *12*, 1545, doi:10.3390/f12111545 . . . . **5**

Larisa Belchinskaya, Konstantin Viktorovich Zhuzhukin, Tatiana Ishchenko and Aleksey Platonov
Impregnation of Wood with Waste Engine Oil to Increase Water- and Bio-Resistance
Reprinted from: *Forests* **2021**, *12*, 1762, doi:10.3390/f12121762 . . . . **17**

Pavlo Bekhta, Tomasz Krystofiak, Barbara Lis and Nataliya Bekhta
The Impact of Sanding and Thermal Compression of Wood, Varnish Type and Artificial Aging in Indoor Conditions on the Varnished Surface Color
Reprinted from: *Forests* **2022**, *13*, 300, doi:10.3390/f13020300 . . . . **31**

Aujchariya Chotikhun, Jitralada Kittijaruwattana, Wa Ode Muliastuty Arsyad, Emilia-Adela Salca, Yusuf Sudo Hadi and Salim Hiziroglu
Some Properties of Wood Plastic Composites Made from Rubberwood, Recycled Plastic and Silica
Reprinted from: *Forests* **2022**, *13*, 427, doi:10.3390/f13030427 . . . . **55**

Lei Han, Andreja Kutnar, José Couceiro and Dick Sandberg
Creep Properties of Densified Wood in Bending
Reprinted from: *Forests* **2022**, *13*, 757, doi:10.3390/f13050757 . . . . **65**

Dennis Jones, Davor Kržišnik, Miha Hočevar, Andreja Zagar, Miha Humar and Carmen-Mihaela Popescu et al.
Evaluation of the Effect of a Combined Chemical and Thermal Modification of Wood though the Use of Bicine and Tricine
Reprinted from: *Forests* **2022**, *13*, 834, doi:10.3390/f13060834 . . . . **77**

Feiyu Ouyang and Wei Wang
Effect of Thermo-Hydro-Mechanical Treatment on Mechanical [-15]Properties of Wood Cellulose: A Molecular Dynamics Simulation
Reprinted from: *Forests* **2022**, *13*, 903, doi:10.3390/f13060903 . . . . **99**

Gaddafi Ismaili, Ellyne Enduat, Nur Syahina Yahya, Fanthy Moola Malek, Noor Azland Jaimudin and Khairul Khuzaimah Abdul Rahim et al.
Physical and Mechanical Properties Performance between Untreated and Treated with CCA Treatment at Different Age Groups of Fast-Growing Acacia Hybrid of Sarawak
Reprinted from: *Forests* **2022**, *13*, 1969, doi:10.3390/f13121969 . . . . **111**

Shaofeng Ru, Can Zhao and Songmei Yang
Multi-Objective Optimization and Analysis of Mechanical Properties of Coir Fiber from Coconut Forest Waste
Reprinted from: *Forests* **2022**, *13*, 2033, doi:10.3390/f13122033 . . . . **129**

František Kačík, Ivan Kubovský, Jiří Bouček, Richard Hrčka, Milan Gaff and Danica Kačíková
Colour and Chemical Changes of Black Locust Wood during Heat Treatment
Reprinted from: *Forests* **2022**, *14*, 73, doi:10.3390/f14010073 . . . . **151**

Pavlo Bekhta, Tomáš Pipíška, Vladimír Gryc, Ján Sedliačik, Pavel Král and Jozef Ráheľ et al.
Properties of Plywood Panels Composed of Thermally Densified and Non-Densified Alder and Birch Veneers
Reprinted from: *Forests* **2023**, *14*, 96, doi:10.3390/f14010096 . . . . **165**

Rasdianah Dahali, Seng Hua Lee, Paridah Md Tahir, Sabiha Salim, Muhammad Syahmi Hishamuddin and Atikah Che Ismail et al.
Influence of *Chrysoporthe deuterocubensis* Canker Disease on the Chemical Properties and Durability of *Eucalyptus urograndis* against Wood Rotting Fungi and Termite Infestation
Reprinted from: *Forests* **2023**, *14*, 350, doi:10.3390/f14020350 . . . . **183**

Preface to "Performance and Modification of Wood and Wood-Based Materials"

Wood remains one of the most attractive building materials. However, its applications are limited because it undergoes biological degradation, is combustible, and changes its properties under the influence of weathering conditions. Many of these disadvantages of solid wood are, to some extent, eliminated in wood-based materials, such as plywood, particleboard, fiberboard, and OSB. However, aside from the fact that it is not possible to eliminate all shortcomings of solid wood, these materials bring new shortcomings, such as toxicity. Therefore, the problem associated with the inadequate properties of wood and wood-based materials still needs a solution. One of the effective ways to improve the properties of wood and wood-based materials is to modify them using conventional processes (thermal, chemical, etc.) or the latest technology (plasma treatment). Concerning wood-based materials, it is possible to treat solid wood's components (particles, fibers, glue, etc.) and use them in wood-based materials, followed by post-treatment of ready-made materials to improve their certain properties.

This reprint contains 12 high-quality original research papers by 74 authors from 17 countries on 3 continents: Asia (China, Indonesia, Malaysia, and Russia), Europe (Bulgaria, the Czech Republic, Germany, Portugal, Romania, Slovakia, Slovenia, Sweden, Ukraine, and the United Kingdom), and North America (USA). These papers were published in a Special Issue, "Performance and Modification of Wood and Wood-Based Materials", of the journal *Forests*. They provide examples concerning conventional and novel modification processes of wood and wood-based materials, their improved/modified properties, and their relevant applications. The Guest Editors would like to thank all authors who contributed to this Special Issue. The Guest Editors would also like to thank the Special Issue Editor, Mila Gao, for her overall professional attitude and kind assistance with the publications.

Tomasz Krystofiak and Pavlo Bekhta
Editors

forests

Editorial

Performance and Modification of Wood and Wood-Based Materials

Pavlo Bekhta [1,2,3,*] and Tomasz Krystofiak [4]

1 Department of Wood-Based Composites, Cellulose and Paper, Ukrainian National Forestry University, 79057 Lviv, Ukraine
2 Department of Wood Science and Technology, Mendel University in Brno, 613 00 Brno, Czech Republic
3 Department of Furniture and Wood Products, Technical University in Zvolen, 960 01 Zvolen, Slovakia
4 Department of Wood Science and Thermal Technics, Poznań University of Life Sciences, 60-627 Poznań, Poland; tomasz.krystofiak@up.poznan.pl
* Correspondence: bekhta@nltu.edu.ua

Citation: Bekhta, P.; Krystofiak, T. Performance and Modification of Wood and Wood-Based Materials. *Forests* **2023**, *14*, 963. https://doi.org/10.3390/f14050963

Received: 2 May 2023
Accepted: 5 May 2023
Published: 7 May 2023

Wood remains one of the most attractive building materials. However, its applications are limited because it undergoes biological degradation, is combustible, and changes its properties under the influence of weathering conditions. Many of these disadvantages of solid wood are to some extent eliminated in wood-based materials such as plywood, particleboard, fiberboard, OSB, etc. However, in addition to the fact that it is not possible to eliminate the shortcomings of solid wood, these materials bring new shortcomings, such as toxicity. Therefore, the problem of improving the properties of wood and wood-based materials still needs a solution. One of the effective ways to improve the properties of wood and wood-based materials is to modify them with conventional processes (thermal, chemical, etc.) or latest ones (plasma treatment). Concerning wood-based materials, it is possible to treat their components (particles, fibers, glue, etc.) and use them in wood-based materials or to use the post-treatment of ready-made materials to improve their certain properties.

This Special Issue, entitled "*Performance and Modification of Wood and Wood-Based Materials*" comprises 12 high-quality original research papers by 74 authors from 17 countries on three continents: Asia (China, Indonesia, Malaysia, and Russia), Europe (Bulgaria, Czech Republic, Germany, Portugal, Romania, Slovakia, Slovenia, Sweden, Ukraine, and United Kingdom), and North America (USA). They provide examples concerning the conventional and novel modification processes of wood and wood-based materials, their improved/modified properties, and their relevant applications.

Usually, the calculation of moisture content (MC) is used to estimate the water absorption of wood. However, the traditional method of determining MC by oven drying does not allow us to accurately obtain the amounts of free and bound water separately. In their work, Gao et al. [1] the absorption of water by heat-treated Chinese fir wood was studied using gravimetric analysis and time-domain nuclear magnetic resonance (TD-NMR). The wood samples were treated in a chamber with steam at several temperature levels, 160, 180, 200, and 220 °C, for 2 h. The results showed that when the temperature of heat treatment increased, the amount of free and bound water decreased. Moreover, the water absorption in the radial direction was faster than that in the tangential direction. Such a trend also existed when the heat treatment process was completed. The authors also stated that the pore structure of wood can seriously changes at a high (over 200 °C) heat treatment temperature causing the closing of the pits in wood cells.

It is well known that the preparation of the surface of the wood substrate is one of the important processes before finishing, and has a significant effect on the surface properties of the coating, as well as on the adhesion of the coating to the substrate. Usually, this process is performed by sanding. Another interesting work [2] confirmed and proved that the thermal compression of wood substrate prior to the varnishing process, instead of the labor-intensive sanding process, makes it possible to apply varnish on the surface of the wood substrate without having an adverse influence on the esthetic characteristics of the samples. The thermally compressed surface still has an attractive "exotic appearance of wood", which allow the use of transparent varnishes that increase the attractiveness of the final product. It was also found that thermally densified surface-varnished veneer is more resistant to color changes during artificial aging compared to sanded surface-varnished veneer. The transparent varnish systems showed better photo-stability when thermally densified wood veneer was used as substrate compared to that of sanded wood. The water-based varnish showed the greatest resistance to discoloration during UVL + IR irradiation, followed by polyurethane and UV-cured varnishes. The findings of this study indicated that replacing the sanding process with the thermal compression of the wood surface before varnishing could be considered an alternative method of producing varnished panels with satisfactory color properties on the surface.

It is a well-known fact that waste plastic material is a serious problem that has a negative impact on the environment. The findings of another study [3] demonstrated that environmentally friendly wood plastic composite (WPC) with different applications can be manufactured using a combination of waste polyethylene terephthalate (PET), silica, and rubberwood particles. The basic properties of this composite were also determined.

In another paper, to make the birch wood more water resistant and resistant to biological degradation, the wood was modified by impregnation with spent engine oil [4]. It was found that this impregnation treatment significantly increased the dimensional stability, water resistance, and biostability of birch wood against brown rot fungi.

Timber is widely used in construction, which is not true of densified timber. This can be explained by our insufficient knowledge on the long-term creep behavior of densified timber. From this point of view, reference [5] is another very interesting and important study. The authors studied the influence of the thermo-hydro-mechanical (THM) densification process on the creep properties of Scots pine sapwood. It was found that the THM densification significantly changed the physical and mechanical properties of wood. The results of this study also demonstrated that the creep of Scots pine timber under a long-term bending load in a constant climate is reduced with the combination of THM densification and impregnation by low-molecular-weight phenol-formaldehyde resin or thermal modification.

To expand our knowledge on the behavioral characteristics of cellulose properties during THM treatment, Ouyang and Wang [6] developed a water and cellulose theoretical model based on molecular dynamics. The authors also analyzed the effects of the hydrogen bond numbers, small molecule diffusion coefficients, end-to-end distances, and mechanical parameters of the water–cellulose model on the mechanical properties of wood cellulose. They found that the pressurized hydrothermal treatment of wood can significantly improve its rigidity and deformation resistance.

In another interesting study, an attempt was made to combine the use of chemical treatments by amines (tricine and bicine) and thermal modification to investigate the properties of the treated spruce and beech wood [7]. However, an improvement of the properties of wood by such combining treatment was not confirmed in these studies. The authors attributed this to the thermal instability of the bicine and tricine.

Another study [8] aimed to expand the fast-growing Acacia hybrid usage in the timber engineering field by using copper chrome arsenic (CCA) treatment. It was found that the treated samples in the tangential direction performed better than the untreated samples, while the radial direction gave a high average strength increment when treated.

Coir fiber has good mechanical properties and has been applied in various fields. Usually, coir fiber is treated with alkali to improve its properties. Ru et al. [9] proposed a different approach to improve the mechanical properties of coir fiber alkali treatment. According to this approach, the properties were optimized and analyzed to find the treatment conditions under which these properties are optimally improved simultaneously. This allowed the authors to find the optimal values of sodium hydroxide (NaOH) concentration, treatment time, and temperature at which the maximum values of mechanical properties are achieved. Such an approach of the optimized combination can be also applied to fiber products when strength and toughness are required.

Another interesting work [10] presents the effects of thermal treatment temperatures (160, 180, and 210 °C) on the color and chemical changes of black locust wood during the ThermoWood process. The authors found that color changes were mainly caused by reducing the content of extractives during thermal treatment. The degradation of wood compounds, such as hemicelluloses and cellulose, also contributes to color changes. It was also confirmed that temperature is a significant factor affecting the color of the wood surface.

It is well known that thermal compression is often used to improve the properties of solid wood. The research conducted by [11] aims to develop plywood panels with two wood species (birch and black alder) and two types of veneer treatments (non-densified and thermally densified) in order to evaluate the influences of different lay-up schemes on the properties of the plywood. The authors used different lay-up schemes to identify opportunities to improve the mechanical and physical properties of the plywood by replacing the birch veneer in the plywood structure with an alternative alder veneer. The results showed that the type of construction, wood species, and applied thermal densification of the veneer affected the examined physical and mechanical properties. The WA, MOR, MOE, and shear strength of plywood were more sensitive to the mixing of wood species in one panel than the mixing of densified and non-densified veneers. Alder veneers can be used to form the inner layers of plywood panels without reducing the shear strength. Increasing the proportion of thermally densified veneer in one panel leads to a higher density, MOR, MOE, shear strength, and TS of plywood panels. It was found that plywood panels manufactured from a mixture of species offered higher bending properties when compared to panels manufactured from alder veneers only. It was also shown that non-treated alder veneer, despite exhibiting somewhat lower strength properties than birch veneer, could be successfully used with proper lay-up schemes in the veneer-based products industry. Therefore, the mixed-species plywood panels allowed an increased use of the lower cost, low-grade, and low-density alder wood veneers as core veneers in panels to reduce production costs and increase the mechanical properties of predominately low-density alder wood plywood.

The findings of another study [12] showed that the Eucalyptus hybrid wood infected by the *Chrysoporthe deuterocubensis* canker disease demonstrates improved biological durability. This opens the possibility of using infected wood in non-structural applications. The authors observed reductions in cellulose and hemicellulose contents and increases in lignin and extractive contents in the infected samples. It was also stated that infected wood demonstrated better durability against fungi and termites than healthy wood.

Author Contributions: Conceptualization, P.B. and T.K.; writing—original draft preparation, P.B.; writing—review and editing, P.B. and T.K.; project administration, T.K. All authors have read and agreed to the published version of the manuscript.

Funding: This research received no external funding.

Acknowledgments: This work was supported by the EU NextGenerationEU through the Recovery and Resilience Plan for Slovakia under project No. 09I03-03-V01-00124.

Conflicts of Interest: The authors declare no conflict of interest.

References

1. Gao, Y.; Zhao, L.; Jiang, J.; Li, Z.; Lyu, J. Water Absorption Properties in Transverse Direction of Heat-Treated Chinese Fir Wood Determined Using TD-NMR. *Forests* **2021**, *12*, 1545. [CrossRef]
2. Bekhta, P.; Krystofiak, T.; Lis, B.; Bekhta, N. The Impact of Sanding and Thermal Compression of Wood, Varnish Type and Artificial Aging in Indoor Conditions on the Varnished Surface Color. *Forests* **2022**, *13*, 300. [CrossRef]
3. Chotikhun, A.; Kittijaruwattana, J.; Arsyad, W.O.M.; Salca, E.-A.; Hadi, Y.S.; Hiziroglu, S. Some Properties of Wood Plastic Composites Made from Rubberwood, Recycled Plastic and Silica. *Forests* **2022**, *13*, 427. [CrossRef]
4. Belchinskaya, L.; Zhuzhukin, K.V.; Ishchenko, T.; Platonov, A. Impregnation of Wood with Waste Engine Oil to Increase Water- and Bio-Resistance. *Forests* **2021**, *12*, 1762. [CrossRef]
5. Han, L.; Kutnar, A.; Couceiro, J.; Sandberg, D. Creep Properties of Densified Wood in Bending. *Forests* **2022**, *13*, 757. [CrossRef]

6. Ouyang, F.; Wang, W. Effect of Thermo-Hydro-Mechanical Treatment on Mechanical Properties of Wood Cellulose: A Molecular Dynamics Simulation. *Forests* **2022**, *13*, 903. [CrossRef]
7. Jones, D.; Kržišnik, D.; Hočevar, M.; Zagar, A.; Humar, M.; Popescu, C.-M.; Popescu, M.-C.; Brischke, C.; Nunes, L.; Curling, S.F.; et al. Evaluation of the Effect of a Combined Chemical and Thermal Modification of Wood though the Use of Bicine and Tricine. *Forests* **2022**, *13*, 834. [CrossRef]
8. Ismaili, G.; Enduat, E.; Yahya, N.S.; Malek, F.M.; Jaimudin, N.A.; Abdul Rahim, K.K.; Wasli, M.E.; Kalu, M.; Openg, I.; Rizalman, A.N.; et al. Physical and Mechanical Properties Performance between Untreated and Treated with CCA Treatment at Different Age Groups of Fast-Growing Acacia Hybrid of Sarawak. *Forests* **2022**, *13*, 1969. [CrossRef]
9. Ru, S.; Zhao, C.; Yang, S. Multi-Objective Optimization and Analysis of Mechanical Properties of Coir Fiber from Coconut Forest Waste. *Forests* **2022**, *13*, 2033. [CrossRef]
10. Kačík, F.; Kubovský, I.; Bouček, J.; Hrčka, R.; Gaff, M.; Kačíková, D. Colour and Chemical Changes of Black Locust Wood during Heat Treatment. *Forests* **2023**, *14*, 73. [CrossRef]
11. Bekhta, P.; Pipíška, T.; Gryc, V.; Sedliačik, J.; Král, P.; Ráheľ, J.; Vaněrek, J. Properties of Plywood Panels Composed of Thermally Densified and Non-Densified Alder and Birch Veneers. *Forests* **2023**, *14*, 96. [CrossRef]
12. Dahali, R.; Lee, S.H.; Md Tahir, P.; Salim, S.; Hishamuddin, M.S.; Che Ismail, A.; Khoo, P.S.; Krystofiak, T.; Antov, P. Influence of Chrysoporthe deuterocubensis Canker Disease on the Chemical Properties and Durability of Eucalyptus urograndis against Wood Rotting Fungi and Termite Infestation. *Forests* **2023**, *14*, 350. [CrossRef]

forests

MDPI

Article

Water Absorption Properties in Transverse Direction of Heat-Treated Chinese Fir Wood Determined Using TD-NMR

Yulei Gao [1], Liyuan Zhao [2], Jinghui Jiang [2], Zhu Li [2] and Jianxiong Lyu [2,*]

[1] Pan Tianshou College of Architecture, Art and Design, Ningbo University, Ningbo 315211, China; gaoyl90@163.com
[2] Research Institute of Wood Industry of Chinese Academy of Forestry, Beijing 100091, China; zhao_li_yuan@126.com (L.Z.); Jiangjh@caf.ac.cn (J.J.); lizhu@caf.ac.cn (Z.L.)
* Correspondence: jianxiong@caf.ac.cn

Abstract: Heat treatment is an environmentally friendly method that improves the moisture-resistant properties and increases the service life of timber. In this work, Chinese fir (*Cunninghamia lanceolata* [Lamb.] Hook.) wood was heat-treated in a chamber with steam at temperatures of 160, 180, 200 and 220 °C for 2 h, and the absorption of water was studied by gravimetric analysis and time domain nuclear magnetic resonance (TD-NMR). The results show that both the amount of bound water and free water decreased with the increasing treatment temperature. The water absorption of wood in the radial direction was faster than that in the tangential direction due to the existence of rays, and this difference remained after the heat treatment. The heat treatment at 220 °C had a significant effect on water absorption in the tangential direction of wood, and the moisture content (MC) was approximately 20% lower than that of samples absorbing water in the radial direction. T_2 (spin–spin relaxation time) distributions showed two main components which were associated with bound water and free water, and for samples absorbing water in the radial or tangential direction, there was only a difference in the amount of free water. The amount of free water significantly decreased for the samples that were heat-treated at 220 °C and absorbed water in the tangential direction, indicating that the high-temperature heat-treated samples tended to close the pits in wood cells.

Keywords: heat treatment; Chinese fir; TD-NMR; free water; bound water; MC

Citation: Gao, Y.; Zhao, L.; Jiang, J.; Li, Z.; Lyu, J. Water Absorption Properties in Transverse Direction of Heat-Treated Chinese Fir Wood Determined Using TD-NMR. *Forests* **2021**, *12*, 1545. https://doi.org/10.3390/f12111545

Academic Editors: Pavlo Bekhta and Tomasz Krystofiak

Received: 13 October 2021
Accepted: 4 November 2021
Published: 9 November 2021

1. Introduction

Wood is a naturally hydrophilic material and its main components are cellulose, hemicellulose and lignin matrix. The existence of hygroscopic groups makes it sensitive to moist conditions and its mechanical properties as well as dimensional stability are deeply influenced by the presence of water [1]. Furthermore, if it is exposed to water, wood is more prone to be decayed or worn and the growth of mold and fungi might be enhanced. Various methods have been developed to make wood a biologically durable material as well as to decrease water absorption and increase the dimensional stability of wood [2,3]. Among these methods, heat treatment is an important approach which can change the relationship between wood and moisture without introducing chemicals, thereby effectively improving the properties of wood.

Heat treatment is a method of exposing wood to an almost oxygen-free environment under high temperature conditions and causing changes in the chemical structure of the wood substance [4,5]. Hemicellulose first degrades during the heat treatment process, including acid hydrolysis and decarboxylation reactions [6]. Lignin is generally considered to be the most thermally stable component; however, with the production of various phenol decomposition products, lignin undergoes thermal degradation at relatively low temperatures [7]. In addition, high temperature will increase the crystallinity of cellulose [8]. Therefore, the amount of hydroxyl sites might be reduced and the pore structure of wood would be changed after heat treatment. All these changes might affect the state and the amount of water adsorbed by wood when being stored in a humid environment or immersed in water; consequently, the water absorption performance of wood would be influenced.

A commonly used method for assessing the water absorption performance of wood is that of calculating the MC after water immersion for a specific time. The traditional method to determine MC is the oven drying method. This method, however, cannot accurately obtain the amounts of different water states (free water, FW; bound water, BW) separately. The NMR technique can be used to investigate the internal structure of wood as well as water state and water content, which makes up for the lack of test methods for characterizing the water absorption performance of timber. Thus, NMR is being more and more widely used in many fields of wood science research for quantitively exploring the characteristics of wood in a quick and noninvasive manner. The NMR signals obtained by the free induction decay (FID) experiments were proven to be a feasible means of calculating the MC of wood [9,10] and the NMR signals of wood substance and water in wood decay at different times could be easily distinguished [11,12]. Furthermore, the peak value of the Carr–Purcell–Meiboom–Gill (CPMG) decay curve was used for fitting the wood MC, and the results indicate that there is a good linear relationship between the signal and MC [13]. Since the mobility and local environment of molecules are connected to the NMR signal decay rate, the water state can be distinguished by the water relaxation characteristics for both hardwood and softwood [14,15]. In addition, the integral peak areas of T_2 curves acquired using the CPMG pulse sequence and the inverse algorithm also have a good linear relationship with the wood MC and could have a good application in calculating the amount of BW and FW.

For wood with large dimensions, water absorption mainly occurs in the transverse direction. In order to see the difference in water absorption in the radial direction and tangential direction and the effect of heat treatment on it, a comparison of the water absorption properties of Chinese fir wood was carried out by the gravimetric analysis and time domain nuclear magnetic resonance (TD-NMR) technique in this study. The wood samples under investigation were immersed in water and weighed as a function of time of water immersion to observe the water absorption progress. The statistical analysis of changes in the average MC of the wood obtained by gravimetric method was conducted and then parts of the samples were selected to perform the TD-NMR experiment. To verify the feasibility of the TD-NMR experiment, the integral peak area of the T_2 curves were firstly used to fit the MC. Thereafter, water states as well as the change of amounts for BW and FW were analyzed according to the T_2 curves.

2. Materials and Methods

2.1. Materials

In this study, the heartwood of Chinese fir (*Cunninghamia lanceolata* [Lamb.] Hook.) was studied. The dimension of the specimen used for heat treatment was 150 (longitudinal, L) $\times$ 15 (radial, R) $\times$ 15 mm^3 (tangential, T). The heat treatments were processed at 160, 180, 200 and 220 °C for 2 h in an airtight chamber. Throughout the heat treatment process, the oxygen content in the thermal treatment chamber was controlled to be no more than 2%. Before the heat treatment, the sample was stored under the air-dry condition, and the MC of the specimens was 9.8%–12.5%. After the heat treatment process, the specimens were cooled to room temperature in the chamber, and the final moisture content was in the range of 3.7%–7.6%. The number of the untreated references was 30 and 50 for each heat treatment group. After heat treatment, the references and each group of the heat-treated specimens were randomly divided into two equal parts: sealed tangential sections and cross-sections or radial sections and cross-sections by epoxy resin, respectively, using the method similar to that described in the previous study [16] and then conduct the experiment of the water absorption in the radial and tangential directions.

2.2. TD-NMR Experiments

The samples used for the TD-NMR experiments were cut in the middle of the heat-treated specimens, as described in their references above, in dimensions of 10(L) × 15(R) × 15(T) mm^3. The NMR instrument used in this study was MesoMR23-060H-I (Niumag Co., Ltd., Suzhou, China) with a 0.52 Tesla permanent magnet (23 MHz proton resonance frequency) operating at 32.00 °C. The spectra were acquired using the CPMG pulse sequence. The number of echoes was 15,000 with 64 scans and the echo time was 0.15 ms. More details were the same as described in the previous study [16]. The simultaneous iterative reconstruction technique (SIRT) algorithm was used to achieve the T_2 distribution curve and the amount of BW and FW was calculated based on Equation (1):

$$MC_i = \frac{(\text{Area}_i/\sum \text{Area}) \times \text{WM}}{\text{OW}} \times 100(\%) \quad (1)$$

where MC_i is the amount of BW or FW; Area_i is the integral area of BW peak or FW peak; ΣArea is the total integral area of the T_2 curve; WM is the water mass of the sample at each water absorption stage which can be calculated on a dry basis; and OW is the oven dry weight of the sample.

In the following analysis, the heat-treated samples are referred to as HT/X, where X represents the heat treatment temperature in degree Celsius, and when necessary, to distinguish the samples of water absorption in the radial direction (R direction) or the tangential direction (T direction), the prefixes R- or T- will be added to indicate the R direction or T direction. For instance, T-HT/220 indicates that the sample was heat-treated at 220 °C and absorbed water in the T direction.

3. Results and Discussion

3.1. MC Change during Water Absorption Process

Measured as a function of water immersion time and analyzed using the oven-drying method, the average MCs (with standard deviations) of the untreated and heat-treated samples are shown in Figure 1a,b. Within the first 30 days, the water absorption of the heat-treated samples was slower than that of the references in both the R direction and T direction. The rate of water absorption was decreased with the increased heat-treated temperature. While compared to the untreated samples, the water absorption rate was gradually increased due to the milder modification of HT/160 and HT/180, both in the R direction and T direction. After immersion for 30 days, the MC change of HT/160 was similar to the untreated samples and the absorption rate was higher than the untreated ones after water immersion for 45 days. Since the hemicellulose in the wood had already begun to degrade when being treated at 160 degrees, more pores might be formed in the wood for holding moisture, which was a possible reason for the higher MC after being immersed in water for long enough, and similar results have been reported in previous research [17]. The absorption rate of HT/180 was also close to the untreated sample after immersion for 45 days. Due to the stronger treatment, the rate of water absorption for HT/200 was lower than their untreated reference throughout the testing process, and the gap was widened along with the immersion time. Heat treatment at 220 °C had the strongest effect on water absorption, especially for samples absorbed water in the T direction, and the MC of the T-HT/220 was approximately 70% of its untreated reference after water absorption for 180 days.

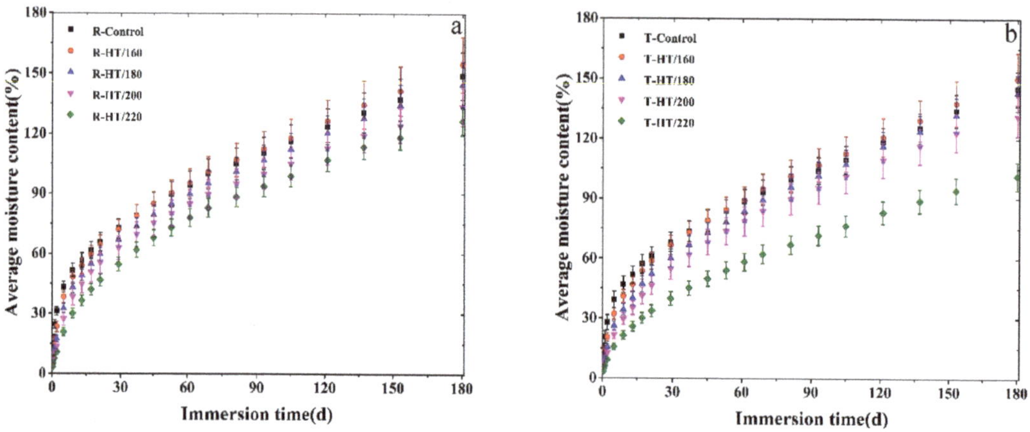

Figure 1. The change of average MC along with the immersion time for the references and heat-treated samples ((**a**) water absorption in the R-direction, (**b**) water absorption in the T-direction).

The decrease in water absorption due to heat treatment within the first 30 days was mainly considered the result of changes in wettability. An increase was observed in the contact angle while the wetting tension was decreased for the heat-treated wood, both indicating that the wood samples were becoming more hydrophobic [18]. Therefore, it was difficult for the water to spread on the wood surface and penetrate the wood sample. However, according to the previous studies, the permeability of wood might increase due to milder heat treatment [19], and owing to the degradation of the courts, cracks appeared along the middle lamella and intercellular cavities [20]. Thus, the internal resistance of water transport was weaker compared with the untreated samples. This might be one of the reasons for the increases in the water absorption rate of HT/160 and HT/180 after prolonged immersion time. However, the degradation of the wood constituents becomes more severe along with the increase in treated temperature, and the micropores would be filled with the degradation products [21]. In addition, many pits were closed because of the heat treatment above 200 °C [14]. These alterations should be the most important reason for the reduction in water absorption of HT/200 and HT/220.

Figure 2 shows the MCs' comparison of water absorption in the R direction and T direction for the samples which were under the same treated temperature as well as the references. As shown in Figure 2a, the water uptakes of R-control and T-control were almost the same after 180 days of water immersion, but the former was slightly higher than the latter. However, the curve revealed that the difference in MC between the two groups of samples was relatively obvious before the 150th day, revealing that the difference gradually decreased. This might indicate that there was a difference in water absorption rates between the R direction and T direction, but it should not affect the water absorption when upon reaching saturation. Although the water absorption of heat-treated samples was decreased compared to the references, no difference of the effect was founded between the R direction and T direction at temperatures below 200 °C. As shown in Figure 2b,c, the decrease in water absorption after the heat treatment was basically the same, and the water absorption was still slightly higher in the R direction. During the heat treatment progress, both the extractives and the products of thermal degradation would move towards the surface of the wood or evaporate from the wood [22]. These substances might partially remain in the surface layer of the wood and in consequence, impeding moisture from penetrating the wood, thereby decreasing the rate of water absorption. After heat-treatment to 220 °C, the water absorption of wood in the T direction significantly decreased and the MC of T-HT/220 was approximately 20% lower than R-HT/220. The possible explanation was the thermoplastic flow of the torus could bring about a filling of the courts after the heat-treated temperature above 200 °C [19], which would block the migration of moisture inside the wood. However, compared with the T direction, rays which exist in the R direction might play an important role during the water uptake process and result in higher water absorption.

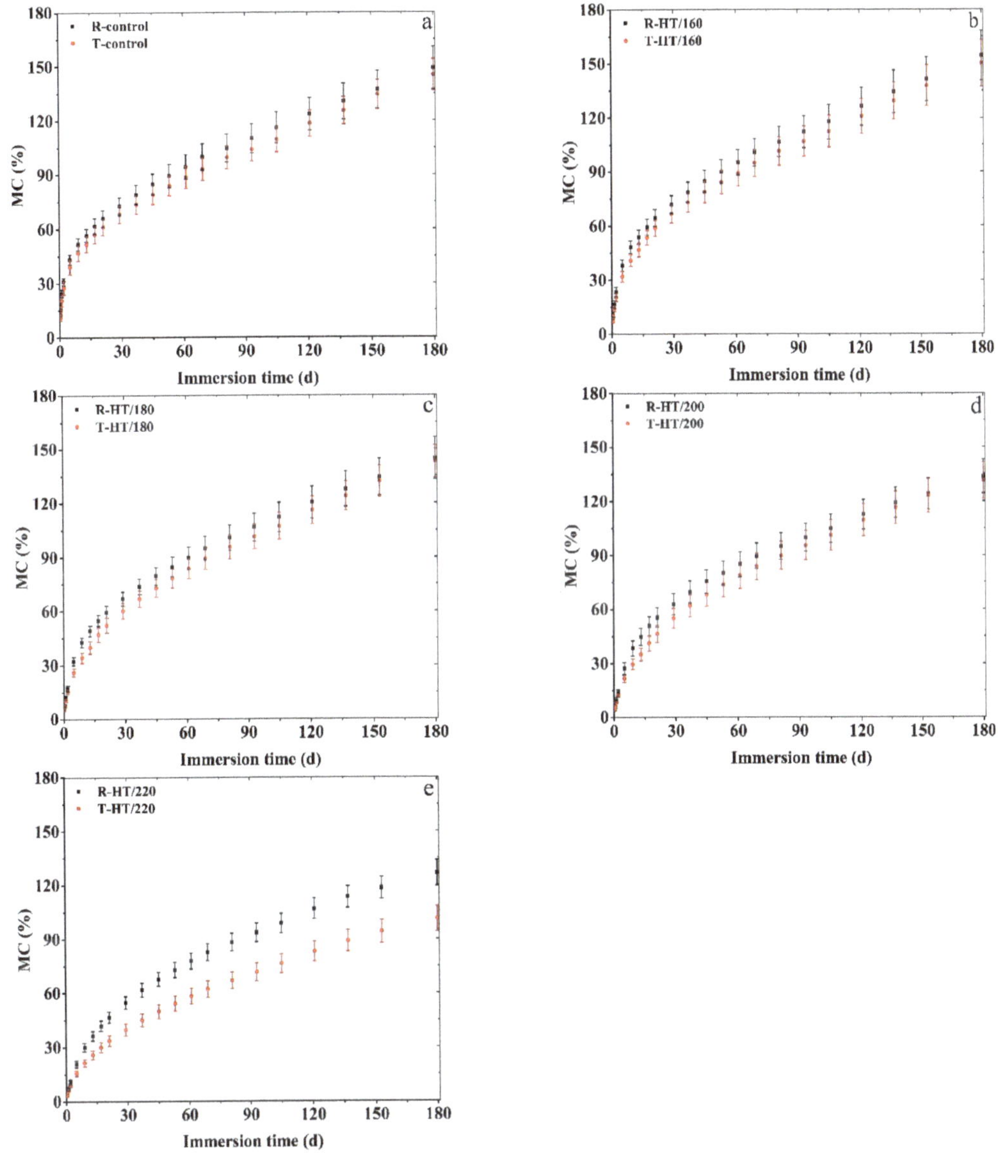

Figure 2. The average MCs' (with standard deviations) comparison of water absorption in the radial and T directions for samples under same heat-treated temperature and the references.

3.2. Determination of the Water States

To evaluate the feasibility of the MC analysis of the wood sample during water absorption process using TD-NMR, the peak area of the T_2 relaxation curves was calculated using the inverse algorithm after conducting each CPMG experiment and then fitted with the MC obtained by the oven-drying method. The response and linear fitting curve of the MC to the peak area are shown in Figure 3a,b. The regression equations and the correlation coefficients (R^2) are listed in Tables 1 and 2. According to the fitting curves and the correlation coefficients, the peak area and the MC has a good linear relationship and the R^2 was higher than 0.99 for all the samples, indicating that the peak areas obtained from the T_2 curves are proportional to the water amounts.

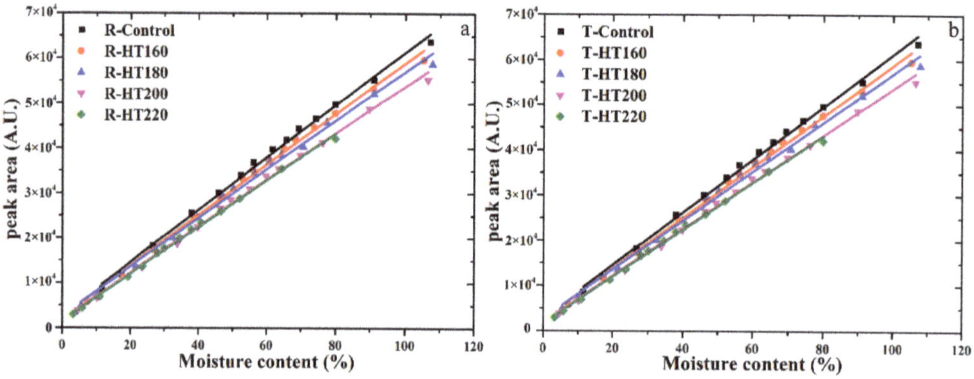

Figure 3. MCs obtained by oven-drying method versus the integral peak areas of the T_2 curves.

Table 1. The regression equations obtained by linear fitting of peak area and MC for samples with water absorption in the R direction.

Samples	Regression Equations	Adj.R^2
R-Control	Y = 583.72x + 2900.81	0.997
R-HT/160	Y = 563.53x + 2386.21	0.996
R-HT/180	Y = 542.72x + 2635.67	0.995
R-HT/200	Y = 518.51x + 1751.49	0.997
R-HT/220	Y = 520.24x + 1520.01	0.998

Table 2. The regression equations obtained by linear fitting of peak area and MC for samples with water absorption in the T direction.

Samples	Regression Equations	Adj.R^2
T-Control	Y = 554.04x + 2263.00	0.998
T-HT/160	Y = 571.94x + 2493.09	0.998
T-HT/180	Y = 482.26x + 2466.21	0.997
T-HT/200	Y = 515.90x + 1849.79	0.997
T-HT/220	Y = 531.70x + 1363.97	0.999

Figure 4 shows the water states of the wood samples determined by TD-NMR which were acquired from a single representative sample. The wood–water system has been widely researched using TD-NMR, and the literature has provided a good summary of relationships between the T_2 value and the states of water in softwood [23]. The T_2 of the BW was a few milliseconds and it was tens of milliseconds or more for FW according to the research results. The anatomical structure of softwood was relatively uncomplicated and only two obvious peaks were found in the water relaxation distribution curves which correspond to different water components, as shown in Figure 4. For all samples, the T_2 of BW was approximately 1 ms, and there was no obvious difference between the heat-treated samples and the untreated references. However, compared with the untreated references, the relaxation time of FW for the heat-treated wood significantly changed, and the change became more obvious with the increase in the heat-treated temperature. According to the relaxation theory of free fluid, the relaxation behavior of water inside wood was mainly relevant to its pore diameter and surface properties [24]. Heat treatment enhanced the surface hydrophobicity of wood due to changes in its chemical composition [25] which might be an important reason of the T_2 of FW becoming greater for the heat-treated wood. Another possible explanation for this was that the wood porosity and average pore diameters might be increased after the heat treatment according to the previous literature [26].

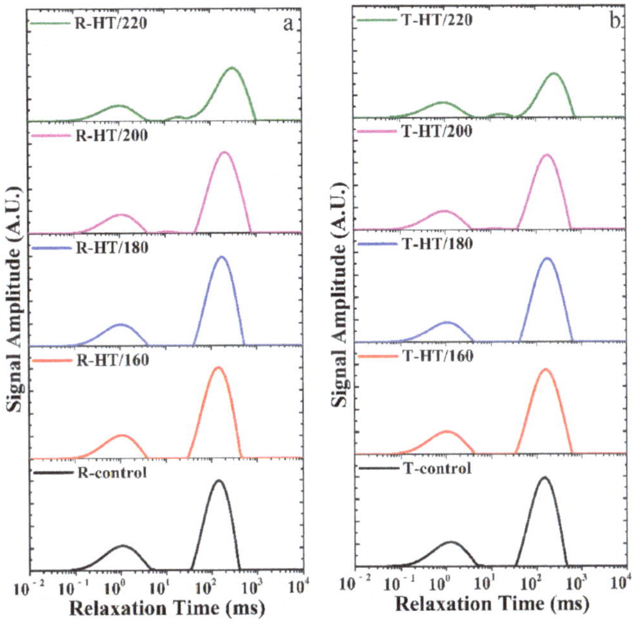

Figure 4. The T_2 curves of the heat-treated samples and the references after 60 days of water absorption.

In addition, as for HT/200 and HT/220, a peak with T_2 of approximately tens of milliseconds appeared on the water relaxation distribution curve, as shown in Figure 4, and which was not found in untreated samples and samples heat-treated below 200 °C. The peak was not obvious on the curve for HT/200; however, it became larger for HT/220 and could be easily observed on the curve. As previously mentioned, the heat treatment might affect the pore structure and therefore affect the relaxation behavior of water inside the wood. Consequently, the above phenomenon indicates that a heat treatment temperature over 200 °C might have a serious effect on the pore structure of wood. According to a previous study [27], a peak with T_2 greater than 100 ms corresponds to water in the lumen of wood cells; by inference, the peak with T_2 of approximately tens of milliseconds might correspond to water in the pores with a diameter less than the cell lumen. During the heat treatment, the cell wall was compressed, the cell lumens became narrow and intercellular cracks may have appeared in the middle lamella [28]. This might be an interpretation of the appearance of a little peak on the water relaxation curve for HT/200 and HT/220.

3.3. *Amount of Bound Water and Free Water*

Over the past decade, the FID signal according to the NMR experiment has been used to determine the MC of wood [29]. The CPMG experiment was employed to investigate the MC of a wood sample also based on the peak area of a T_2 relaxation curve [30]. In this study, both the amounts of BW and FW were calculated using MCs acquired by the oven-drying method and the peak area of the T_2 curves. The change in the amount of BW and FW along with the water immersion time are illustrated in Figure 5. For a more direct comparison, the values of the BW content and FW content after 60 days of water immersion are listed in Tables 3 and 4.

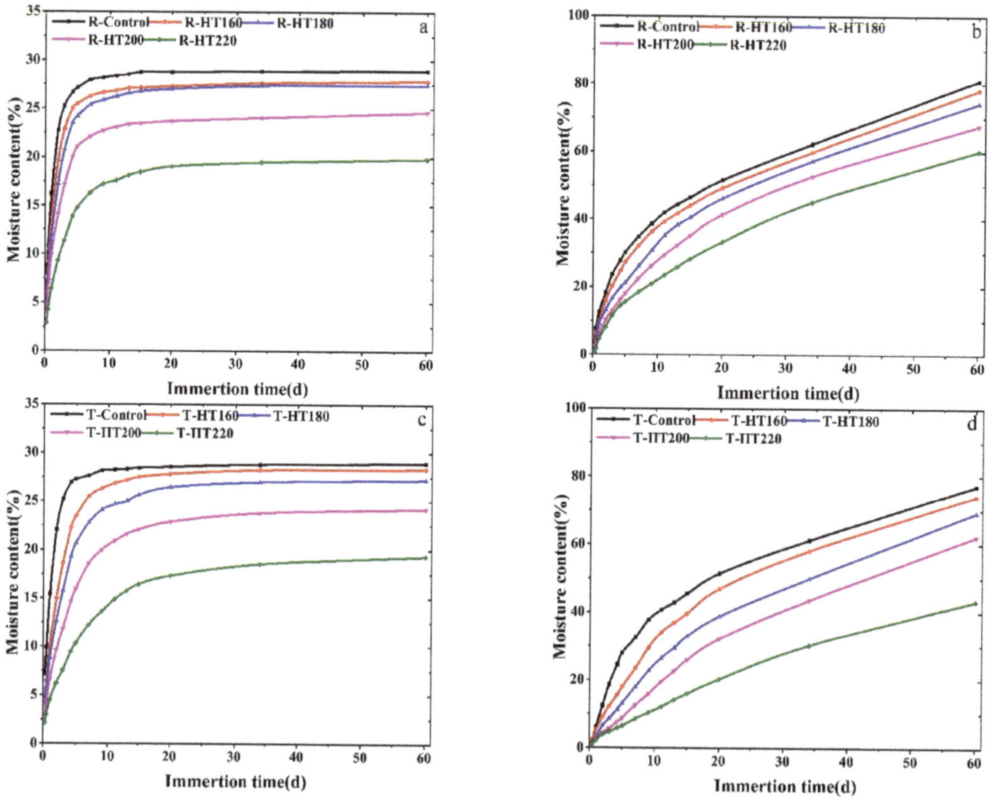

Figure 5. The dynamic change in BW (**a**,**c**) amount and FW amount (**b**,**d**) for the wood samples during 60 days of water absorption.

Table 3. The content of BW and FW of the wood samples after 60 days of water absorption in the R direction.

Samples	R-Control	R-HT/160	R-HT/180	R-HT/200	R-HT/220
BW Content	28.92%	27.84%	27.43%	24.64%	19.83%
FW Content	81.00%	78.10%	74.20%	67.56%	60.17%

Table 4. The content of BW and FW of the wood samples after 60 days of water absorption in the T direction.

Samples	T-Control	T-HT/160	T-HT/180	T-HT/200	T-HT/220
BW Content	28.93%	28.32%	27.23%	24.21%	19.33%
FW Content	77.20%	74.12%	69.25%	62.20%	43.23%

According to the data in Tables 3 and 4, no significant difference of the BW content was found between HT/160 and HT/180, but the value was slightly lower compared with the references. After the treatment temperature reached 200 °C, the BW content obviously decreased, and when the temperature rose to 220 °C, the value dropped by more than 30% compared to the untreated samples. As we all know, changes in wood chemistry and mass loss will occur after heat treatment [31]. The opinion of most researchers is that the hygroscopicity change of heat-treated wood is related to the reduction in available hydroxyl groups in the cell wall because of the degradation of chemicals [32]. At lower temperatures, wood chemical composition, especially that of cellulose and lignin, is less affected. Thus, the BW content is only slightly reduced for HT/160 and HT/180. At higher temperatures, the rearrangement or reorientation of cellulose molecules inside quasicrystalline amorphous regions and the crosslinking condensation reaction of lignin occur, whilst the wood chemical composition as well as wood pore structure are changed. The increase in matrix stiffness and lignin cross-linking would decrease the capacity of the expansion of the cell wall and lead to the reduction in the polylayer sorption [33]; as a result, the BW content markedly decreases as the heat treatment temperature rises.

The amount of FW in the heat-treated samples and untreated reference were measured as a function of the time of water immersion and the results are shown in Figure 5b,d. The values increased with prolonging water immersion time, but did not reach a saturation value after immersion in water for 60 days. For untreated samples, the absorption of FW was slightly faster in the R direction than that in the T direction, which can be inferred from Tables 3 and 4. This may be caused by rays, which are oriented in the R direction. After the heat treatment, the difference in water transport speed between the R direction and the T direction still existed, and the gap became more obvious after heat treatment at 220 °C. The amount of FW for T-HT/220 was approximately 28% lower than R-HT/220 after water immersion for 60 days. Based on the above analysis, after 60 days of immersion, water absorption in the R direction or the T direction almost did not influence the BW content for the untreated samples or samples treated at the same temperatures, and the difference in MC was mainly caused due to the different amounts of FW. This might indicate that the heat treatment changes the wood chemical composition and pore structure, which affects the migration of liquid water but has no influence on the diffusion of water vapor inside it.

4. Conclusions

In this study, the water absorption performance of the heat-treated wood at different temperatures, i.e., 160, 180, 200 and 220 °C, as well as that of untreated samples was investigated as a function of the water immersion time. The MC analysis by the oven-drying method showed that after the heat treatment below 200 °C, the rate of water absorption slightly decreased, while higher treatment temperatures (200 °C and 220 °C) had an obvious effect on the water absorption rate. The comparison of MCs revealed that the water absorption in the R direction was faster than that in the T direction, and the difference still existed after the heat treatment. It is worth mentioning that the 220 °C-heat treatment had a significant effect on water absorption in the T direction, and the MC was significantly lower than that in the R direction (R-HT/220) after the same time of water immersion. T_2 distributions showed two main components corresponding to FW and BW, both in the case of the heat-treated samples and the references. Whereas there was also a component with T_2 of tens of milliseconds which belonged to FW for HT/200 and HT/220, the water amount was small. This might be because of changes in the pore structure caused by the chemical degradation of wood. For the heat-treated samples, the analysis of MCs obtained by the oven-drying method and the integral peak area of the T_2 curves showed that both the content of FW and BW decreased, which became more obvious after heat treatment at higher temperature. For the samples which were under the same heat-treatment conditions but in different water absorbing directions, the difference in MC was mainly due to the different amount of FW. Moreover, for the T-HT/220 samples, the significant reduction in FW content might be because the closing of pits reduces the access of water to the interior of wood.

Author Contributions: Conceptualization, J.L., Y.G. and J.J.; methodology, Y.G. and L.Z.; investigation, Y.G. and Z.L.; writing—original draft preparation, Y.G.; writing—review and editing, J.L., Y.G. and L.Z. All authors have read and agreed to the published version of the manuscript.

Funding: This research was funded by the National Natural Science Foundation of China (No.31971591).

Institutional Review Board Statement: Not applicable.

Informed Consent Statement: Not applicable.

Data Availability Statement: Not applicable.

Conflicts of Interest: The authors declare no conflict of interest.

References

1. Kamperidou, V. The Biological Durability of Thermally- and Chemically-Modified Black Pine and Poplar Wood Against Basidiomycetes and Mold Action. *Forests* **2019**, *10*, 1111. [CrossRef]
2. Kartal, S.N.; Hwang, W.J.; Imamura, Y. Water absorption of boron-treated and heat-modified wood. *J. Wood Sci.* **2007**, *53*, 454–457. [CrossRef]
3. Lu, J.; Xu, K.; Liu, Y.; Wu, Y.; Li, X. Research progresses on reinforced modifi cation of poplar wood from fastgrowing plantation. *J. Cent. South Univ. For. Technol.* **2014**, *34*, 99–103. (In Chinese)
4. Herrera-Builes, J.F.; Sepúlveda-Villarroel, V.; Osorio, J.A.; Salvo-Sepúlveda, L.; Ananías, R.A. Effect of Thermal Modification Treatment on Some Physical and Mechanical Properties of Pinus oocarpa Wood. *Forests* **2021**, *12*, 249. [CrossRef]
5. Gu, L.; Ding, T.; Jiang, N. Development of wood heat treatment research and industrialization. *J. For. Eng.* **2019**, *22*, 9–19. (In Chinese)
6. Altgen, M.; Hofmann, T.; Militz, H. Wood moisture content during the thermal modification process affects the improvement in hygroscopicity of Scots pine sapwood. *Wood Sci. Technol.* **2016**, *50*, 1181–1195. [CrossRef]
7. Hill, C.A.S. *Wood Modification: Chemical, Thermal and Other Processes*; John Wiley & Sons: Hoboken, NJ, USA, 2006; pp. 99–127.
8. Akgül, M.; Korkut, S. The effect of heat treatment on some chemical properties and colour in Scots pine and Uludağ fir wood. *Int. J. Phys. Sci.* **2012**, *7*, 2854–2859.
9. Chao, L.; Minghui, Z.; Jianfang, Y. Determining wood moisture content by free induction decay of nuclear magnetic resonance. *J. Beijing For. Univ.* **2012**, *34*, 142–145.
10. Gao, Y.; Zhang, M. Moisture Sorption in Wood Studied by Time Domain Nuclear Magnetic Resonance. *Chin. J. Magn. Reson.* **2016**, *33*, 295–304. (In Chinese)
11. Nanassy, A.J. Use of wide line NMR for measurement of moisture content in wood. *Wood Sci. Technol.* **1973**, *5*, 187–193.
12. Xu, Y.; Araujo, C.D.; Mackay, A.L.; Whittall, K.P. Proton Spin–Lattice Relaxation in Wood— T 1 Related to Local Specific Gravity Using a Fast-Exchange Model. *J. Magn. Reson.* **1996**, *110*, 55–64. [CrossRef]
13. Xu, K.; Lv, J.; Gao, Y.; Wu, Y.; Li, X. Determination of Moisture Content and Moisture Content Profiles in Wood during Drying by Low Field Nuclear Magnetic Resonance. *Dry. Technol.* **2017**, *35*, 1909–1918. [CrossRef]
14. Kekkonen, P.M.; Ylisassi, A.; Telkki, V.V. Absorption of Water in Thermally Modified Pine Wood As Studied by Nuclear Magnetic Resonance. *J. Phys. Chem. C* **2014**, *118*, 2146–2153. [CrossRef]
15. Gao, Y.; Li, X.; Lei, P.; Zhang, M. Water Distribution in Poplar during High-Temperature Drying Process Studied by Time-Domain Nuclear Magnetic Resonance. *J. Magn. Reson.* **2016**, *33*, 479–490. (In Chinese)
16. Gao, Y.; Xu, K.; Peng, H.; Jiang, J.; Lu, J. Effect of Heat Treatment on Water Absorption of Chinese fir Using TD-NMR. *Appl. Sci.* **2018**, *9*, 78. [CrossRef]
17. Ahmed, S.A.; Morén, T. Moisture properties of heat-treated Scots pine and Norway spruce sapwood impregnated with wood preservatives. *Wood Fiber Sci.* **2012**, *44*, 85–93.
18. Metsä-Kortelainen, S.; Viitanen, H. Wettability of sapwood and heartwood of thermally modifi ed Norway spruce and Scots pine. *Eur. J. Wood Prod.* **2011**, *70*, 135–139. [CrossRef]
19. Scheiding, W.; Direske, M.; Zauer, M. Water absorption of untreated and thermally modified sapwood and heartwood of Pinus sylvestris L. *Eur. J. Wood Wood Prod.* **2016**, *74*, 585–589. [CrossRef]
20. Biziks, V.; Beļkova, Ļ.; Kapača, E.; Militz, H. Changes in the microstructure of birch wood after hydrothermal treatment. *Wood Sci. Technol.* **2013**, *47*, 717–735. [CrossRef]
21. Biziks, V.; Andersons, B.; Sansonetti, E.; Andersone, I.; Militz, H.; Grinins, J. One-stage thermo-hydro treatment (THT) of hardwoods: An analysis of form stability after five soaking-drying cycles. *Holzforschung* **2015**, *69*, 563–571. [CrossRef]
22. Popper, R.; Niemz, P.; Eberle, G. Investigations on the sorption and swelling properties of thermally treated wood. *Holz Roh Werkst.* **2005**, *63*, 135–148. [CrossRef]
23. Zhang, M.; Li, X.; Zhou, Y.; Gao, Y. Water Status Change in Wood Drying Studied By Time-Domain NMR. *Sci. Silvae Sin.* **2014**, *50*, 109–113. (In Chinese)
24. Brownstein, K.R.; Tarr, C.E. Importance of classical diffusion in NMR studies of water in biological cells. *Phys. Rev. A* **1979**, *19*, 2446–2453. [CrossRef]
25. Yan, Y.; Li, F.; Zhang, K.; Gu, X.; Guo, M. Dimensional Stability and finishing performance of heat-treated Pinus koraiensis and Rubber Wood. *J. Northeast For. Univ.* **2018**, *46*, 45–48. (In Chinese)
26. Pfriem, A.; Zauer, M.; Wagenführ, A. Alteration of the pore structure of spruce (Picea abies (L.) Karst.) and maple (Acer pseudo-platanus L.) due to thermal treatment as determined by helium pycnometry and mercury intrusion porosimetry. *Holzforschung* **2009**, *63*, 94–98. [CrossRef]

27. Zhang, X.; Zhang, J.; Wang, Q.; Mu, J.; Chu, D. Study on Water Status and Distribution in 2 Fast-growing Woods During Drying Based on LF-NMR and MRI. *J. Southwest For. Univ.* **2020**, *40*, 143–150. (In Chinese)
28. Boonstra, M.J.; Acker, J.V.; Tjeerdsma, B.F.; Kegel, E.V. Strength properties of thermally modified softwoods and its relation to polymeric structural wood constituents. *Ann. For. Sci.* **2007**, *64*, 679–690. [CrossRef]
29. Araujo, C.D.; Avramidis, S.; Mackay, A.L. Behaviour of Solid Wood and Bound Water as a Function of Moisture Content. A Proton Magnetic Resonance Study. *Holzforschung* **1994**, *48*, 69–74. [CrossRef]
30. Labbé, N.; Jéso, B.D.; Lartigue, J.-C.; Daudé, G.; Pétraud, M.; Ratier, M. Time-domain 1H NMR characterization of the liquid phase in greenwood. *Holzforschung* **2006**, *60*, 265–270. [CrossRef]
31. Stamm, A.J. Thermal Degradation of Wood and Cellulose. *Ind. Eng. Chem.* **1956**, *48*, 413–417. [CrossRef]
32. Källbom, S.; Rautkari, L.; Wålinder, M.; Johansson, L.S.; Campbell, J.M.; Segerholm, K.; Jones, D.; Laine, K. Water vapour sorption characteristics and surface chemical composition of thermally modified spruce (Picea abies karst). *Int. Wood Prod. J.* **2016**, *7*, 116–123. [CrossRef]
33. Jalaludin, Z.; Hill, C.A.S.; Xie, Y.; Samsi, H.W.; Husain, H.; Awang, K.; Curling, S.F. Analysis of the water vapour sorption isotherms of thermally modified acacia and sesendok. *Wood Mater. Sci. Eng.* **2010**, *5*, 194–203. [CrossRef]

forests

Article

Impregnation of Wood with Waste Engine Oil to Increase Water- and Bio-Resistance

Larisa Belchinskaya, Konstantin Viktorovich Zhuzhukin *, Tatiana Ishchenko and Aleksey Platonov

Department of Chemistry, Voronezh State University of Forestry and Technologies named after G.F. Morozov, 8, Timiryazeva, 394087 Voronezh, Russia; belbom@mail.ru (L.B.); tl_ischenko@mail.ru (T.I.); aleksey66@yandex.ru (A.P.)
* Correspondence: kinkon18@yandex.ru

Abstract: Impregnation is a common method of protecting wood from external influences. This study proposes the use of spent engine oil as an impregnating composition for modifying birch wood to make it resistant to biological degradation and water. The indicators of water resistance and dimensional stability of wood such as wetting contact angle, thermogravimetric analysis, Fourier transform infrared spectroscopy (FTIR), and biodegradation tests have been determined. It has been found that treatment with spent engine oil significantly increases the dimensional stability (56.8% and 45.7% in tangential and radial directions) and water-resistant indicators of wood. Thermogravimetric analysis has showed that the curves for the impregnated specimens were different from the control group and had two sharp peaks at 302 and 357 °C. However, FTIR indicated that no clear chemical reactions occur between spent engine oil and wood. A study on wood resistance to biological degradation has showed a significant increase in resistance against brown rot (*Poria placenta* fungi) in the treated specimens, in contrast to the control group. Thus, impregnation of wood with spent engine oil makes it possible to increase wood resistance to water and biological degradation.

Keywords: birch wood; spent engine oil; water resistance; biostability; dimensional stability

Citation: Belchinskaya, L.; Zhuzhukin, K.V.; Ishchenko, T.; Platonov, A. Impregnation of Wood with Waste Engine Oil to Increase Water- and Bio-Resistance. *Forests* **2021**, *12*, 1762. https://doi.org/10.3390/f12121762

Academic Editors: Pavlo Bekhta and Tomasz Krystofiak

Received: 27 November 2021
Accepted: 10 December 2021
Published: 13 December 2021

1. Introduction

Wood is the most widespread environmentally friendly, renewable natural polymer. It possesses a complex structure, the main components of which are cellulose, hemicellulose, lignin and extractives. This material is important in a number of industrial sectors of the economy and areas of economic activity such as construction, furniture production, and the chemical industry [1].

Wood has significant advantages over other materials, for example, a high ratio of strength indicators to its weight, high impact resistance, the possibility of using it in many technological processes, etc. [2]. However, due to the presence of a large number of hydroxyl groups in its chemical structure, wood is susceptible to atmospheric influences, under the impact of which there is a change in its size and decrease in performance, a significant reduction in the service life of products, and biological decomposition [3–6]. Increase in dimensional stability, strength, hydrophobic properties and resistance to biological degradation of wood can be achieved by reducing its hygroscopicity using many modification methods, such as steam thermal treatment [7–9], cell wall modification with methyltrimetosiloxane [10,11], modification using styrene [12] and phenol-containing resins [13,14], modification using boron and compatibilizers [15], high-density polyethylene [16] and siloxanes [17], modification using 1,3-dimethylol-4,5-dihydroxyethylene urea [18], thermal processing [19], modification using waxes, paraffins [20,21], vegetable oils [22,23], etc. However, these methods have their disadvantages: heat treatment can reduce the strength properties of wood [24,25], and chemical modification is characterized by the complexity of the process and high energy consumption.

In accordance with the current Russian Federation standards, wood used in high humidity environments and constant contact with water, for example, railway sleepers, wooden bridge structures, power transmission poles, must be impregnated [26]. The main impregnating compounds for such wood are creosote oil, coal oil and other oily compounds [27]. These compositions provide high protection of wood from biological destruction and give it sufficient water resistance. However, they show high toxicity and are hazardous to the environment [28]. Waste engine oil also includes some hazardous substances, but according to the current standards of the Russian Federation is considered a moderately hazardous waste, while creosote and coal oils are highly hazardous substances.

In this work, it is proposed to use waste gasoline engine oil as an alternative to a highly toxic antiseptic hydrophobizator (creosote, which is still used in Russia in significant quantities) to protect wood used in non-residential construction from water, changes in size and biodegradation. Spent motor oil is a waste product from the automotive industry; it shows water-repellent and antiseptic qualities [29]. There is a small amount of research on the use of waste engine oil as an anticorrosive and preservative agent [30], as well as a component of a stabilizer to obtain hydrophobizing composition used in railway sleeper impregnation [31]. In remote regions (where it is not possible to ensure constant delivery of goods) the use of spent engine oil as a water repellent capable of protecting wood from water and making it resistant to biological degradation could be an alternative when using wood products in non-residential construction.

The purpose of this work was to study the effect of birch wood impregnation on the indicators of its water and moisture resistance, dimensional stability, as well as susceptibility to biological degradation. The impregnation was made by heating the specimens in a cold bath with waste engine oil.

2. Materials and Methods

2.1. Materials

Wood of silver birch (*Betula pendula*) was obtained from the educational and experimental forestry of Voronezh State University of Forestry and Technologies (Voronezh region, Russia). The specimens had dimensions of 20 × 20 × 20 mm (length × width × thickness) with an initial moisture content of 80 ± 5% (according to GB/T 1931–2009 [32]). All the specimens were dried at 103 °C to constant weight. Spent engine oil GULF Formula GX Powermax SAE 5W-40 was purchased from Avrora Avto (Voronezh, Russian Federation); the manufacturer was LLK International (Voronezh, Russian Federation). The oil was drained from the gasoline engine of a Granta car (Lada, Togliatti, Russia); the engine capacity was 1.6 L. The total mileage of the car was 149,000 km, and the mileage on purchased oil was 7900 km. The oil was poured into a disposable polyethylene terephthalate (PET) bottle. Indicators of spent engine oil were determined by the supplier at the OEM-OIL laboratory (Moscow, Russia), test report No. 000113u (Table 1).

Table 1. Physical and chemical indicators of spent engine oil provided by Avrora Avto.

Indicator	Unit	Test Method	Measured Value
Kinematic viscosity at 40 °C	mm^2/s	ASTM D 445	82.6
Kinematic viscosity at 100 °C	mm^2/s	ASTM D 445	13.7
Viscosity index	-	ASTM D 2270	170
TBN (Total Base Number)	mg, KOH	ASTM D 4739	8.28
TAN (Total Acid Number)	mg, KOH	ASTM D 664	2.08
pH	-	ASTM D 664	5.8
Water content	IR Units	ASTM E 2412	<0.1
Colloidal carbon content			0.1
Oxidation product content			11
Oil content			<0.1
Content of nitration products			8

2.2. Wood Impregnation Using Spent Engine Oil

Impregnation of wood specimens was carried out by the hot-cold baths method at atmospheric pressure. At least 30 specimens were used in each experiment. The control group of specimens was not impregnated. The impregnation was carried out in several stages. After weighing the specimens, they were immersed in a container with spent engine oil for 60 min at a temperature of 120 °C. Then the specimens were moved into a container with cold spent engine oil at a temperature of 30 °C and held for 60 min, so as not to allow them to come into contact with air. The wood was impregnated as a result of a sharp temperature drop, leading to the creation of negative pressure in the wood. These conditions contributed to the intensification of penetration of impregnating composition into the wood structure. The next step was drying the wood specimens, first at room temperature under ambient conditions for 3 days, and then drying in an oven at a constant temperature of 60 °C for 72 h (the temperature was controlled to avoid oil escaping from the specimens).

2.3. Measuring Weight Gain as a Percentage (WPG)

The *WPG* was determined based on the variations in the weight before and after impregnation. It was calculated by Equation (1):

$$WPG = \frac{W_w - W_o}{W_o} \cdot 100\ (\%) \quad (1)$$

where W_o the oven-dried weight of specimens before treatment and W_w is the oven-dried weight of specimens after treatment.

2.4. Water Absorption and Dimensional Stability

To determine water absorption (*WA*) and volumetric swelling (*S*), pre-dried specimens (impregnated, control ones) were placed in a dessicator in distilled water at a temperature of 20 °C for 1, 2, 3, 6, 9, 13, 20, and 30 days. After each measurement, excessive water amount was removed from the specimens with filter paper. And distilled water in a dessicator was replaced with a new portion. *WA* was determined by Equation (2):

$$WA = \frac{m_i - m_0}{m_0} \cdot 100\ (\%) \quad (2)$$

where m_i—mass of the specimen after being in distilled water for a certain period of time, and m_0 is the mass of the pre-dried specimen before placing in water.

Volumetric swelling (S) and swelling (a) in the radial and tangential directions were calculated after holding the specimens in distilled water for 40 days using Equations (3) and (4), respectively:

$$S = \frac{V_\omega - V_d}{V_d} \cdot 100\ (\%) \tag{3}$$

where V_ω—volume of specimens after impregnation in distilled water for a certain time, V_d—volume of pre-dried specimen before placing it in water:

$$a = \frac{l_w - l_0}{l_0} \cdot 100\ (\%) \tag{4}$$

where l_w—size of the specimens after being in distilled water for a certain time, l_0—initial size of the specimen.

2.5. Moisture Absorption

To determine moisture absorption (MA), the specimens were placed in a climatic chamber (TYP KBF 240, Binder, Tuttlingen, Germany) under constant environmental conditions (temperature was 20 °C, humidity was 95%) for 1, 2, 3, 6, 9, 13, 20, 30, and 40 days. After certain intervals of conditioning in a climatic chamber, MA was calculated using Equation (5):

$$MA = \frac{W_a - W_b}{W_b} \cdot 100\ (\%) \tag{5}$$

where W_a—weight of specimens after being in a climatic chamber for a certain time, and W_b—weight of specimens before conditioning in a climatic chamber.

2.6. Anti-Swelling Efficiency (ASE)

ASE was calculated from the change in volumetric swelling of natural and impregnated wood before and after being in distilled water for 1, 10, or 30 days using Equation (6):

$$ASE = \frac{S_u - S_t}{S_u} \cdot 100\ (\%) \tag{6}$$

where S_u—volumetric swelling of impregnated specimens, and S_t—volumetric swelling of non-impregnated specimens.

2.7. Thermogravimetric Analysis

Thermogravimetric analysis (TGA) was carried out on a STA449F3 analyzer (Netzsch, Weimar, Germany) at a heating rate of 5 °C/min to a temperature of 450 °C, in a nitrogen gas atmosphere to study the change in mass and heat effects of impregnated and non-impregnated specimens.

2.8. Determination of the Contact Angle

The wetting contact angle of wood with distilled water was measured by the sessile drop method in the laboratory on a goniometer using the HIview 10 program. The liquid was applied to the wood surface with a 0.01 mL microsyringe. The image was recorded using a portable digital microscope camera (Ruihoge, Nanchang, China) and recorded for 90 s. Then the obtained data were analyzed using the KOMPAS 3D software package (Askon, Russian Federation). Measurements were carried out at four different points on the surface of each specimen.

2.9. IR Spectroscopic Studies

FTIR analysis was performed to study the intermolecular interaction of functional groups in treated and untreated wood specimens. The studies were carried out on a VERTEX 70 spectrometer (Bruker, city, Germany) by the method of disturbed complete internal reflection using a diamond prism in the frequency range from 400–4000 cm^{-1} with a resolution of 2 cm^{-1} in the transmission mode.

2.10. Biodegradation Resistance Test

Sterile culture medium with a volume of 25 mL, obtained from barley malt (40 g) and bacteriological agar (20 g) (HiMedia Laboratories, Moscow, Russian Federation) dissolved in 1 L of distilled water was placed in Petri dishes with a diameter of 9 cm. Then a small amount of freshly grown culture mycelium was inoculated with *Poria placenta* incubated for 2 weeks to ensure complete colonization of the medium with mycelium at 22 °C and 65% relative humidity. The treated and control specimens (dimensions of 20 × 20 × 20 mm (length × width × thickness)) were placed in a Petri dish under sterile conditions. Incubation was carried out for 16 weeks at 22 °C and 65% relative humidity in a Binder TYP KBF 240 climatic chamber. At the end of the test (after 16 weeks), the mycelium was removed with a dry brush and the specimens were placed in an oven to dry to a constant weight at 103 °C. Weight loss (WL) was determined by Equation (7):

$$WL = \frac{m_0 - m_1}{m_0} \cdot 100\ (\%) \tag{7}$$

where m_0 and m_1—initial and final weights of dried wood specimens before and after exposure to the fungus, respectively.

3. Results and Discussion

3.1. Measurements of Weight Percent Gain (WPG)

The WPG indicating net spent engine oil uptake is presented in Table 2. The average weight of the dried specimens before impregnation with spent engine oil was 2.12 g, after impregnation it was 3.36 g. WPG after impregnation is 58.5%, which makes it possible to estimate a fairly high penetration of spent engine oil into the wood structure.

Table 2. Weight of the specimen before and after spent motor oil impregnation.

	Before Impregnation, g.	After Impregnation, g.	WPG,%
Weight of the specimen	2.12 ± 0.14	3.36 ± 0.16	58.5

3.2. Dimensional Stability and Water Resistance of Wood

3.2.1. Water and Moisture Absorption

Kinetic dependences of water absorption (WA) and moisture absorption (MA) of treated and untreated wood (Figure 1a,b) were plotted to study the hydrophobic properties of wood impregnated with spent engine oil. Figure 1a shows MA after conditioning the specimens in a climate chamber at 20 °C and 95% humidity. After 1 day in the climatic chamber, an increase in moisture content in all the specimens by 12% was observed (Figure 1a). In the next 10 days, MA on the wood was 24.69% and 10.44% for untreated and treated wood, respectively. After conditioning the specimens for 40 days, their MA was 25.31%. It was 12.87% for the treated ones, which is two times lower than for untreated wood specimens.

Figure 1b shows the change in water absorption of treated and untreated wood over 30 days. Water absorption increased to 14.2% and 43.9%, respectively, after 1 day of keeping impregnated and non-impregnated specimens in distilled water. A significant increase in WA for both groups of specimens was observed after 13 days of the experiment and amounted to 31.4% for treated and 132.0% for untreated specimens. After keeping the specimens for 30 days in distilled water, the WA of the treated specimens was 3.7 times less than for the untreated ones and amounted to 37.5% and 141.2%, respectively.

The water absorption of wood impregnated with creosote and coal oil [33,34] turned out to be similar in results, but on average, depending on the type of wood, it is 10%–15%

higher. The MA of beech wood impregnated with hemp oil after 40 days of conditioning was 17% [35]. When impregnated with wax emulsions of European spruce, WA turned out to be significantly worse than the results obtained, so after 500 h of the experiment, WA was 80% [36]. This can be explained by the relatively small size of spent motor oil molecules formed during operation in the engine as a result of thermal degradation, which is indirectly confirmed by the high depth of impregnation of wood (Figure 1c) and the high value of WPG (58.5%). Used engine oil, like any other oil, is hydrophobic, while allowing additional protection of wood from water and moisture.

Spent engine oil impregnation significantly reduced water and moisture absorption properties of wood. Destruction of intermolecular van der Waals and hydrogen bonds between adsorption centers of wood and water molecules occurs at the first, hot stage of impregnation. At the second, cold stage of impregnation, free adsorption centers form intermolecular bonds with the functional groups of the engine oil chemical components. It can significantly increase water resistance of wood and stability of its dimensions.

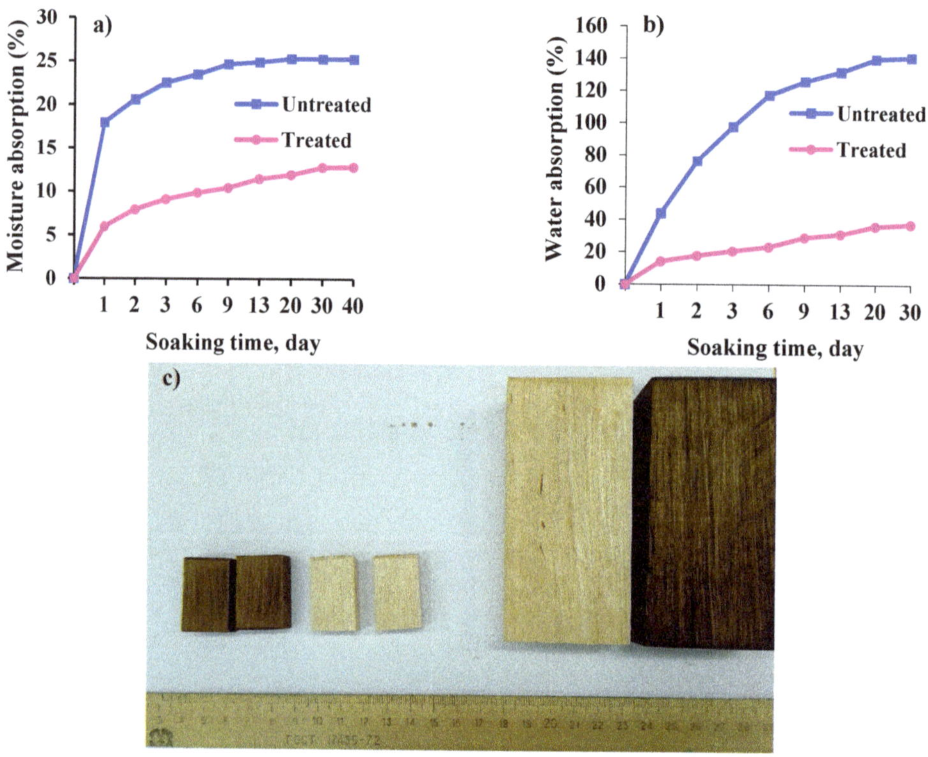

Figure 1. Moisture absorption (**a**) and water absorption (**b**) of treated and untreated birch wood, split birch wood sample (**c**) before and after impregnation.

3.2.2. Swelling and Anti-Swelling Efficiency

Dimensional stability has a significant impact on the quality and service life during the operation of wood products. Quantitative assessment of wood dimensional stability was carried out according to the values of wood swelling in tangential and radial directions, volumetric swelling and anti-swelling efficiency. Figure 2a shows the results of determination of radial, tangential, and volumetric wood swelling after soaking in distilled water for 40 days. Mean tangential swelling (%) for the control group was 12.7 and swelling decreased by 56.8% for the treated group (relative to the control one) and it was 8.1. Mean radial swelling (%) was 10.2 for the control group of specimens. This indicator for the processed specimens decreased by 45.7% and amounted to 7.0. Volumetric swelling (S, %) of the treated specimens decreased by 25.23% in comparison with the untreated specimens. It was 16.09% for the treated group of specimens and 20.15% for the untreated one. Anti-swelling efficiency was 73.20% in one day of soaking. This indicator has significantly decreased in ten days and amounted to 38.61%. ASE decreased by only 2.43% in 30 days, and it was equal to 36.18% (Figure 2b).

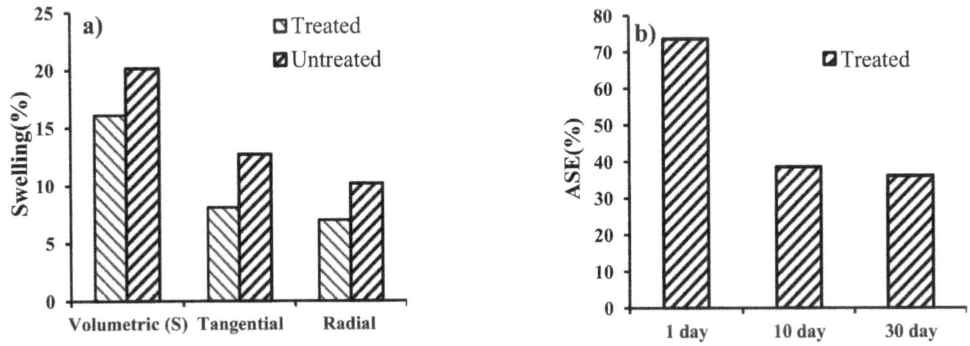

Figure 2. Swelling (**a**) and anti-swelling efficiency (**b**) of treated and untreated birch wood.

3.3. Thermogravimetric Analysis

Thermogravimetric (TG) and differential thermal (DTG) curves of untreated and treated wood specimens were obtained to assess the change in the thermal properties of wood before and after impregnation (Figure 3). A small weight loss (about 2–4 wt.%) can be observed in the area marked with the number "1" of the TG curve, in the temperature range from 30 to 95 °C. It is associated with the endothermic process of dehydration of unbound water from wood without decomposition of the main wood components [37,38]. On the obtained TG curves, no significant dehydration is observed due to preliminary drying of the specimens in the oven. In the area 2, where the temperature ranges from 95 to 230 °C, some wood components (for example, hemicellulose) undergo degradation [39–42], while bond breakage can occur between other structural components. The most intense area of specimen weight loss is area 3, when the temperature changes from 230 to 385 °C. In this area, all structural components of the cell wall underwent thermal destruction with a loss of 87.29% of the total weight for the treated specimens and 89.67% for the control group. In addition, in this area, specimens soaked in used engine oil have two sharp endothermic peaks at 302 and 357 °C. At these temperatures, oxidation and decomposition of engine oil hydrocarbon components occurs. The curves for the control group show a peak at 352 °C. It corresponds to the thermal decomposition of cellulose [43]. In the area marked 4, weight loss of the specimens is practically absent. The residual mass for the specimens (control group) was 4.32%, and it was 10.64% for the impregnated specimen. The difference in residual weight is caused by the presence of heat-resistant components in the impregnated specimens [44].

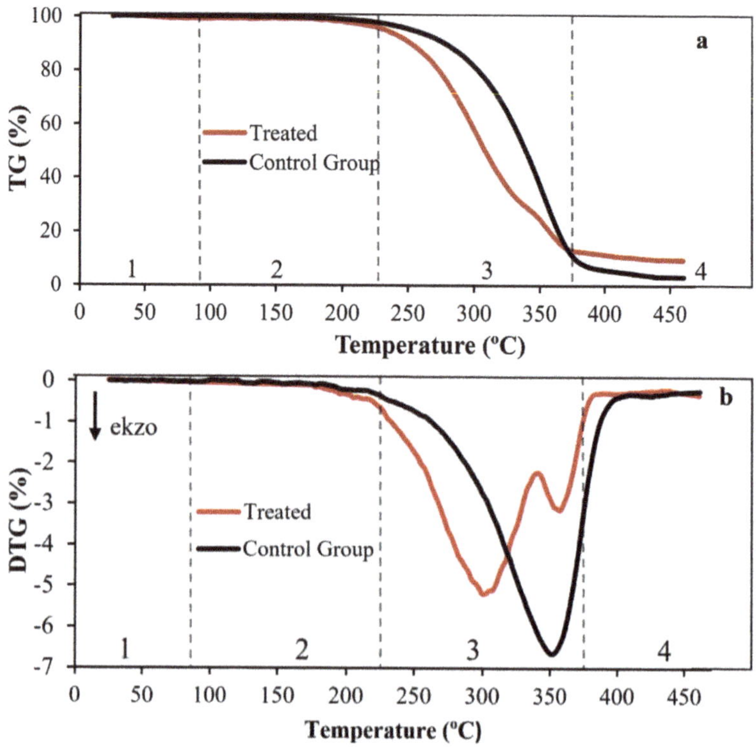

Figure 3. TG (**a**) and DTG (**b**) curves of wood specimens impregnated and not impregnated with spent engine oil.

3.4. Angle of Contact

The degree of hydrophobicity of wood is determined by the possibility of wetting it with water. The main method for determining the surface wettability is the angle of contact between the surface and the liquid [45]. The value of this indicator determines water resistance of wood and its dimensional stability, and therefore directly affects the quality and durability of wood products. Figure 4a–c show the results of determining the contact angle of treated and untreated wood in three mutually perpendicular directions (transverse, radial and tangential one).

The angle of contact for the control specimen during the first 20 s in the transverse direction decreased from 90.60° to 26.43°. This indicates a high degree of surface interaction of wood with further penetration of distilled water through the anatomical structures. The angle of contact was 61.21° for the treated wood in 20 s in the transverse direction. It is more than three times higher than for the control group. After 60 s, the lateral contact angle for treated and untreated wood was 55.30° and 3°, respectively. With a further increase of time when water is on the wood surface, the angle of contact changed insignificantly. It was 52.30° for treated wood and 0° for untreated wood.

In the radial direction (Figure 4b) the angle of contact for the treated wood decreased to 71.21° in the first 20 s and further, in contrast to the contact angle in the transverse direction, decreased less intensively. It was equal to 59.30° after 60 s. The contact angle for untreated wood in 60 s was 62% less relative to impregnated wood and amounted to 36.39°. The change in the angle of contact in the tangential direction is insignificant and comparable to the transverse direction. The contact angle after 90 s is 52.30° for treated wood and 0° for untreated wood.

The given change in the angles of contact enables to conclude that impregnated wood is significantly hydrophobic in three directions.

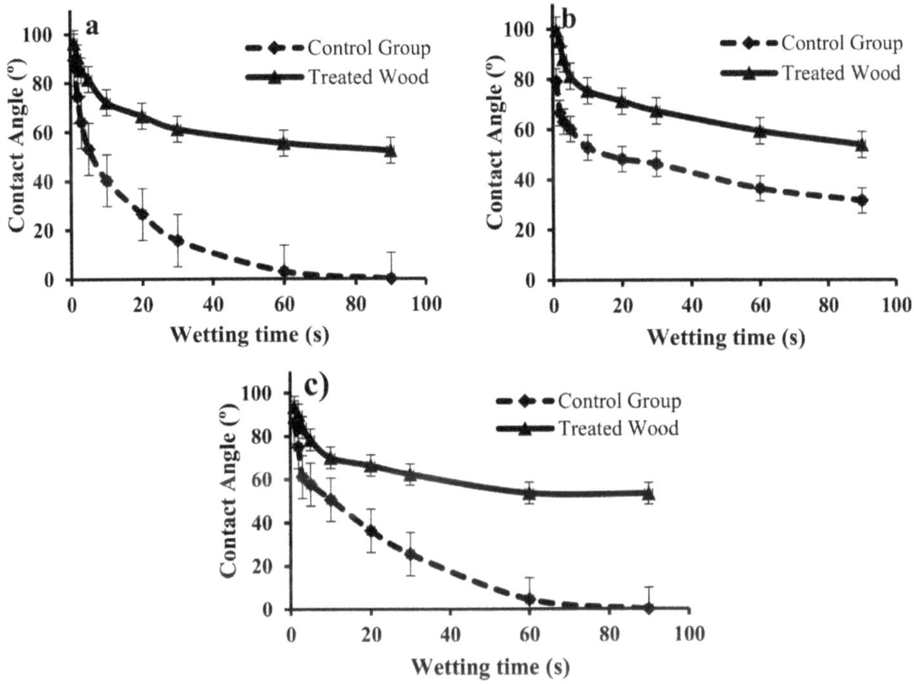

Figure 4. The angle of contact of treated and untreated wood with distilled water in the transverse (**a**), radial (**b**) and tangential (**c**) directions. Vertical lines indicate the standard error of measurements.

3.5. Investigation of Possibility of Intermolecular Interaction Formation (Method of IR-Fourier Spectroscopy) during Wood Impregnation

The process of wood impregnation is accompanied by the formation of intermolecular bonds between the functional groups of wood and the impregnating composition. Hydroxyl groups play an essential role in this process. Dimensional stability and water-resistant characteristics of wood are determined by the presence of a large number of hydroxyl groups (-OH) in it, a decrease in access to which makes it possible to improve its performance [46–49]. The FTIR spectra of spent engine oil, impregnated and untreated wood are shown in Figure 5. There is an absorption band in the region of 3300–3400 cm^{-1} in the obtained spectra. This is a characteristic of the stretching vibration of O–H groups of water in treated and untreated wood [50] and for ethylene glycol contamination of used motor oil during operation in the engine [51,52]. As a result of impregnation of birch wood with used engine oil, the spectra show clear absorption bands in the 2925 cm^{-1} and 2850 cm^{-1} region. These bands characterize the symmetric and asymmetric vibrations of methylene ($-CH_2$) and methyl groups ($-CH_3$) in aliphatic chains [44] in large quantities in used engine oil and preserved in treated wood. Peaks are preserved for untreated wood and used engine oil in the area of 1735 cm^{-1}. They are characteristic of stretching vibrations of the carbon skeleton in wood [53] and vibrations of the carbonyl group in spent engine oil [44] with an increase in the intensity of this band after wood impregnation. The absorption bands at 1380 cm^{-1} and 1460 cm^{-1}, present in used engine oil and appearing in wood after impregnation, represent asymmetric deformation of CH, CH_2 and CH_3 groups. The peak of 1240 cm^{-1}, which is present in wood before and after treatment, corresponds to the stretching vibrations of CO bonds in combination with vibrations of the aromatic ring in lignin, the value of which increases in impregnated wood [39]. In the spectrum of treated and untreated wood, the absorption band at 1030 cm^{-1} is characteristic of symmetric stretching of C–O–C dialkyl ethers, as well as deformation of CH bonds and β–O–4 bonds in lignin [39]. During the operation of the internal combustion engine, engine oils are exposed to high temperatures and pressure, contact with oxygen and various metals, and as a result, oil hydrocarbons undergo oxidation, condensation and decomposition processes [54]. In the composition of spent motor oils, absorption bands were found at a frequency of 720 cm^{-1} and 1150 cm^{-1}, which correspond to the valence vibrations of the peroxide group (-C-O-O-) with increased chemical activity [55]. These groups are formed during the oxidation of hydrocarbons as a result of the operation of car engines. When impregnating birch wood with used engine oil, the absorption bands 720 and 1150 cm^{-1} practically disappear, probably as a result of the interaction between the groups -OH of wood and the peroxide groups of spent engine oil. The absorption band of 3350 cm^{-1} in natural wood corresponds to the oscillation frequency of the O-H groups [39]. This band in the impregnated sample shifts to the lower frequency range (by 50 cm^{-1}) due to the possible participation of these groups in the formation of a hydrogen bond. No other changes in functional groups after treatment with spent engine oil were found according to FTIR data. Thus, the impregnation of wood with spent engine oil (to some extent) changed the chemical groups of the wood, but without altering the wood structure.

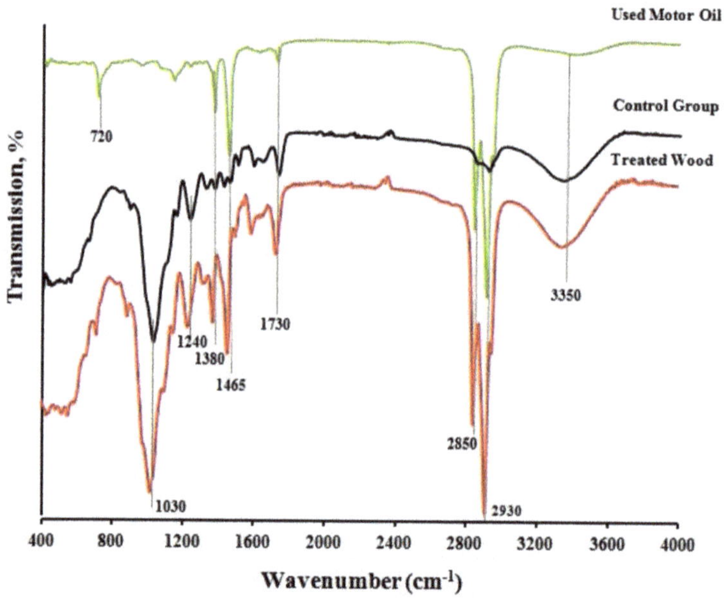

Figure 5. FTIR study of used engine oil, treated and untreated wood.

3.6. Biodurability Test

Figure 6 shows the results of determining the biological resistance of impregnated and untreated birch wood to *Poria placenta*. After 16 weeks, the untreated specimens were completely covered with brown rot fungus mycelium, in contrast to the specimens soaked in spent engine oil, which showed a high resistance to *Poria placenta*. After 10 weeks of incubation in a climatic chamber, the weight loss of untreated specimens was 34.25%, and it was 2.57% for the treated ones. It indicates a high resistance of wood impregnated with spent engine oil to biodegradation. After 16 weeks, the percentage of weight loss for untreated specimens increased by 13.07% and amounted to 47.32%. Treated specimens showed 3.61%

weight loss. Thus, the impregnation of wood with spent engine oil significantly increased the resistance of birch wood to biodegradation.

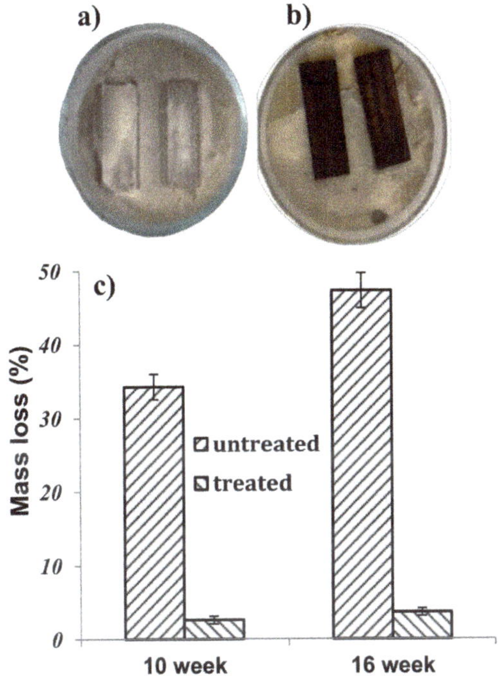

Figure 6. Biostability test against *Poria placenta* of untreated (**a**) and treated wood (**b**) after 16 weeks; (**c**) weight loss of specimens after 10 and 16 weeks of incubation with brown rot.

4. Conclusions

When birch wood was impregnated with spent engine oil, the weight percent gain (WPG) was 58.6%, which indicates a high degree of impregnation and easy penetration of the composition into the wood. Impregnation with spent engine oil significantly reduced WA and MA (by 3.7-fold and 97.6%, respectively). The treatment of wood with spent engine oil reduced swelling in tangential and radial directions by 56.8% and 45.7%, respectively. Therefore, the treatment with this composition improves dimensional stability of the wood. The increase in the hydrophobicity of the wood surface after impregnation is confirmed by the data on the determination of the contact angle before and after impregnation. After impregnation of wood specimens, a significant increase in the contact angle was observed in all three directions (radial, tangential and transverse) relative to untreated wood. This indicates hydrophobization of the wood surface associated with the adsorption coating of the specimen surface by impregnating composition and its diffusion penetration into the volumetric space. After evaluating thermal stability, the specimens treated with spent engine oil, differed from the control group by having two sharp endothermic peaks at 302 °C and 357 °C, while the DTG of the untreated specimens appeared as a regular curve with sharp peaks at 352 °C during TGA. After treating the wood with spent engine oil, FTIR spectra showed the presence of absorption bands in the spent engine oil, although no significant structural changes were observed. In addition, the treated specimens showed a significant increase in biostability against brown rot fungi (Poria placenta) and a weight loss was only 3.61% after 16 weeks of incubation.

Author Contributions: Conceptualization, L.B.; methodology, K.V.Z. and A.P.; validation, L.B. and K.V.Z.; formal analysis, K.V.Z.; resources, L.B., A.P., T.I.; data curation, L.B. and K.V.Z.; writing—original draft preparation, K.V.Z. and L.B.; writing—review and editing, L.B. and K.V.Z.; visualization, K.V.Z.; supervision, L.B., A.P., T.I. and K.V.Z. All authors have read and agreed to the published version of the manuscript.

Funding: This research was funded by FOUNDATION FOR THE PROMOTION OF INNOVATION, grant number 15368GU/2020.

Acknowledgments: Voronezh State University of Forestry and Technologies named after G.F. Morozov (VSUFT) for the opportunity to conduct research.

Conflicts of Interest: The authors declare no conflict of interest.

References

1. Rowell, R.; Pettersen, R.; Tshabalala, M. Cell Wall Chemistry. In *Handbook of Wood Chemistry and Wood Composites*, 2nd ed.; Taylor & Francis Group: Tokyo, Japan, 2012.
2. Popescu, C.-M.; Popescu, M.-C. A near Infrared Spectroscopic Study of the Structural Modifications of Lime (Tilia Cordata Mill.) Wood during Hydro-Thermal Treatment. *Spectrochim. Acta Part A Mol. Biomol. Spectrosc.* **2013**, *115*, 227–233. [CrossRef] [PubMed]
3. Korkut, D.S.; Hiziroglu, S. Experimental Test of Heat Treatment Effect on Physical Properties of Red Oak (Quercus Falcate Michx.) and Southern Pine (*Pinus Taeda* L.). *Materials* **2014**, *7*, 7314–7323. [CrossRef]
4. Okon, K.E.; Lin, F.; Chen, Y.; Huang, B. Effect of Silicone Oil Heat Treatment on the Chemical Composition, Cellulose Crystalline Structure and Contact Angle of Chinese Parasol Wood. *Carbohydr. Polym.* **2017**, *164*, 179–185. [CrossRef] [PubMed]
5. Li, W.; Wang, H.; Ren, D.; Yu, Y.; Yu, Y. Wood Modification with Furfuryl Alcohol Catalysed by a New Composite Acidic Catalyst. *Wood Sci. Technol.* **2015**, *49*, 845–856. [CrossRef]
6. Kasemsiri, P.; Hiziroglu, S.; Rimdusit, S. Characterization of Heat Treated Eastern Redcedar (*Juniperus Virginiana* L.). *J. Mater. Process. Technol.* **2012**, *212*, 1324–1330. [CrossRef]
7. Cao, Y.; Lu, J.; Huang, R.; Jiang, J. Increased Dimensional Stability of Chinese Fir through Steam-Heat Treatment. *Eur. J. Wood Wood Prod.* **2012**, *70*, 441–444. [CrossRef]
8. Chung, M.J.; Wang, S.Y. Effects of Peeling and Steam-Heating Treatment on Mechanical Properties and Dimensional Stability of Oriented Phyllostachys Makinoi and Phyllostachys Pubescens Scrimber Boards. *J. Wood Sci.* **2018**, *64*, 625–634. [CrossRef]
9. Hill, C.; Altgen, M.; Rautkari, L. Thermal Modification of Wood—A Review: Chemical Changes and Hygroscopicity. *J. Mater. Sci.* **2021**, *56*, 1–34. [CrossRef]
10. Broda, M.; Mazela, B. Application of Methyltrimethoxysilane to Increase Dimensional Stability of Waterlogged Wood. *J. Cult. Herit.* **2017**, *25*, 149–156. [CrossRef]
11. Reinprecht, L.; Vacek, V.; Grznárik, T. Enhanced Fungal Resistance of Scots Pine (*Pinus Sylvestris* L.) Sapwood by Treatment with Methyltrimethoxysilane and Benzalkoniumchloride. *Eur. J. Wood Wood Prod.* **2017**, *75*, 17–31. [CrossRef]
12. Nguyen, T.T.; Xiao, Z.; Che, W.; Trinh, H.M.; Xie, Y. Effects of Modification with a Combination of Styrene-Acrylic Copolymer Dispersion and Sodium Silicate on the Mechanical Properties of Wood. *J. Wood Sci.* **2019**, *65*, 1–11. [CrossRef]
13. Lukowsky, D. Influence of the Formaldehyde Content of Waterbased Melamine Formaldehyde Resins on Physical Properties of Scots Pine Impregnated Therewith. *Holz Als Roh-Und Werkst.* **2002**, *60*, 349–355. [CrossRef]
14. Kajita, H.; Furuno, T.; Imamura, Y. The Modification of Wood by Treatment with Low Molecular Weight Phenol-Formaldehyde Resin: A Properties Enhancement with Neutralized Phenolic-Resin and Resin Penetration into Wood Cell Walls. *Wood Sci. Technol.* **2004**, *37*, 349–361. [CrossRef]
15. Ayrilmis, N. Combined Effects of Boron and Compatibilizer on Dimensional Stability and Mechanical Properties of Wood/HDPE Composites. *Compos. Part B Eng.* **2013**, *44*, 745–749. [CrossRef]
16. Adhikary, K.B.; Pang, S.; Staiger, M.P. Dimensional Stability and Mechanical Behaviour of Wood–Plastic Composites Based on Recycled and Virgin High-Density Polyethylene (HDPE). *Compos. Part B Eng.* **2008**, *39*, 807–815. [CrossRef]
17. Giudice, C.A.; Alfieri, P.V.; Canosa, G. Decay Resistance and Dimensional Stability of Araucaria Angustifolia Using Siloxanes Synthesized by Sol–Gel Process. *Int. Biodeterior. Biodegrad.* **2013**, *83*, 166–170. [CrossRef]
18. Pfeffer, A.; Mai, C.; Militz, H. Weathering Characteristics of Wood Treated with Water Glass, Siloxane or DMDHEU. *Eur. J. Wood Wood Prod.* **2012**, *70*, 165–176. [CrossRef]
19. Candan, Z.; Korkut, S.; Unsal, O. Effect of Thermal Modification by Hot Pressing on Performance Properties of Paulownia Wood Boards. *Ind. Crop. Prod.* **2013**, *45*, 461–464. [CrossRef]
20. Kocaefe, D.; Huang, X.; Kocaefe, Y. Dimensional Stabilization of Wood. *Curr. For. Rep.* **2015**, *1*, 151–161. [CrossRef]
21. Lesar, B.; Straže, A.; Humar, M. Sorption Properties of Wood Impregnated with Aqueous Solution of Boric Acid and Montan Wax Emulsion. *J. Appl. Polym. Sci.* **2011**, *120*, 1337–1345. [CrossRef]
22. Varganici, C.-D.; Rosu, L.; Rosu, D.; Mustata, F.; Rusu, T. Sustainable Wood Coatings Made of Epoxidized Vegetable Oils for Ultraviolet Protection. *Environ. Chem. Lett.* **2021**, *19*, 307–328. [CrossRef]

23. Ahmed, S.A.; Morén, T.; Sehlstedt-Persson, M.; Blom, Å. Effect of Oil Impregnation on Water Repellency, Dimensional Stability and Mold Susceptibility of Thermally Modified European Aspen and Downy Birch Wood. *J. Wood Sci.* **2017**, *63*, 74–82. [CrossRef]
24. Dündar, T.; Büyüksarı, Ü.; Avcı, E.; Akkılıç, H. Effect of heat treament on the physical and mechanical properties of compression and opposite wood of black pine. *BioResources* **2012**, *7*, 5009–5018. [CrossRef]
25. Esteves, B.; Videira, R.; Pereira, H. Chemistry and Ecotoxicity of Heat-Treated Pine Wood Extractives. *Wood Sci. Technol.* **2011**, *45*, 661–676. [CrossRef]
26. Johnson, M.G.; Luxton, T.P.; Rygiewicz, P.T.; Reichman, J.R.; Bollman, M.A.; King, G.A.; Storm, M.J.; Nash, M.S.; Andersen, C.P. Transformation and Release of Micronized Cu Used as a Wood Preservative in Treated Wood in Wetland Soil. *Environ. Pollut.* **2021**, *287*, 117189. [CrossRef] [PubMed]
27. Hebisch, R.; Karmann, J.; Schäferhenrich, A.; Göen, T.; Berger, M.; Poppek, U.; Roitzsch, M. Inhalation and Dermal Exposure of Workers during Timber Impregnation with Creosote and Subsequent Processing of Impregnated Wood. *Environ. Res.* **2020**, *181*, 108877. [CrossRef] [PubMed]
28. Dolmatov, L.V.; Akhmetov, A.F.; Kutukov, I.E. Aggregative Stability and Thermal Stability of the Petroleum-Based Biocide ZhTK. *Chem. Technol. Fuels Oils* **2001**, *37*, 274–277. [CrossRef]
29. Unnisa, S.A.; Hassanpour, M. Development Circumstances of Four Recycling Industries (Used Motor Oil, Acidic Sludge, Plastic Wastes and Blown Bitumen) in the World. *Renew. Sustain. Energy Rev.* **2017**, *72*, 605–624. [CrossRef]
30. Animpong, M.A.B.; Oduro, W.O.; Koranteng, J.; Ampomah-Benefo, K.; Boafo-Mensah, G.; Akufo-Kumi, K.; Tottimeh, G.O.; Amoah, J.Y. Coupling Effect of Waste Automotive Engine Oil in the Preparation of Wood Reinforced LDPE Plastic Composites for Panels. *S. Afr. J. Chem. Eng.* **2017**, *24*, 55–61. [CrossRef]
31. Belchinskaya, L.; Zhuzhukin, K.; Dmitrenkov, A.; Roessner, F. Studying and Imparting Moisture Absorption Qualities of the New Wood Based Bio-Composite Material. In *IOP Conference Series: Earth and Environmental Science*; IOP Publishing: Bristol, UK, 2020; Volume 595. [CrossRef]
32. Zhao, R.J.; Fei, B.H.; Lv, J.X.; Yu, H.Q.; Huang, R.F.; Zhao, Y.Q.; Huang, A.M.; Cui, Y.Z. Wood Moisture Content Measuring Method. Wood—Determination of Moisture Content for Physical and Mechanical Test. MOD CN-GB GB/T. 2009. Available online: https://wenku.baidu.com/view/df6e7b14d35abe23482fb4daa58da0116d171feb.html (accessed on 9 December 2021).
33. Cheremisinoff, N.P.; Rosenfeld, P.E. Wood-Preserving Technology. In *Handbook of Pollution Prevention and Cleaner Production*; Elsevier: Amsterdam, The Netherlands, 2010; pp. 27–41.
34. Betts, W.D. The Properties and Performance of Coal-Tar Creosote as a Wood Preservative. In *The Chemistry of Wood Preservation*; Elsevier: Amsterdam, The Netherlands, 2005; pp. 136–160.
35. Baar, J.; Brabec, M.; Slávik, R.; Čermák, P. Effect of Hemp Oil Impregnation and Thermal Modification on European Beech Wood Properties. *Eur. J. Wood Wood Prod.* **2021**, *79*, 161–175. [CrossRef]
36. Lesar, B.; Humar, M. Use of Wax Emulsions for Improvement of Wood Durability and Sorption Properties. *Eur. J. Wood Wood Prod.* **2011**, *69*, 231–238. [CrossRef]
37. Chen, W.; Yu, H.; Liu, Y.; Chen, P.; Zhang, M.; Hai, Y. Individualization of Cellulose Nanofibers from Wood Using High-Intensity Ultrasonication Combined with Chemical Pretreatments. *Carbohydr. Polym.* **2011**, *83*, 1804–1811. [CrossRef]
38. Lin, Y.-Y.; Chen, W.-H.; Colin, B.; Pétrissans, A.; Quirino, R.L.; Pétrissans, M. Thermodegradation Characterization of Hardwoods and Softwoods in Torrefaction and Transition Zone between Torrefaction and Pyrolysis. *Fuel* **2022**, *310*, 122281. [CrossRef]
39. Lin, B.-J.; Colin, B.; Chen, W.-H.; Pétrissans, A.; Rousset, P.; Pétrissans, M. Thermal Degradation and Compositional Changes of Wood Treated in a Semi-Industrial Scale Reactor in Vacuum. *J. Anal. Appl. Pyrolysis* **2018**, *130*, 8–18. [CrossRef]
40. Candelier, K.; Chaouch, M.; Dumarçay, S.; Pétrissans, A.; Pétrissans, M.; Gérardin, P. Utilization of Thermodesorption Coupled to GC–MS to Study Stability of Different Wood Species to Thermodegradation. *J. Anal. Appl. Pyrolysis* **2011**, *92*, 376–383. [CrossRef]
41. Mastouri, A.; Efhamisisi, D.; Shirmohammadli, Y.; Oladi, R. Physicochemical Properties of Thermally Treated Poplar Wood in Silicone and Rapeseed Oils: A Comparative Study. *J. Build. Eng.* **2021**, *43*, 102511. [CrossRef]
42. Pétrissans, A.; Younsi, R.; Chaouch, M.; Gérardin, P.; Pétrissans, M. Experimental and Numerical Analysis of Wood Thermodegradation. *J. Therm. Anal. Calorim.* **2012**, *109*, 907–914. [CrossRef]
43. Giuntoli, J.; de Jong, W.; Arvelakis, S.; Spliethoff, H.; Verkooijen, A.H.M. Quantitative and Kinetic TG-FTIR Study of Biomass Residue Pyrolysis: Dry Distiller's Grains with Solubles (DDGS) and Chicken Manure. *J. Anal. Appl. Pyrolysis* **2009**, *85*, 301–312. [CrossRef]
44. Mishra, A.; Kumari, U.; Turlapati, V.Y.; Siddiqi, H.; Meikap, B.C. Extensive Thermogravimetric and Thermo-Kinetic Study of Waste Motor Oil Based on Iso-Conversional Methods. *Energy Convers. Manag.* **2020**, *221*, 113194. [CrossRef]
45. Rossi, D.; Rossi, S.; Morin, H.; Bettero, A. Within-Tree Variations in the Surface Free Energy of Wood Assessed by Contact Angle Analysis. *Wood Sci. Technol.* **2012**, *46*, 287–298. [CrossRef]
46. Mitani, A.; Barboutis, I. Changes Caused by Heat Treatment in Color and Dimensional Stability of Beech (*Fagus Sylvatica* L.) Wood. *Drv. Ind.* **2014**, *65*, 225–232. [CrossRef]
47. Engelund, E.T.; Thygesen, L.G.; Svensson, S.; Hill, C.A.S. A Critical Discussion of the Physics of Wood–Water Interactions. *Wood Sci. Technol.* **2013**, *47*, 141–161. [CrossRef]
48. Altgen, M.; Willems, W.; Hosseinpourpia, R.; Rautkari, L. Hydroxyl Accessibility and Dimensional Changes of Scots Pine Sapwood Affected by Alterations in the Cell Wall Ultrastructure during Heat-Treatment. *Polym. Degrad. Stab.* **2018**, *152*, 244–252. [CrossRef]

49. Li, J.-Z.; Furuno, T.; Katoh, S. Dimensional Stability and Flame Resistance of Silicate-Acetylated and -Propionylated Wood Composites. *J. Wood Chem. Technol.* **2000**, *20*, 441–453. [CrossRef]
50. Wu, Z.; Deng, X.; Li, L.; Xi, X.; Tian, M.; Yu, L.; Zhang, B. Effects of Heat Treatment on Interfacial Properties of Pinus Massoniana Wood. *Coatings* **2021**, *11*, 543. [CrossRef]
51. Jia, X.; Huang, B.; Bowers, B.F.; Zhao, S. Infrared Spectra and Rheological Properties of Asphalt Cement Containing Waste Engine Oil Residues. *Constr. Build. Mater.* **2014**, *50*, 683–691. [CrossRef]
52. Liu, S.; Peng, A.; Wu, J.; Zhou, S.B. Waste Engine Oil Influences on Chemical and Rheological Properties of Different Asphalt Binders. *Constr. Build. Mater.* **2018**, *191*, 1210–1220. [CrossRef]
53. Esteves, B.; Velez Marques, A.; Domingos, I.; Pereira, H. Chemical Changes of Heat Treated Pine and Eucalypt Wood Monitored by FTIR. *Maderas. Cienc. y Tecnol.* **2013**, *15*, 245–258. [CrossRef]
54. Wei, L.; Duan, H.; Jin, Y.; Jia, D.; Cheng, B.; Liu, J.; Li, J. Motor Oil Degradation during Urban Cycle Road Tests. *Friction* **2021**, *9*, 1002–1011. [CrossRef]
55. Yang, Z.; Liao, Y.; Ren, H.; Hao, X.; Song, X.; Liu, Z. A Novel Co-Treatment Scheme for Waste Motor Oil and Low Rank Coal Slime: Waste Dispose Waste. *Fuel* **2021**, *292*, 120275. [CrossRef]

forests

Article

The Impact of Sanding and Thermal Compression of Wood, Varnish Type and Artificial Aging in Indoor Conditions on the Varnished Surface Color

Pavlo Bekhta [1,*], Tomasz Krystofiak [2], Barbara Lis [2] and Nataliya Bekhta [3]

1 Department of Wood-Based Composites, Cellulose and Paper, Ukrainian National Forestry University, 79057 Lviv, Ukraine
2 Department of Wood Science and Thermal Technics, Poznań University of Life Sciences, 60-627 Poznań, Poland; tomasz.krystofiak@up.poznan.pl (T.K.); barbara.lis@up.poznan.pl (B.L.)
3 Department of Design, Ukrainian National Forestry University, 79057 Lviv, Ukraine; n.bekhta@nltu.edu.ua
* Correspondence: bekhta@nltu.edu.ua

Abstract: The aim of this pioneering study was to compare and evaluate two different pre-treatment processes of wooden surfaces prior to varnishing by sanding or thermal compression in terms of the impact on the color characteristics of the varnished surface. In the experiment, a wood veneer of black alder and birch before varnishing was subjected to sanding with a sandpaper of 180 grit size, or thermal compression at temperatures of 180 and 210 °C. Three different kinds of commercially manufactured transparent varnish (water-based (WB) varnish, polyurethane (PUR) varnish and UV-cured (UV) varnish) were applied to the prepared veneer surfaces. The samples prepared in this way were also subjected to an artificial aging process in indoor conditions by ultraviolet light and infrared irradiation (UVL + IR). The colors of the surfaces in the CIE $L^*a^*b^*$ system were measured and color differences (ΔL^*, Δa^*, Δb^* and ΔE) were determined for sanded and thermally densified, unvarnished and varnished, as well as subjected to accelerated aging surfaces. It was found that thermally densified surface-varnished veneer is more resistant to color changes during artificial aging compared to when sanded and surface-varnished. The transparent varnish systems showed better photo-stability, when thermally densified wood veneer was used as substrate than that of sanded wood. The WB varnish showed the greatest resistance to discoloration during UVL + IR irradiation, followed by PUR and UV. The preliminary findings obtained in this study indicated that replacing the sanding process with thermal compression of wood surface before varnishing could be considered as an alternative method of producing varnished panels with satisfactory color properties of surface.

Keywords: black alder; birch; color; varnish system; surface pre-treatment; sanding; thermal compression; artificial aging

Citation: Bekhta, P.; Krystofiak, T.; Lis, B.; Bekhta, N. The Impact of Sanding and Thermal Compression of Wood, Varnish Type and Artificial Aging in Indoor Conditions on the Varnished Surface Color. *Forests* **2022**, *13*, 300. https://doi.org/10.3390/f13020300

Academic Editor: Luigi Todaro

Received: 26 January 2022
Accepted: 11 February 2022
Published: 12 February 2022

1. Introduction

Wood-based materials are very popular and widespread materials that have been used in diverse fields, including furniture production, for interior decoration materials, for interior panels in cars and many others. Sliced decorative veneer with a thickness of 0.2–0.9 mm [1] produced from naturally beautiful precious woods is usually used to overlay on the wood panels' surface to improve their appearance and color. Common precious wood species are oak, maple, ash, walnut, cherry, mahogany and others. However, due to the low thickness, the decorative veneer easily cracks during storage, transportation and use [1]. In addition, part of the precious wood is irretrievably lost in the sanding process of veneered wood panels before their varnishing [2]. However, the resources of valuable wood species, which can be used as a material for veneering, are sharply depleted and have become more expensive. Less valuable wood species are characterized by various defects, mechanical damage and less decorative appearance than valuable wood species. This necessitates the improvement of the aesthetic properties of such wood veneer.

The preparation of the surface of the wood substrate is one of the important processes before finishing and has a significant effect on the surface properties of the coating (color, gloss, and roughness) as well as on the adhesion of the coating to the substrate. Usually, this process can be performed by helical planing, face milling and sanding, etc. [3]. Sanding is the most common pre-treatment process of a wooden surface before finishing. Sanding makes the surface more homogeneous and reduces the influence of wood anatomy on the roughness profile [3,4]. In addition, a smooth surface is a prerequisite for a quality interaction between the coating and the substrate and is also necessary for a good appearance [5]. During the sanding process a layer of wood, usually of a valuable species, is removed and this portion of wood irreversibly goes to waste, namely, generates economic losses. Moreover, sander dust creates unfavorable conditions for workers and pollutes the environment. However, sanding is one of the most skilled, time-consuming and expensive operations in the woodworking industry [6].

On the other hand, in our previous studies [7–12] it was found that the combination of heat treatment and compression enhanced the overall surface quality of the samples making it denser, smooth and homogeneous. After such treatment, sanding of the densified wood surface before finishing is no longer required [2,9]. This process is called thermomechanical densification and is the simplest and most ecological method of wood modification, which does not use chemicals [13]. Thermal compression differs from the sanding process in that the wood is compressed between heated press plates or rolled between heated or cold polishing drums to smooth the wood surface [13,14]. In addition, the thermomechanical densification makes it possible to provide new useful characteristics of wood species with low quality and technical characteristics [7–12]. In particular, after such treatment the wood species with lower decorative characteristics are made similar in color to exotic wood [7]. In addition, transparent coatings allow to keep the wood to retain its good aesthetic properties. Together, this can increase the demand for such wood/wood-based panels, as well as to determine their value and hence price.

The results obtained in our previous studies indicate that the replacement of the sanding by thermal compression pre-treatment of wooden substrate can be considered as an alternative way to prepare the wood surface before finishing in the production of veneered panels with satisfactory aesthetic properties and reduce the cost of varnish by ≈50% [2,15]. Therefore, it was proposed to replace the time-consuming sanding process of wooden substrate with the thermal compression process [16]. However, little or no information is available on the possibility of using thermally compressed wood as a substrate for finishing [2,15,17,18]. More information can be found in these studies on the finishing of heat-treated wood [19–23].

One of the main parameters that is crucial for the final decision of the consumer are the aesthetic and decorative properties of wood. Wood color belongs among the most decisive purchasing criteria [24]. Moreover, wood color is a very important characteristic of wood, which often determines its final price, especially when it will be used for furniture production [25]. The color is determined by the chemical components of the wood, environmental conditions (humidity, solar radiation, pollution and wind), as well as the area where the trees were grown [26]. CIE $L^*a^*b^*$ is the most commonly used color space for measuring the color of wood surface [27].

It should be noted that the literature analyzes the effect of thermal compression on wood color [7], on the anatomical structure [11,28], physical and mechanical properties of densified wood [7–12], on the bondability [29–31] and coatability [2,12,15] of wood. Wood, like most natural and synthetic polymers, is sensitive to abiotic and biotic factors that cause various changes in its properties, including changes in its decorative characteristics [32,33]. In addition to a number of environmental parameters (various types of radiation, including UVL radiation, moisture, temperature, etc.) that significantly contribute to wood degradation, the rate of degradation may also be affected by wood properties [34] and the wood species (softwood or hardwood) [35]. Kubovský and Kačík [36] note that the changes associated with the aging process of wood are color changes due to photodegradation of lignin, extractives and to some extent, hemicellulose. At the same time, there is no comprehensive data on the influence of abiotic factors, ie UVL radiation, on the color and color stability of densified wood, in particular wood used as flooring or furniture materials. One of the first works on this issue is the work of Laskowska [37], which examined the influence of ultraviolet radiation on the color of thermo-mechanically modified beech and oak wood. The author found that the greatest changes in the color of non-densified and densified beech and oak wood occurred after 20 h of light irradiation.

Wood varnishes add beauty and provide powerful protection of wooden surfaces against the adverse effects of external factors, namely: heat, light, temperature, moisture, wind, abrasion, etc [5,17,22,23,38–40]. However, varnish materials applied to wood, as well as wood itself, due to external factors are also subject to aging, which leads to loss of their useful potential [41–45]. This is manifested primarily in the change of aesthetic and decorative characteristics of coatings, which affects the aesthetic appeal in the interior. The UVL-induced photo-degradation of wood products surface-treated with varnishes takes place also in indoor conditions [43,46,47]. The measured and calculated color change values by Kúdela and Kubovský [43] demonstrated that the most color changes are generated during the first 100 h of accelerated ageing process.

However, there is no information on the effect of artificial UVL radiation on the color of alder and birch wood, pre-treated with sanding or thermomechanical compression, as well as varnished with various varnish systems. Thus, the main aim of this study was to compare and evaluate two different pre-treatment processes of wooden surfaces prior to varnishing by sanding or thermal compression in terms of the impact on the color characteristics of the varnished surface. Sanding was used as a conventional surface preparation process for the comparison.

The objectives of the current experiments are as follows: first, to determine the color change in the CIE $L^*a^*b^*$ color space of selected wood species, namely birch and black alder, under the influence of thermal compression at 180 and 210 °C; secondly, to assess the influence of sanding on the color of the wood species; thirdly, to compare the color changes of the wood species after sanding and thermal compression; fourthly, to determine the color stability of the wood species after sanding and thermal compression, using artificial aging; fifthly, to assess the color changes of the wood species varnished by different varnish systems on a sanded and thermally densified surface; and sixthly, to evaluate the effect of artificial aging on the color change of the varnished wood surface on the sanded and thermally densified surface. The obtained information will make it possible to determine whether it is expedient to replace sanding with thermal compression in terms of the aesthetic properties of the varnished surface.

2. Materials and Methods

2.1. Materials

Rotary-peeled black alder (*Alnus glutinosa* Gaertn.) and birch (*Betula verrucosa* Ehrh.) wood veneers with the nominal thickness of 1.5 mm and moisture content of 6 ± 2% were used as cladding materials. The alder and birch peeled veneers are the most produced and most commonly used for plywood production in Ukraine. Commercially manufactured (SWISS KRONO Sp. z o.o., Żary, Poland) medium density fibreboard (MDF), with thickness 16 mm and density 750 ± 10 kg/m^3 was used as the substrate material. Single-component waterborne Jowacoll® 148.00 adhesive supplied by Jowat SE Corporation, Detmold based on EVA copolymer with density 1.35 g/cm^3, apparent viscosity of 13,000 mPa·s (Brookfield), solids content 70%, and pH value 7.0 were used in the veneering process of MDF panels. Waterborne lacquer (IQ-HY1330-15), polyurethane lacquer (R533-2-15 + 10% RLH6110 (hardener) + 10% V721 (solvent)) and UV-hardened lacquer (UV120-45) were used for varnishing. The parameters of the varnish products are presented in Table 1.

Table 1. Parameters of the varnish products.

Varnish Type	Density (g/cm^3)	Solid Content (%)	Viscosity (s)	pH
Water-borne (WB)	1.03 ± 0.15	29 ± 3	55–75	7.0–9.0
Polyurethane (PUR)	0.88 ± 0.05	17 ± 2	15–20	-
UV acrylic (UV)	1.29 ± 0.15	95 ± 3	5–70	-

2.2. Pre-Treatment Process of Wood Veneer

Two different pre-treatment processes of wood veneer before varnishing were used, sanding and thermal compression. Half of the veneer samples were thermally compressed between the smooth and carefully cleaned heated plates of a laboratory press at temperatures of 180 and 210 °C. After thermal compression, the MDF panels were veneered with densified veneer and were used as wooden substrate before varnishing.

For the other batch of veneer samples, the conventional sanding process using sandpaper of 180 grit size was performed for the comparison. Before that, the MDF panels were veneered with non-sanded veneers and were used as wooden substrate before varnishing.

2.3. Surface Varnishing Process

After thermal compression or sanding process of the substrate surface, a three various varnishes system were applied to the surface of the samples at various layers (1, 3, 2NS - two layers of varnish without intermediate sanding and 2S - two layers of varnish with intermediate sanding). The varnish with three layers was applied, in accordance with the lacquer producer's recommendations, only on the sanded surface for comparison.

The lacquer product was applied to every surface of the test samples using a roller system or by spraying. Finished panels were finally conditioned at a temperature of 23 ± 2 °C and a relative humidity of 50 ± 5% in time 168 h. After the varnish application and technological exposure, the varnished surfaces were subjected to accelerated aging (simultaneous action of UVL and IR irradiation) in indoors' conditions.

2.4. Artificial Aging of the Samples

The sanded and thermally compressed samples before and after varnishing, positioned at an angle of 45° were exposed to intensive light and infrared radiation (UVL + IR). The artificial aging test was carried out with a special quartz lamp (VT-800, FAMED Lodz S.A., Lodz, Poland) having radiation of 740 W. The radiation was applied from a distance of 40 cm to the unvarnished samples 1, 2, 4, 6 and 8 h and to the varnished samples 15, 30, 45 and 60 min exposure time. The temperature of 65 °C at the surface of the samples was determined with the help of a temperature detector (DT 8662 Dual Infrared Thermometer, CEM, Shenzhen, China).

2.5. Color Measurement

The surface color of natural alder and birch veneer was determined, after thermal compression and sanding, after varnishing with various varnish systems, as well as the color of sanded, thermally densified and varnished surfaces after UVL and IR irradiation (Figure 1). The color measurements of all specimens were recorded on the surface of veneer specimens before and after relevant treatment with a colorimeter Testan DT-145 (Anticorr, Gdansk, Poland). The sensor head was 6 mm in diameter. Measurements were made using a D65 illuminant and 10-degree standard observer. Color measurements were performed always on the same marked locations on the sample.

Percentage of reflectance, collected at 10 nm intervals over the visible spectrum (from 400 to 700 nm) was converted into the CIE $L^*a^*b^*$ color system, where L^* describes the lightness, a^* and b^* describe the chromatic coordinates on the green-red and blue-yellow axis, respectively. From the $L^*a^*b^*$ values, the color uniformity were calculated as a difference in the lightness (ΔL^*) and chromaticity parameters (Δa^* and Δb^*) between treated and non-treated veneer samples using the following formulae:

$$\Delta L^* = \Delta L_t^* - \Delta L_n^* \quad (1)$$

$$\Delta a^* = \Delta a_t^* - \Delta a_n^* \quad (2)$$

$$\Delta b^* = \Delta b_t^* - \Delta b_n^* \quad (3)$$

where "t" refers to treated and "n" to non-treated samples.

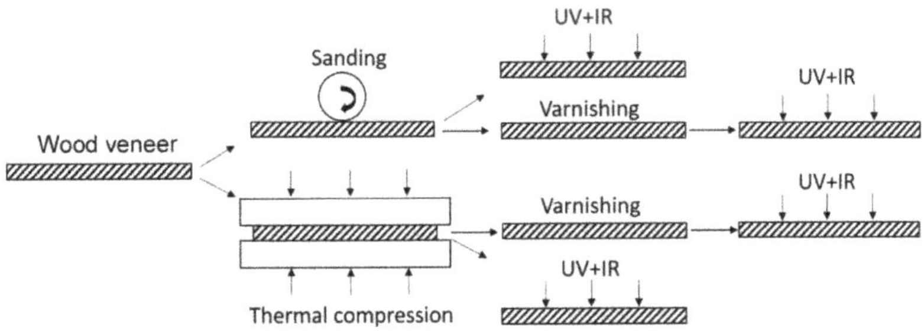

Figure 1. The diagram of sequence of samples treatment.

In addition, total color difference (ΔE) were calculated as:

$$\Delta E = \sqrt{(\Delta L^{*2} + \Delta a^{*2} + \Delta b^{*2})} \quad (4)$$

For each sample, ten random measurements of surface color were taken.

Positive values of ΔL^* show whitening, and negative values of ΔL^* indicate a shift toward gray. Positive values of Δa^* indicate reddening, and negative values of Δa^* show a shift towards green. Positive values of Δb^* represent yellowing, and negative values of Δb^* represent a shift towards blue.

The classification of the overall color change was carried out based on assessment guidelines by Cividini et al. [48] (Table 2).

Table 2. The Criteria to Assess the Color Change.

Variability Range	Color Difference
$\Delta E < 0.2$	Invisible changes
$0.2 < \Delta E < 2$	Smart changes
$2 < \Delta E < 3$	Color changes visible by high quality filter
$3 < \Delta E < 6$	Color changes visible by medium quality filter
$6 < \Delta E < 12$	Distinct color changes
$\Delta E > 12$	A different color

2.6. Statistical Analysis

Statistical analysis was conducted using SPSS software program version 22 (IBM Corp., Armonk, NY, USA). Analysis of variance (ANOVA) was performed on the data to evaluate the effect of the surface pre-treatment, type of varnish, number of varnish layers and duration of artificial aging on the color changes of the samples. The effects were considered not to be statistically significant when the p-value was higher than 0.05 at the 95% confidence level. Duncan's multiple range test was used to determine the significant difference between and among the groups.

3. Results

3.1. Influence of Pre-Treatment Process on Color Changes of Wood Surface

The pre-treatment processes of the wood substrate surface before varnishing, such as sanding and thermal compression, on the color changes were investigated and compared. In addition, the effect of artificial aging on the color changes of the sanded and thermally compressed surfaces was also investigated.

Figure 2 shows the comparison of color changes in alder and birch surfaces of native (non-treated), sanded and thermally compressed veneer. The surface color of thermally compressed veneer samples varied appreciably from lighter to darker with increasing temperature. However, it is difficult to analyse the color in samples based only on their visual observation. Detailed analysis of the color changes was performed using CIE $L^*a^*b^*$ systems and is presented below. Initial colorimetric parameters of wood veneers native surfaces, as well as surfaces after sanding and thermal compression process are given in the Table 3. The alder wood was less resistant to a temperature of 210 °C than birch wood. The L^* values for the alder and birch wood densified at a temperature of 210 °C were, respectively, 22.5% and 13.3% lower than the L^* values for non-densified alder and birch wood. According to the previous studies [7,27,49,50], the darkening of wood as a result of thermo-mechanical treatment may be caused by the degradation of hemicelluloses. Hydrolysis of the hemicelluloses produces changes in parameter L^*, while reduction of lignin produces changes mainly in parameter a^* [51].

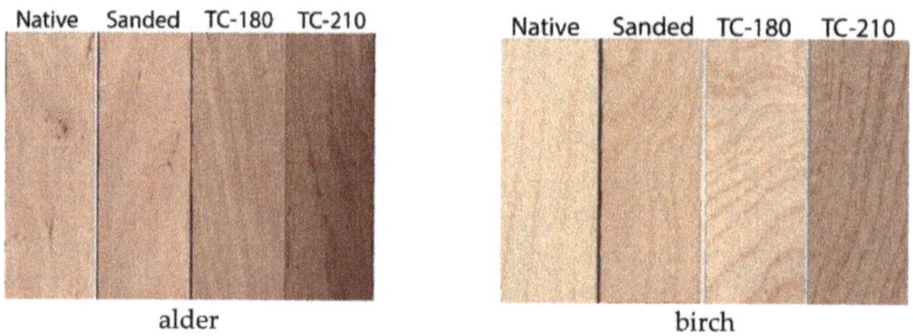

Figure 2. Visual appearance of the surface of alder and birch veneers, native, sanded and thermally compressed at different temperatures.

Table 3. Initial colorimetric parameters of wood veneers.

Wood Species	Wood Treatment	Lightness L^*	Redness a^*	Yellowness b^*	Hue Angle h_{ab}	Chroma C^*_{ab}
Alder	Native	73.8 (±1.5)	11.0 (±0.8)	23.0 (±1.7)	72.8 (±2.3)	24.1 (±1.6)
	Sanded	75.8 (±2.1)	11.2 (±0.3)	21.8 (±0.6)	62.5 (±0.6)	24.5 (±0.6)
	TC-180	67 (±2)	12.6 (±0.9)	22.1 (±2.3)	79 (±3)	22.6 (±2.2)
	TC-210	57 (±4)	13.7 (±1.7)	19.4 (±1.9)	71 (±6)	20.6 (±1.5)
Birch	Native	79.3 (±1.2)	9.0 (±0.6)	23.9 (±1.2)	80.8 (±1.1)	24.2 (±1.2)
	Sanded	79 (±2)	10.2 (±1.2)	23 (±3)	65.9 (±1.7)	24.1 (±2.9)
	TC-180	75.1 (±1.2)	12.6 (±1.5)	24.8 (±0.9)	84v (±3)	25 (±1)
	TC-210	68.7 (±2.6)	13.7 (±1.4)	23.9 (±0.8)	79.1 (±2.9)	24.4 (±0.9)

Changes in the color of the veneer surface after sanding and thermal compression processes are shown in Figure 3. As the compression temperature increases, the color becomes darker (L^* decreases). Compared with the native alder veneer surface the ΔL changed to −6.0 and −16.6 units shifting toward gray with increasing the compression temperature to 180 °C and 210 °C, respectively. A similar trend is observed for birch veneer: ΔL amounted −4.1 and −10.5 units, respectively, for compression temperatures of 180 °C and 210 °C. After sanding, ΔL amounted 2.0 (indicate whitening) for alder and −0.3 (indicate darkening) for birch. The changes mainly concern the parameters L^* and a^*, both for sanded and thermally compressed wood. The reddening of both wood species was observed, the smallest for sanded surface ($\Delta a^* = 0.3$ for alder and $\Delta a^* = 1.2$ for birch wood), the highest for TC-210 pre-treatment ($\Delta a^* = 2.7$ for alder and $\Delta a^* = 4.7$ for birch wood). The birch wood reddened more than alder wood. The reddening is due to the presence of extractives in the wood [7,52]. In addition, the blueish pigmentation became more intense for alder wood, especially after treatment TC-210 ($\Delta b^* = -3.6$); after sanding $\Delta b^* = -1.2$. On the other hand, birch surface shifts to yellow after sanding ($\Delta b^* = -0.6$) and shifts to blue after thermal compression ($\Delta b^* = 0.9$ and $\Delta b^* = 0.1$ for TC-180 and TC-210, respectively). For both alder and birch veneers, the smallest total color differences correspond to pre-treatment by sanding, with a ΔE of 2.4 and 1.4, respectively, followed by TC-180 with a ΔE of 6.3 and 5.6, respectively, and TC-210, with a ΔE of 17.2 and 11.5, respectively. Some researchers have also observed that sanding of marupa wood with different grits (280 and 320) did not significantly alter the surface color [25].

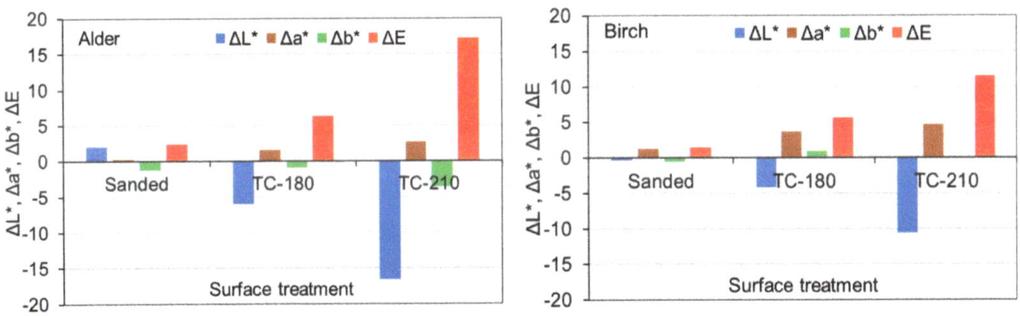

Figure 3. Color differences on sanded and thermally compressed surfaces.

There is a significant difference in the total color between the sanded surface and the surface thermally compressed at temperatures of 180 °C and 210 °C. This is natural, because during thermal compression at temperature of 210 °C there is a significant darkening of the surface. The total color change for birch is $\Delta E = 11.5$, distinct color changes (see Table 2), and for alder $\Delta E = 17.2$, different color (see Table 2). Despite the darkening, the color is attractive with a pronounced wood texture and often resembles the color of a tropical wood species (Figure 2).

It is natural that the color of the sanded and thermally compressed surfaces is different. The sanded surface is formed due to the short-term mechanical action of the cutting tool on the wood surface, which does not lead to significant changes in surface color. Thermal compression, especially at high temperatures, is the effect of heat and pressure on the wood, because of which chemical changes occur in the surface layers of wood [49,50], which lead to changes in the surface color, especially its darkening. This may be caused by the degradation of hemicelluloses and increasing the lignin content in wood [27,49,50].

3.2. *Influence of Artificial Aging on Color Changes of Sanded and Thermally Compressed Surfaces*

There is some time between the pre-treatment process of wood surface by sanding or thermal compression and varnishing. Within this period, wood surface may be exposed to various factors, including UVL + IR radiation. This can cause color changes. Very often, such color changes in wood surface are undesirable, especially when valuable wood species are finished with transparent varnishes. Therefore, one of the objectives of this work was to find out what happens to the sanded or thermally compressed surface before varnishing. Do these surfaces change color under the action of UVL + IR radiation before varnishing?

Figures 4 and 5 show the effect of artificial aging on the color changes of sanded and thermally compressed alder and birch wood surfaces. Artificial aging affects the color of alder and birch surface in different ways. This can be explained by the different anatomical structure and chemical composition of these wood species. ANOVA analysis confirmed that the susceptibility of alder and birch wood to photodegradation is different since the effects of wood species, temperature of thermal compression and their interaction on the color changes were significant ($p \leq 0.05$). This is in good agreement with the results by Laskowska [37] who also found that the total changes in the color of thermally densified beech that took place under the influence of UVL radiation were twice as high as in the case of densified oak wood.

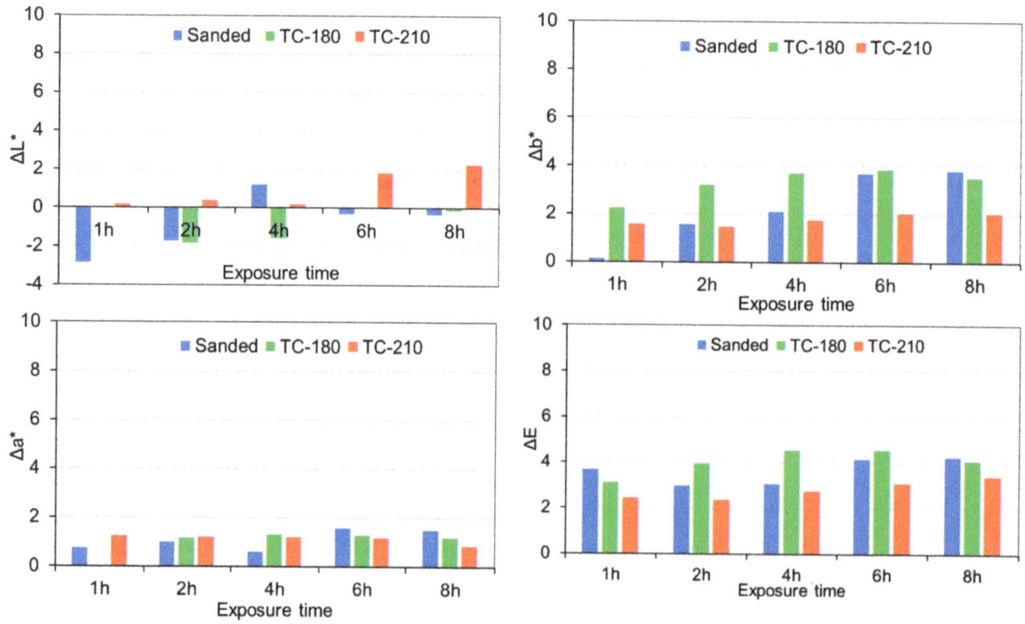

Figure 4. Color differences on sanded and thermally compressed alder surfaces after the artificial aging.

ANOVA analysis showed that the duration of artificial aging significantly affects the changes ΔL^* and Δb^*, but insignificantly affects the change Δa^* and total color change ΔE of the sanded alder surface before varnishing. Figure 4 shows that the maximum differences in brightness occurred during the first hour of artificial aging ($\Delta L^* = -2.9$). After that, during the next exposure time, this difference was reduced and was $\Delta L^* = -0.3$ in 8 h of exposure. At the same time, the smallest ($\Delta b^* = 0.2$) and highest ($\Delta b^* = 3.8$) yellowish of sanded alder surface was observed in 1 h and 8 h of exposure time, respectively. Despite the fact that artificial aging affects ΔL^* and Δb^* significantly, according to the Duncan test, the differences in ΔL^* values between 1 and 2 h, 2, 6 and 8 h, and 4, 6 and 8 h of exposure time were insignificant; and the differences in Δb^* values between 1 and 2 h, 2 and 4 h, 6 and 8 h of exposure time were also insignificant.

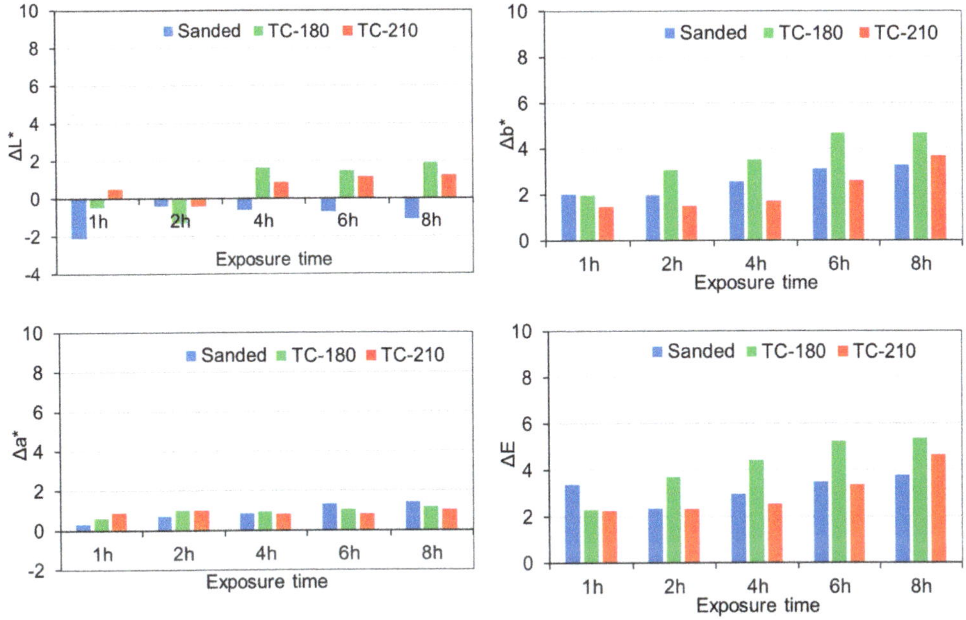

Figure 5. Color differences on sanded and thermally compressed birch surfaces after the artificial aging.

Quite different results were found for the thermally compressed surfaces of alder veneer. The duration of artificial aging effects insignificantly on the surface color changes ΔL^*, Δa^*, Δb^* and ΔE after treatments TC-180 and TC-210 (except Δa^* for TC-180 treatment). Total color change values were ΔE = 3.0–4.3 for the sanded surface, ΔE = 3.1–4.6 for the TC-180 treatment and ΔE = 2.4–3.4 for the TC-210 treatment. According to Table 2, the total color variation of alder wood surface was classified as visible by medium quality filter after applying the two pre-treatment processes: sanding or thermal compression. In addition, according to the Duncan test, the difference of ΔE values between the sanded surface and the surface after treatment TC-180 is insignificant ($p > 0.05$). There is a significant difference of ΔE values between the sanded surface and surface after treatment TC-210, as well as between surfaces after treatments TC-180 and TC-210.

In contrast, the mechanism of the effect of artificial aging on the color change of birch wood was more complex than for alder wood. It is established that the duration of artificial aging significantly affects the change:

- Δa^*, Δb^* and ΔE for sanded surface. As the duration of exposure time of artificial aging increases, along with insignificant ($p > 0.05$) darkening (ΔL^* varies from −2.1 to −1.1), the surface becomes significantly ($p \leq 0.05$) redder (Δa^* varies from +0.3 to +1.4) and yellower (Δb^* varies from +2.0 to +3.3);
- ΔL^*, Δb^* and ΔE for treatment TC-180. The surface after treatment of TC-180 significantly ($p \leq 0.05$) brightens (ΔL^* varies from −0.5 to +1.9) and there is a shift towards yellowing (Δb^* varies from +2.0 to +4.7) with insignificant ($p > 0.05$) reddening (Δa^* varies from +0.6 to +1.2);
- Δb^* and ΔE for treatment TC-210. After treatment TC-210, the surface becomes yellower (Δb^* varies from +1.5 to +3.7) with insignificant ($p \leq 0.05$) brightness (ΔL^* varies from −0.4 to +1.2) and redness (Δa^* varies from +0.9 to +1.1).

Thus, birch wood surfaces pre-treated by sanding or thermal compression are unstable to UVL + IR radiation and are subject to color changes during artificial aging. As in the case of alder wood, the total color variation was classified as visible by medium quality filter.

The ΔE values vary from 2.4 to 3.8 for treatment by sanding, from 2.3 to 5.4 for treatment TC-180 and from 2.3 to 4.7 for treatment TC-210. The difference of ΔE values between the sanded surface and the surface after treatment TC-210 is insignificant ($p > 0.05$).

The yellow and red tones of wood are primarily governed by the photochemistry of the essential wood components, particularly lignin and extractives, respectively [27,52]. All the chemical constituents of wood are susceptible to degradation by sunlight. Cellulose and hemicellulose are not good light absorbers and, therefore, to be degraded by light to a lesser extent [53]. Among the wood constituent polymers, lignin is the most sensitive to light [54]. According to Pandey, "UVL + IR irradiation modified physical and chemical characteristics of wood surfaces and resulted in rapid colour changes, degradation of lignin and increased concentration of chromophoric groups on the wood surfaces" [54]. Moreover, the lignin and its derivatives are factors accountable mainly for the process of wood yellowing [52,55]. The smaller the share of extractives in wood, the lower is its susceptibility to a change in color towards red [37]. These correlations are confirmed by the studies conducted by Tolvaj et al. [56].

It can be concluded that wood thermo-mechanically densified exhibits' acceptable photostability. The higher the temperatures of the thermal compression treatment, the smaller were the total changes in the color of alder and birch wood under the influence of UVL + IR radiation. Laskowska [37] also found that with increasing the temperature of thermo-mechanical treatment, the beech and oak wood had a darker color and, consequently, were least susceptible to a color change under the influence of UVL irradiation. Some other experiments [57,58] also showed that the color stability for heat-treated wood is better after exposure to UVL when compared to untreated wood due to a condensed lignin structure and the antioxidants formed during treatment.

Therefore, artificial aging has an insignificant effect on the surface color of sanded or thermally compressed alder wood and significantly alters the surface color of sanded and thermally compressed birch wood. This fact must be taken into account in the technological process, providing appropriate time for exposure of wood before varnishing.

3.3. Color Changes in the Varnished Surface

Figures 6 and 7 show the graphical interpretation of the effects of a pre-treatment process of wooden substrate, the kind of varnish and the number/type of varnish layers on the color changes in the varnished surface of alder and birch wood.

Varnishing of wood changes the color of the surface. ANOVA analysis showed that the pre-treatment process of the wood substrate, the kind of varnish and the number/type of varnish layers significantly affect the changes in ΔL^*, Δa^*, Δb^* (except the number of varnish layers) and ΔE (except the kind of varnish for birch wood) for both wood species. When analysing the impact of factors on the color changes, it is worth noting that there were considerable differences in the influence of these factors. For both wood species the number of varnish layers has the weakest effect on the change ΔL^*, Δa^* and ΔE and does not affect the change Δb^* at all. The pre-treatment process influenced the changes in ΔL^* (except alder wood), Δa^*, Δb^* and ΔE of varnished surface to a greater extent than the kind of varnish and number of varnish layers.

Figure 6 shows that the maximum darkening (mean value $\Delta L^* = -5.17$), reddening (mean value $\Delta a^* = 3.3$) and yellowing (mean value $\Delta b^* = 5.2$) are observed for the alder wood surface after treatment TC-210, then for the surface after sanding (mean values $\Delta L^* = -4.4$, $\Delta a^* = 1.5$, $\Delta b^* = 1.3$) and for the surface after treatment TC-180 (mean values $\Delta L^* = -2.5$, $\Delta a^* = 1.2$, $\Delta b^* = 1.7$). According to the Duncan test, it was found that there is an insignificant difference between the values of ΔL^* for the sanded surface and the surface thermally compressed at 210 °C (treatment TC-210), and between the values of Δa^* and Δb^* for the sanded surface and the surface thermally compressed at 180 °C (treatment TC-180). Among the investigated varnishes, the surface with UV varnish darkens the most (mean value $\Delta L^* = -5.6$), followed with PUR varnish (mean value $\Delta L^* = -4.2$) and the weakest with WB varnish (mean value $\Delta L^* = -1.2$). The smallest mean value $\Delta a^* = 1.0$

corresponds to WB varnish, followed by UV varnish with a mean Δa^* = 2.3 and PUR varnish with a mean value Δa^* = 2.6. However, the difference between the values of Δa^* for surfaces varnished with UV and PUR is insignificant ($p > 0.05$). The surface becomes the most yellow after varnishing with PUR (mean value Δb^* = 4.3), then the surfaces that are varnished with UV (mean value Δb^* = 2.7) and WB (mean value Δb^* = 1.3). Vidholdová and Slabejová [40], studying the influence of the various types of transparent surface finishes (acrylic-polyurethane, polyacrylic and aldehyde resin, and alkyd resin) on the change of surface color, also found that the discoloration of all the transparent varnish systems resulted mainly from the photoyellowing of the underlying native or thermally modified wood.

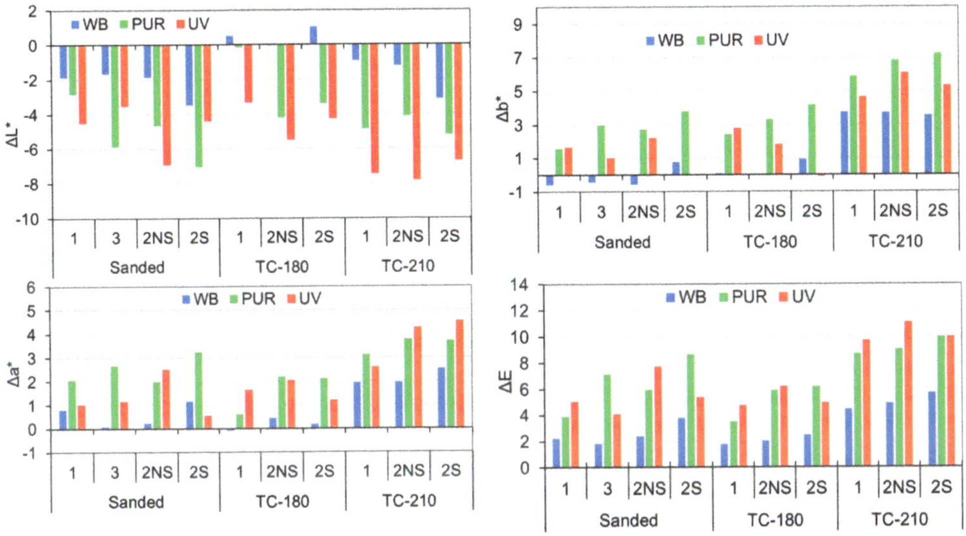

Figure 6. Color differences on sanded and thermally compressed alder surfaces after varnishing.

The maximum total color differences correspond to treatment TC-210 with a mean value ΔE = 8.5, followed by sanded surface with a mean value ΔE = 5.2 and TC-180 with a mean value ΔE = 4.4. The difference in values ΔE is significant between the sanded surface and thermally compressed surfaces, as well as between both treatments (TC-180 and TC-210). The lowest mean value of ΔE = 3.3 appears for WB varnish, followed by PUR varnish with a mean value of ΔE = 7.0 and by UV varnish with a mean value ΔE = 7.1. The difference between the ΔE values for PUR and UV varnished surfaces is insignificant ($p > 0.05$). The lowest value of ΔE = 4.3 is observed for a sanded surface with three layers of varnish, the highest value of ΔE = 6.5 is found for a surface with two layers of varnish without intermediate sanding (2NS). The mean values ΔE = 5.4 and ΔE = 6.4 were for 1 and 2S layers, respectively. In addition, the difference between the values of ΔE for surfaces with the number of layers of varnish 2S and 2NS is insignificant ($p > 0.05$).

If for alder wood there is a darkening of the varnished surface, treated by sanding or thermal compression, then a slightly different picture is observed for birch wood (Figure 7). The greatest darkening is manifested for the treatment TC-180 (average value ΔL^* = −4.5) and for the treatment TC-210 (average value ΔL^* = −3.2). The sanded birch surface becomes lighter after varnishing (average value ΔL^* = 0.2). In addition, the surfaces treated with TC-180 and TC-210 are reddened with average values of Δa^* = 1.0 and Δa^* = 1.1, respectively; and for the sanded surface there is a shift to green (average value Δa^* = −1.5). The greatest yellowing is manifested for the surface treated by TC-210 (average value Δb^* = 2.8), then for the surface treated by TC-180 (average value Δb^* = 2.2), and for the sanded surface there is a shift to the blue (average value Δb^* = −2.1).

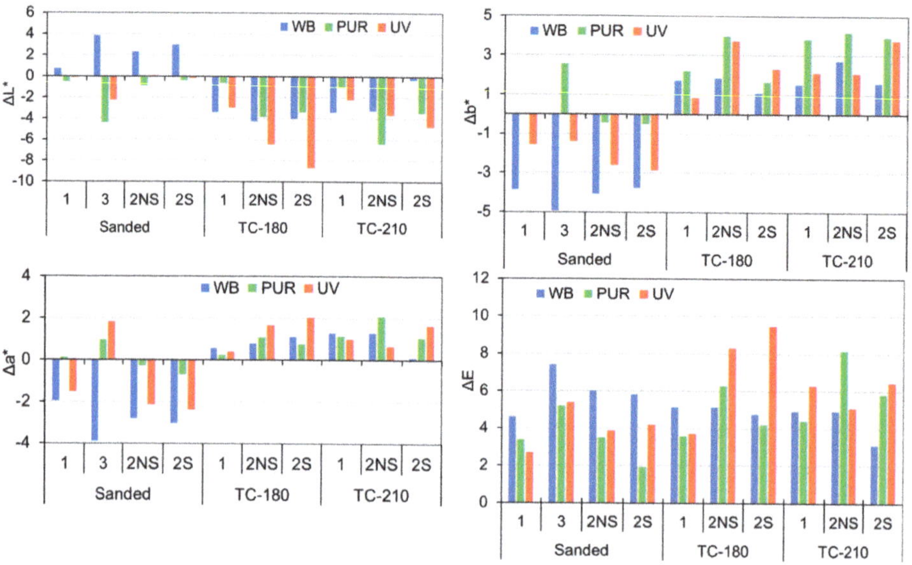

Figure 7. Color differences on sanded and thermally compressed birch surfaces after varnishing.

The difference in the lightness ΔL^* for the birch varnished surface is slightly lower than for the alder varnished surface. As for alder wood, the darkest surface is varnished with UV (average value $\Delta L^* = -3.3$), followed by PUR (average value $\Delta L^* = -2.5$) and WB (average value $\Delta L^* = -1.2$). However, the difference between the values of ΔL^* for surfaces varnished with UV and PUR is insignificant ($p > 0.05$). In addition, if the surface varnished with PUR and UV becomes redder (with average values of $\Delta a^* = 0.7$ and $\Delta a^* = 0.3$, respectively), then for the surface varnished with WB there is a shift to green (average value of $\Delta a^* = -0.4$). The difference between the values of Δa^* for surfaces varnished with UV and PUR is insignificant ($p > 0.05$). Similar changes of Δb^* occured for these varnishes. If yellowing is observed for the surface varnished with UV and PUR (with average values of $\Delta b^* = 0.9$ and $\Delta b^* = 2.2$, respectively), then for WB there is a shift to blue (average value of $\Delta b^* = -0.2$).

The minimum total color change with an average value of $\Delta E = 4.3$ was found for the sanded surface, a larger one with an average value of $\Delta E = 5.6$ and $\Delta E = 5.9$ for surfaces thermally compressed at 210 °C and 180 °C, respectively. According to the Duncan test, the difference between ΔE values for surfaces thermally compressed at different temperatures is insignificant ($p > 0.05$). The smallest color changes were demonstrated by PUR (average value $\Delta E = 4.7$), followed by WB (average value $\Delta E = 5.1$) and UV (average value $\Delta E = 5.7$). However, the difference between the ΔE values for surfaces varnished by PUR, WB and UV is insignificant. The lowest value of $\Delta E = 4.5$ is observed for a surface with one varnish layer, the highest value of $\Delta E = 5.9$ is found for a sanded surface with three varnish layers. The difference between ΔE values for surfaces with the number of varnish layers 1, 2S and 2NS, or 3, 2S and 2NS is insignificant.

The different behavior of the color values of wood color after varnishing could be explained by changes in the chemical composition of the wood samples caused by pre-treatment process by sanding or thermal compression as well as by the different nature and chemical composition of the investigated varnishes. Therefore, not only the pre-treatment of the substrate, but also the type of varnish affects the color changes and determines the aesthetic characteristics of the varnished surface. As noted by Vidholdová and Slabejová [40] the change in color due to the surface finish is an interaction of the changed wood color and the color of the coating film itself. Among the investigated varnishes for both wood species, the WB varnish showed the greatest resistance to discoloration, followed by PUR and UV. Because of such color changes, the surface varnished with WB varnish is the lightest, and the surfaces varnished with PUR and UV varnishes are darker, redder and yellower.

3.4. *Influence of Artificial Aging on Color Changes of Varnished Alder Surfaces*

3.4.1. WB Varnish

Figure 8 shows the effect of the pre-treatment process, the number of varnish layers and the duration of artificial aging on the change in color parameters ΔL^*, Δa^*, Δb^* and ΔE for the surface varnished by WB. It is established that the duration of artificial aging significantly affects the change of ΔL^*, Δa^*, Δb^* and ΔE; as the duration increased, the values of color parameters increased too.

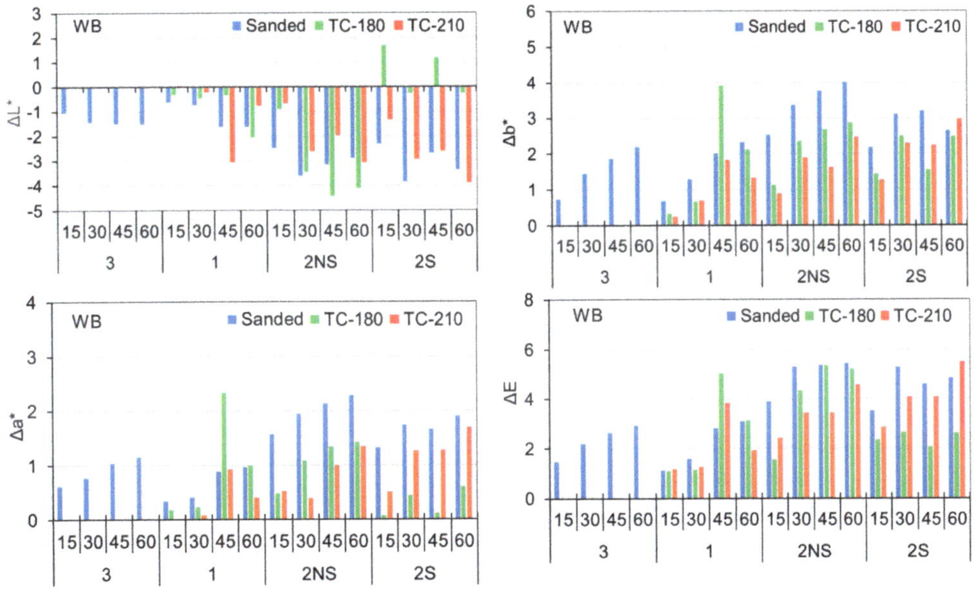

Figure 8. Color differences on sanded and thermally compressed alder varnished surfaces with WB varnish after artificial aging.

If we analyze the percentage influence of the studied factors on the change of color parameters of varnished alder wood, it should be noted that there were significant differences in the influence of these factors. The number of varnish layers has the strongest effect on the change of ΔL^*, Δb^* and ΔE, while the pre-treatment process has the strongest effect on the change of Δa^*.

A result of artificial aging was that the sanded surface darkens (average value $\Delta L^* = -2.1$), reddens (average value $\Delta a^* = 1.3$) and turns yellow (average value $\Delta b^* = 2.3$) more than surfaces after treatments TC-180 and TC-210. Laskowska [37] also observed that light irradiation had the greatest effect on the change in the lightness of unmodified beech and oak wood, and the smallest effect on the lightness of both wood densified at a temperature of 200 °C. However, it should be noted that according to the Duncan test, the difference in the values of ΔL^* between the sanded surface and the surface after treatment TC-210, as well as between surfaces after treatments TC-180 (average value $\Delta L^* = -1.2$) and TC-210 $\Delta L^* = -1.7$) was insignificant ($p > 0.05$). Similarly, the differences in the average values of Δa^* and Δb^* between the surfaces after treatments TC-180 and TC-210 are insignificant ($p > 0.05$). The smallest shifts towards to red were found for the surface after treatments TC-180 (mean value $\Delta a^* = 0.6$) and TC-210 (mean value $\Delta a^* = 0.7$), and the smallest yellowing for the surface after treatments

TC-180 (mean value $\Delta b^* = 1.7$) and TC-210 (mean value $\Delta b^* = 1.5$). For the surface with two varnish layers without intermediate sanding (2NS) the highest darkening (average value $\Delta L^* = -2.8$), reddening (average value $\Delta a^* = 1.3$) and yellowing (average value $\Delta b^* = 2.5$) was observed. Surfaces with one and three varnish layers, as well as with two varnish layers with intermediate sanding (2S) are characterized by a tendency to reduce shades of gray, red and yellow. The smallest darkening (average value $\Delta L^* = -1.0$), redness (average value $\Delta a^* = 0.5$) and yellowing (average value $\Delta b^* = 1.2$) was observed for the surface with one varnish layer. In addition, the difference in the values of ΔL^*, Δa^* and Δb^* between the surfaces with 1, 3 and 2S; 3 and 2S; 2S and 2NS number of varnish layers, respectively, was insignificant ($p > 0.05$).

Regarding the total color change, the smallest ΔE with an average value of $\Delta E = 2.6$ was experienced by the surface after treatment TC-180, and the largest with an average value of $\Delta E = 3.5$, sanded surface. In addition, it should be noted that the surfaces after treatments TC-180 and TC-210 (average value $\Delta E = 2.9$) on the values of ΔE differ insignificantly ($p > 0.05$). The lowest average value of $\Delta E = 2.0$ was observed for the surface with one varnish layer, the highest average value of $\Delta E = 4.2$ was found for the surface with two varnish layers without intermediate sanding (2NS). The difference in the values of ΔE between the surfaces with one and three, 2S and 2NS number of varnish layers was insignificant ($p > 0.05$).

3.4.2. PUR Varnish

Figure 9 shows the effect of the pre-treatment process, the number of varnish layers and the duration of artificial aging on the change in color parameters ΔL^*, Δa^*, Δb^* and ΔE for the surface varnished by PUR. It was found that the duration of artificial aging, as well as for WB varnish, significantly affects the change in ΔL^*, Δa^*, Δb^* and ΔE. As the duration of artificial aging increased, the values in color parameters increased too. The findings in this research agree with other authors [40,59] who also noted that the total color difference ΔE shows a systematic trend towards higher values with increasing irradiation time for the surfaces of thermally modified wood coated by transparent varnishes.

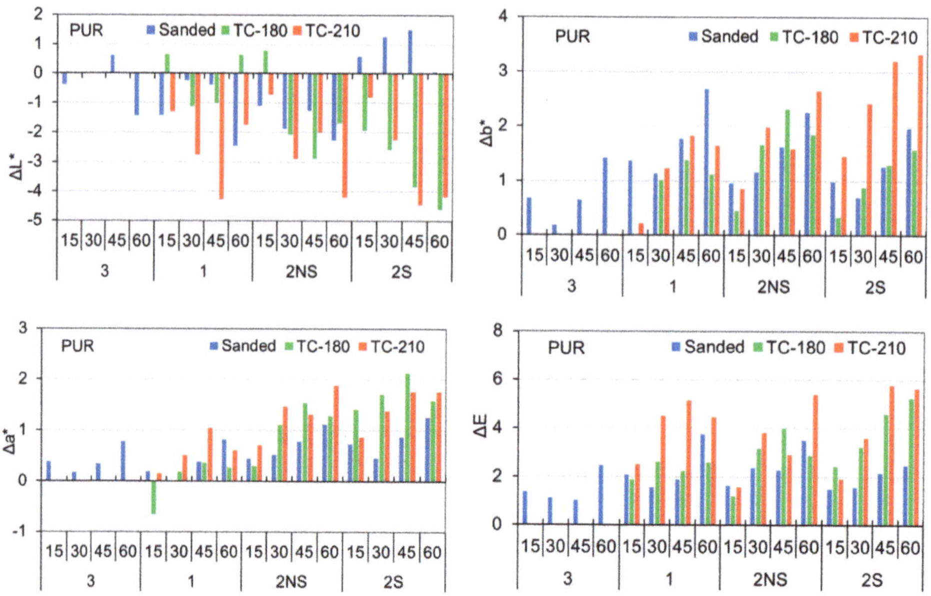

Figure 9. Color differences on sanded and thermally compressed alder varnished surfaces with PUR varnish after artificial aging.

However, if we compare these results with the percentage influence of investigated factors, we found some differences in the color change of the surface varnished with PUR. In particular, it was found that the number of varnish layers has a negligible effect on the change of ΔL^* and ΔE. The most strongly influenced ΔL^* and ΔE was the duration of artificial aging, followed by the pre-treatment process. The change in Δa^* was most strongly influenced by the pre-treatment process, then the duration of artificial aging and the number of varnish layers. In contrast, the change in Δb^* was most strongly influenced by the number of varnish layers, then the pre-treatment process and the duration of artificial aging.

During artificial aging, the surface varnished with PUR, in contrast to WB varnished surface, darkened (average value $\Delta L^* = -2.6$), reddened (average value $\Delta a^* = 1.0$) and yellowed (average value $\Delta b^* = 1.7$) the most after treatment TC-210. The smallest average values of $\Delta L^* = -0.6$ and $\Delta a^* = 0.6$ were found for the sanded surface, while the smallest average values of $\Delta b^* = 1.1$, for the surface after treatment TC-180. However, it should be noted that according to the Duncan test, the difference in the values of ΔL^*, Δa^* and Δb^* between the sanded surface and the surface after treatment TC-180 was insignificant ($p > 0.05$). Similarly, the difference in the average values of Δa^* between the surfaces after treatments TC-180 and TC-210 was also insignificant ($p > 0.05$). The control sanded surface with three varnish layers was characterized by a tendency to reduce shades of gray (with an average value of $\Delta L^* = -0.3$) and yellow (with an average value of $\Delta b^* = 0.7$). The surface varnished with two varnish layers without intermediate sanding (2NS) was the darkest (with an average value of $\Delta L^* = -1.8$), while the surface varnished with two varnished layers with intermediate sanding (2S) was the reddest (average $\Delta a^* = 1.3$) and the yellowest (with the average value of $\Delta b^* = 1.6$). The smallest redness was observed for surfaces with one and three varnish layers with average values of $\Delta a = 0.3$ and $\Delta a = 0.4$, respectively. In addition, the difference in the values of ΔL^*, Δa^* and Δb^* between the surfaces with 1, 2S and 2NS, 3 and 1; 1 and 3, 2NS and 2S; 1, 2S and 2NS number of varnish layers, respectively, was insignificant ($p > 0.05$).

Regarding the total color change, the smallest ΔE with an average value of ΔE = 2.1 corresponded to the sanded surface, and the largest with an average value of ΔE = 4.0 to the surface after treatment TC-210. In addition, it should be noted that the surfaces after sanding and treatment TC-180 (average value $\Delta E = 2.8$) and TC-210 on the values of ΔE differed significantly ($p < 0.05$). Moreover, the smallest total color change with an average value of $\Delta E = 1.5$ was found for the control sanded surface with three varnish layers, while the largest with an average value of $\Delta E = 3.4$ for a surface with two varnish layers with intermediate sanding (2S). The difference in the values of ΔE between surfaces with 1, 2S and 2NS number of varnish layers was insignificant ($p > 0.05$).

3.4.3. UV Varnish

The Figure 10 shows the effect of pre-treatment process, the number of varnish layers and the duration of artificial aging on the change in color parameters ΔL^*, Δa^*, Δb^* and ΔE for the surface varnished with UV. It was found that the duration of artificial aging, in contrast to WB and PUR varnishes, had little effect on the change in ΔL^*, Δa^* and ΔE of the surface varnished with UV. As the duration of artificial aging increased, only the values Δb^* changed significantly. In addition to the duration of artificial aging, the number of varnish layers had also insignificant effect on the change of ΔL^*, Δa^* and ΔE. The pre-treatment process had a significant effect on the change in ΔL^*, Δa^* and ΔE, but an insignificant effect on the change in Δb^*.

It was found that the surface treated with TC-210 was the lightest (with an average value of $\Delta L^* = 0.3$) compared to the surface treated with TC-180 (with an average value of $\Delta L^* = 0.5$) and the sanded surface, which was the darkest (with an average value of $\Delta L^* = -1.3$). However, it should be noted that the difference in the average values of ΔL^* and Δa* between the surfaces after treatments with TC-180 and TC-210 was insignificant. The sanded surface, in addition to the highest darkening, was also characterized by the highest reddnening (with an average value $\Delta a^* = 0.8$) and yellowing (average value of

Δb^* = 1.1). The smallest reddening (with an average value Δa^* = 0.1) and yellowing (with an average value Δb^* = 0.8) was the surface after the treatment TC-180. The lightest (with an average value ΔL^* = 0.2), and at the same time the yellowest (with an average value Δb^* = 1.5), was the surface with two varnish layers with intermediate sanding (2S). The control sanded surface with three varnish layers was the darkest (with an average value ΔL^* = −1.3) and the reddest (average value Δa^* = 0.7). At the same time, the surface with one varnish layer had the lowest value Δa^* = 0.1 (was the least red) and Δb^* = 0.7 (was the least yellow). In addition, the difference between the values Δb^* for surfaces with 1, 3 and 2NS number of varnish layers was insignificant ($p > 0.05$).

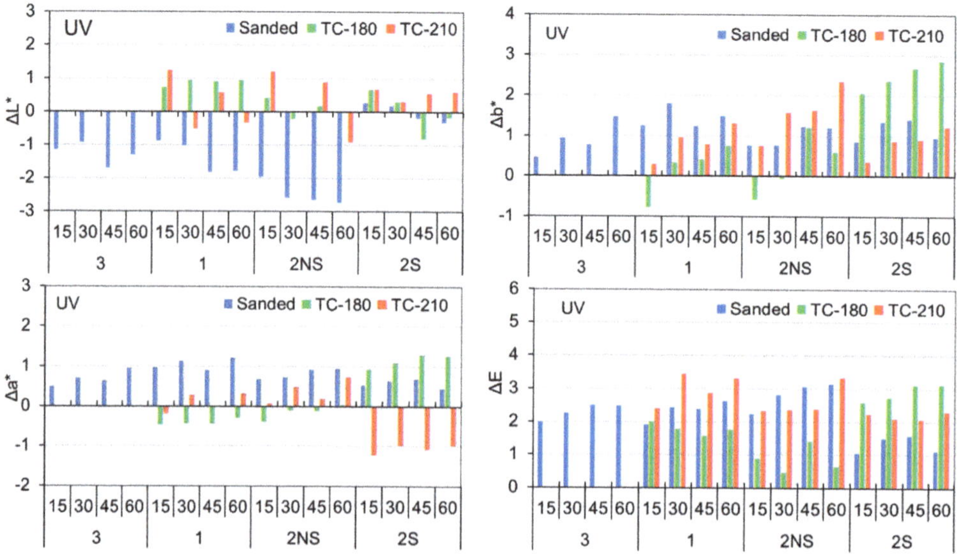

Figure 10. Color differences on sanded and thermally compressed alder varnished surfaces with UV varnish after artificial aging.

Regarding the total color change, the smallest color changes with an average value of ΔE = 1.8 were experienced by the surface after treatment TC-180, and the highest with an average value of ΔE = 2.7, the surface after treatment TC-210. In addition, it should be noted that the surfaces after sanding (average value ΔE = 2.2) and treatment TC-180 in the values of ΔE differ insignificantly ($p > 0.05$). The smallest total color changes with an average value of ΔE = 2.1 and the highest with an average value of ΔE = 2.4 were found for the surfaces with two varnish layers without intermediate sanding (2NS) and one varnish layer, respectively. The difference between the values of ΔE for surfaces with 1, 3, 2S and 2NS number of varnish layers was insignificant ($p > 0.05$).

As a result, it can be argued that UVL + IR-induced slight photo-degradation of alder wood-varnished surface pre-treated by sanding or thermal compression takes place in indoor conditions. Similar results are presented in [59]. The authors came to the conclusion that transparent varnishes cannot also prevent photo-degradation of the wood underneath during exposure to UVL in the case of thermally modified wood. Thermally densified surface-varnished wood is more resistant to color changes compared to sanded surface-varnished wood. This agrees with the results of other authors [37] who noted that wood that densified at a higher temperature was less susceptible to color changes under the influence of light irradiation. It could be due to an increase in lignin stability by condensation at the thermal compression process at 210 °C [59,60]. The better photo-stability of the thermally modified wood compared to the non-modified wood for the transparent varnishes was also reported by Deka and Petric [59], which supports our present results. Some authors [58] indicate acceptable photostability also of thermally treated wood, when the wood is coated with PUR and UV-hardened coatings. Sanded surface-varnished wood is darker, redder and yellower than a thermally densified surface. It is known that clear transparent varnishes allow transmission of sunlight, and surface degradation can take place in the underlying wood [61]. However, the findings in this research showed that various transparent varnishes produce different colors in wood surfaces. Among the studied varnishes, the darkest, reddest and yellowest surface was found for WB varnish; UV varnish showed the lightest and least red and least yellow surfaces. The least darkening, rednening and yellowing was observed for varnished surfaces with one and three layers of varnish, the most for varnished surfaces with 2NS and 2S layers.

3.5. Influence of Artificial Aging on Color Changes of Varnished Birch Surfaces

3.5.1. WB Varnish

Figure 11 shows the effect of the pre-treatment process, the number of varnish layers and the duration of artificial aging on the changes in color parameters ΔL^*, Δa^*, Δb^* and ΔE for the surface of birch wood varnished with WB. With an increasing duration of artificial aging, the values in color parameters increased. The maximum changes in color parameters are observed after 60 min of artificial aging, the minimum, after 15 min of exposure time. However, ANOVA analysis showed that the duration of artificial aging significantly affects the changes in Δa^* and Δb^*, but insignificantly affect ΔL^* and ΔE. In addition, the pre-treatment process significantly affects the changes in Δa^* and Δb^*, but insignificantly affects the changes in ΔL^* and ΔE. The number of varnish layers significantly affects the changes of all parameters ΔL^*, Δa^*, Δb^* and ΔE.

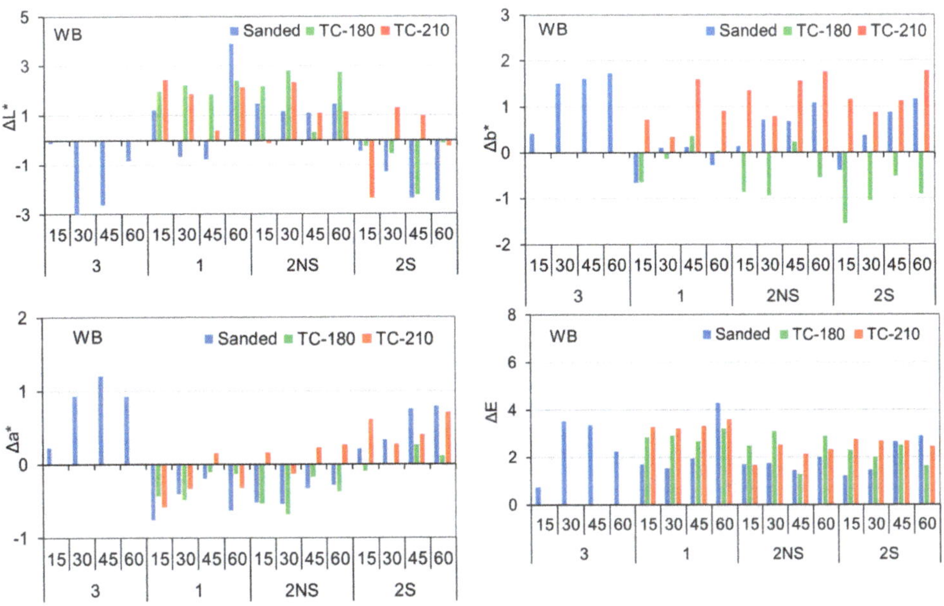

Figure 11. Color differences on sanded and thermally compressed birch varnished surfaces with WB varnish after artificial aging.

During artificial aging, the sanded surface remains the darkest (with an average value of $\Delta L^* = -0.3$), the surfaces are lighter after the treatments TC-210 and TC-180 with average values of $\Delta L^* = 1.1$ and $\Delta L^* = 1.4$, respectively. However, the difference between the values of ΔL^* for these two surfaces treated with TC-180 and TC-210 is insignificant ($p > 0.05$).

Regarding the reddening of the surface, the smallest, with average values of $\Delta a^* = 0.1$ and $\Delta a^* = 0.0$, respectively, was recorded for the sanded surface and the surface after the treatment TC-210. The surface after the treatment TC-180 demonstared a shift towards green (with an average value of $\Delta a^* = -0.2$). The difference between the values of Δa^* for the sanded surface and the surface after the treatment TC-210 was insignificant ($p > 0.05$). The yellowest surface was detected after the treatment TC-210 with an average value of $\Delta b^* = 1.1$. The surface after treatment TC-180 and the sanded surface showed a shift towards blue and yellow with average values of $\Delta b^* = -0.4$ and $\Delta b^* = 0.6$, respectively.

The surfaces with two varnish layers without intermediate sanding (2NS) and one varnish layer were lighter, with average values of $\Delta L^* = 1.5$ and $\Delta L^* = 1.7$, respectively. The darker surface was the one with two varnish layers with intermediate sanding (2S) with an average value of $\Delta L^* = -0.8$, and the darkest surface was the control sanded surface with three varnish layers with an average value of $\Delta L^* = -1.6$. However, according to the Duncan test, the difference between the values of ΔL^* for surfaces with 3 and 2S, 2NS and 1 number of varnish layers was insignificant ($p > 0.05$).

Among the surfaces, the control sanded surface withthree varnish layers was also characterized by the highest redness (average value $\Delta a^* = 0.8$). For the surfaces with two varnish layers without intermediate sanding (2NS) and one varnish layer there was a shift towards green with average values of $\Delta a^* = -0.2$ and $\Delta a^* = -0.3$, respectively. The difference between the values of Δa^* for surfaces with 1 and 2NS number of varnish layers is insignificant ($p > 0.05$).

The highest yellowing with an average value of $\Delta b^* = 1.3$ was found for the sanded surface with three varnish layers. The smallest yellowing with average values of $\Delta b^* = 0.2$, $\Delta b^* = 0.3$ and $\Delta b^* = 0.5$ was observed for the surfaces with two varnish layers with intermediate sanding (2S), with one varnish layer and with two varnish layers without intermediate sanding (2NS), respectively. The difference between the values of Δb^* for surfaces with 2S, 1 and 2NS number of varnish layers is insignificant ($p > 0.05$).

Regarding the total color change, the smallest change with an average value of $\Delta E = 2.2$ was experienced by the sanded surface, higher total changes were experienced by the surfaces after treatments TC-180 and TC-210 with average values $\Delta E = 2.6$ and $\Delta E = 2.9$, respectively. In addition, it should be noted that the sanded surface and surfaces after treatments TC-180 and TC-210 in the values of ΔE differ insignificantly ($p > 0.05$). The smallest and highest average values of $\Delta E = 2.1$ and $\Delta E = 3.0$ were observed for the surfaces with two varnish layers without intermediate sanding (2NS) and with one varnish layer, respectively. The difference between the values in ΔE for surfaces with 2NS, 2S and 3; 2S, 3 and 1 number of varnish layers is insignificant ($p > 0.05$).

3.5.2. PUR Varnish

Figure 12 shows the effect of pre-treatment process, the number of varnish layers and the duration of artificial aging on the changes in color parameters ΔL^*, Δa^*, Δb^* and ΔE for the surface varnished with PUR. With an increasing duration of artificial aging, the values of the color parameters increased; the maximum changes in color parameters are observed after 45–60 min of artificial aging, the minimum in 15 min of exposure time. The ANOVA analysis showed that the duration of artificial aging, as well as for WB varnish, significantly affects the change in Δa^* and Δb^*, but effects insignificantly ($p > 0.05$) on ΔL^* and ΔE.

However, if we compare these results with the percentage influence of factors on the color change of the surface varnished with PUR, some differences were found. In particular, it was found that the pre-treatment process has the greatest effect and the number of varnish layers has the lowest effect on ΔL^*, Δa^*, Δb^* and ΔE.

During artificial aging, the surface varnished with the PUR, as well as for the WB varnish, darkens (average value $\Delta L^* = -1.5$), reddens (average value $\Delta a^* = 0.7$) and turns yellow (average value $\Delta b^* = 0.9$) the most after the treatment TC-210. The surface after treatment TC-180 was the lightest (average value $\Delta L^* = 0.9$), the least red (average value $\Delta a^* = 0.0$) and the least yellow (average value $\Delta b^* = 0.5$). For the sanded surface, there is a shift towards gray (average value $\Delta L^* = -0.1$), towards green (average value $\Delta a^* = -0.1$) and towards blue (average value $\Delta b^* = -0.1$). However, according to the Duncan test, the difference in the values of Δa^* for the sanded surface and the surface after the treatment TC-180 was insignificant ($p > 0.05$).

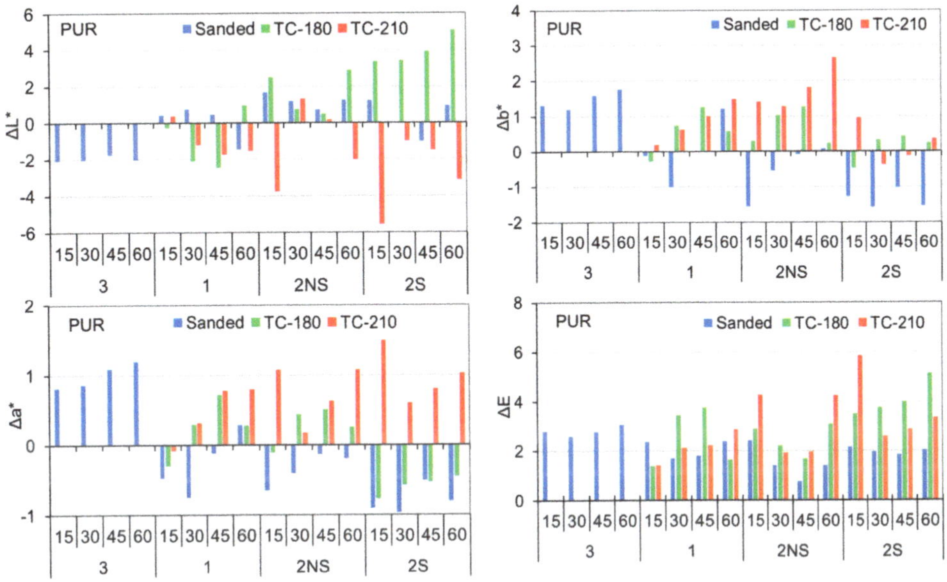

Figure 12. Color differences on sanded and thermally compressed birch varnished surfaces with PUR varnish after artificial aging.

The brightest were surfaces with two layers of varnish with intermediate sanding (2S) and without intermediate sanding (2NS) with average values $\Delta L^* = 0.5$ and $\Delta L^* = 0.6$, respectively. The difference in the values of ΔL^* between these surfaces was insignificant. For the surface with one varnish layer there was a shift towards gray (with an average value of $\Delta L^* = -0.8$). The control sanded surface with three layers of varnish was the darkest (with an average value of $\Delta L^* = -2.0$). In addition, the control sanded surface with three varnish layers was the reddest (average value $\Delta a^* = 1.0$) and the yellowest (with average value $\Delta b^* = 1.5$). For the surface with two varnish layers with intermediate sanding (2S) there was a shift towards green (average value $\Delta a^* = -0.1$) and towards blue (with an average value of $\Delta b^* = -0.3$). In addition, it should be noted that surfaces with one varnish layer and two varnish layers without intermediate sanding (2NS) on the values of Δa^* and Δb^* differ insignificantly ($p > 0.05$).

Regarding the total color change, the smallest change with an average value of $\Delta E = 2.1$ was the sanded surface, and the highest was the surfaces after the treatments TC-210 and TC-180 with average values $\Delta E = 2.8$ and $\Delta E = 2.9$, respectively. In addition, the surfaces after the treatments TC-180 and TC-210 on the values of ΔE differ insignificantly ($p > 0.05$). The surface with one varnish layer undergoes the smallest total color change with an average value of $\Delta E = 2.3$, then the surface with two varnish layers without intermediate sanding (2NS) with an average value of $\Delta E = 2.3$ and the control sanded surface with three varnish layers with an average value of $\Delta E = 2.8$. The highest color changes (with an average value of $\Delta E = 3.2$) were observed for the surface with two varnish layers with intermediate sanding (2S). The difference between the values of ΔE for surfaces with 1, 2NS and 3; 3 and 2S number of varnish layers was insignificant ($p > 0.05$).

3.5.3. UV Varnish

Figure 13 shows the effect of the pre-treatment process, the number of varnish layers and the duration of artificial aging on the changes in color parameters ΔL^*, Δa^*, Δb^* and ΔE for the birch surface varnished with UV. The ANOVA analysis showed that the duration of artificial aging has an insignificant effect on the changes in color parameters. In addition to the duration of artificial aging, the number of varnish layers also has an insignificant effect on the changes in ΔL^* and Δb^*, while the pre-treatment process significantly affects the changes in ΔL^*, Δb^* and ΔE, but has an insignificant effect on the change in Δa^*.

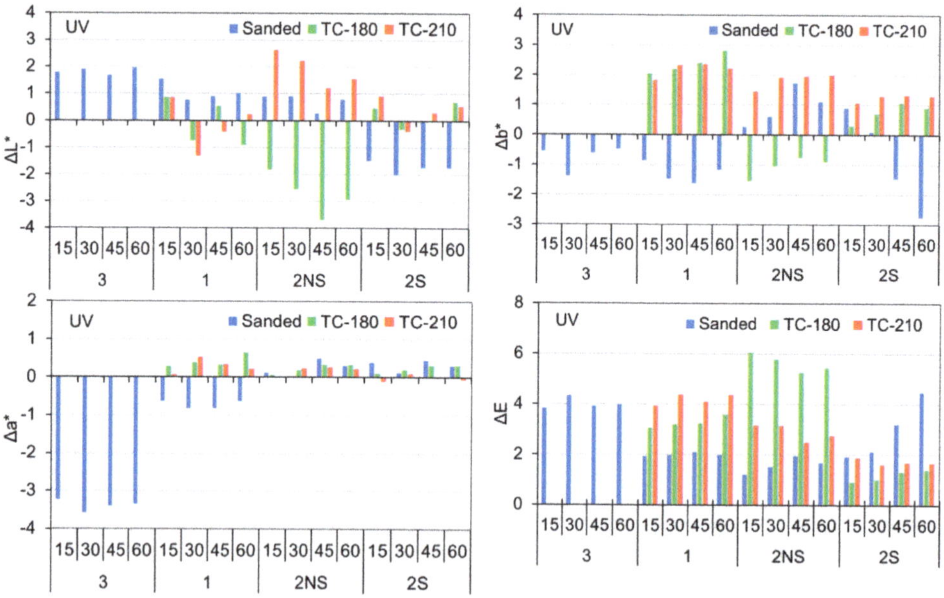

Figure 13. Color differences on sanded and thermally compressed birch varnished surfaces with UV varnish after artificial aging.

It was found that the surface treated with TC-210 and the sanded surface were the brightest (with an average value of $\Delta L^* = 0.5$) compared to the surface treated with TC-180 (with an average value of $\Delta L^* = -0.7$), which was the darkest. However, the difference between the mean values of ΔL^* for all three surfaces was insignificant ($p > 0.05$). Redder surfaces were after treatments TC-210 and TC-180 with average values of $\Delta a^* = 0.2$ and $\Delta a^* = 0.3$, respectively. For the sanded surface, a shift towards green with an average value of $\Delta a^* = -0.9$ was observed. The difference between the values of Δa^* for the surfaces after treatments TC-180 and TC-210 was insignificant ($p > 0.05$). The surfaces after treatments of TC-180 and TC-210 were yellower, with average values of $\Delta b^* = 1.1$ and $\Delta b^* = 1.9$, respectively. For the sanded surface, a shift towards blue with an average value of $\Delta b^* = -0.5$ was observed. The difference between the values in Δa^* and Δb^* for surfaces after treatments TC-180 and TC-210 was insignificant ($p > 0.05$).

The control sanded surface with three varnish layers and the surface with one varnish layer with average values of $\Delta L^* = 1.8$ and $\Delta L^* = 0.1$, respectively, were lighter. For surfaces with two varnish layers without intermediate sanding (2NS) and with intermediate sanding (2S), a shift towards gray was observed with average values of $\Delta L^* = 0.0$ and $\Delta L^* = -0.4$, respectively. There is no difference in ΔL^* between surfaces with 1, 2S and 2NS numbers of layers. At the same time, surfaces with the number of layers 1, 2S and 2NS were characterized by a slight reddening with average values of $\Delta a^* = 0.1$, $\Delta a^* = 0.2$ and $\Delta a^* = 0.2$, respectively. Moreover, there is no difference in Δa^* between these surfaces. In addition to a slight reddening of surfaces with the number of layers 2S, 2NS and 1, they were also characterized by yellowing with average values of Δb^* = 0.4, Δb^* = 0.6 and Δb^* = 1.6, respectively. The opposite trend was observed for the control sanded surface with three layers of varnish, for which there was a significant shift towards green and blue with average values of $\Delta a^* = -3.4$ and $\Delta b^* = -0.7$, respectively. In addition, the difference between the values in Δb^* for surfaces with 2S and 2NS number of varnish layers was insignificant ($p > 0.05$).

Regarding the total color changes, the smallest color changes with an average value of ΔE = 2.6 corresponded to the sanded surface and the highest, the surface after treatments TC-210 and TC-180 with average values ΔE = 3.3 and ΔE = 3.3, respectively. In addition, according to the Duncan test, the surfaces after sanding and treatments TC-180 and TC-210 on the values of ΔE differed insignificantly ($p > 0.05$). The smallest total color changes with an average value of ΔE = 1.9 were found for the surface with two varnish layers with intermediate sanding (2S). The highest total color changes with average values of ΔE = 3.4, ΔE = 3.4 and ΔE = 4.0 were found for surfaces with two varnish layers without intermediate sanding (2NS), with one varnish layer and for the control sanded surface with three varnish layers, respectively. The difference between the values in ΔE for surfaces with 2NS, 1 and 3 number of varnish layers was insignificant ($p > 0.05$).

Therefore, we can conclude that the studied transparent varnishes cannot prevent the photo-degradation of the birch wood underneath them, during exposure to UVL + IR radiation in indoor conditions. There was still a slight degree of color change for sanded and thermally densified wood surfaces during UVL + IR exposure. This might be due to the fact that the absorption of UVL by lignin leads to the formation of free radicals which react with oxygen, producing carbonyl and carbonyl chromophoric groups, which are responsible for wood's color changes [54,57,59,60]. Free radical reactions were proved by the detection of free radical signals in EPR spectra after exposure to light [59]. It is clear that different transparent varnishes behaved differently on both wood species. Therefore, the photo-degradation of the birch wood's varnished surface had a slightly different behavior than the alder wood-varnished surface. A sanded surface that is varnished is darker, greener and less yellow than the thermally densified surface, which is lighter, redder and more yellow. Among the studied varnishes, the darkest and reddest surface was found for the PUR varnish; UV varnish showed the yellowest surface and WB varnish provided the lightest and least yellow surface. The lightest varnished surfaces were with one and 2NS layers of varnish, the darkest—with 3 and 2S layers of varnish. Moreover, the surfaces were redder and yellower with 1, 2NS and 2S layers of varnish, and were greener with 3 layers of varnish.

4. Conclusions

The results of this work confirmed and proved that thermal compression of wood substrate prior to the varnishing process, instead of the labor-intensive sanding process, makes it possible to apply varnish on the surface of wood substrate without having an adverse influence on the esthetic characteristics of the samples. The thermally compressed surface still has an attractive "exotic appearance of wood", which allow the use of transparent varnishes that increase the attractiveness of the final product. The alder wood was less resistant to a temperature of 210 °C than birch wood. Sanding of both alder and birch wood did not significantly alter the surface color.

Despite the improved esthetic and decorative properties compared to sanded wood, thermally compressed wood, both varnished and unvarnished, is still prone to surface degradation during UVL + IR radiation, even in indoor conditions. However, it should be noted that alder and birch wood, thermally compressed at higher temperatures, was less susceptible to discoloration under the influence of UVL + IR radiation than sanded wood. It was found that by increasing the duration of artificial aging the values in color differences (ΔL^*, Δa^*, Δb^* and ΔE) increased too. The alder and birch wood that was surface-varnished, both sanded and thermally densified, showed a similar susceptibility to color changes after

UVL + IR irradiation. The WB varnish showed the greatest resistance to discoloration during UVL + IR irradiation, followed by the PUR and the UV. The surface varnished with WB was the lightest, and the surfaces varnished with PUR and UV varnishes were darker, redder and yellower. In all cases, the ΔE values were lower than six, which indicates slight color changes in surface-varnished after artificial aging compared to the original surface color and these color changes are visible by medium quality filter. Therefore, thermal compression of wood veneer followed by an appropriate transparent varnish system, could be considered as an industrially acceptable method to protect wood against photodegradation in indoor conditions with simultaneous improvement of aesthetic surface properties and preservation of wood in the absence of sanding process.

Author Contributions: Conceptualization, P.B.; methodology, P.B., T.K., B.L. and N.B.; investigation, T.K., B.L. and N.B.; writing—original draft preparation, P.B.; writing—review and editing, P.B., T.K., B.L. and N.B.; project administration, P.B. All authors have read and agreed to the published version of the manuscript.

Funding: This research was supported by the Polish National Agency for Academic Exchange (NAWA) under contract No. PPN/ULM/2020/1/00188/U/00001, to implement the project "Development of a Novel Wood Surface Preparation Method before Varnishing" by Prof. Pavlo Bekhta at the Poznan University of Life Science, Poland.

Institutional Review Board Statement: Not applicable.

Informed Consent Statement: Not applicable.

Data Availability Statement: The data that support the findings of this study are available upon reasonable request from the authors.

Acknowledgments: Bekhta acknowledges the Polish National Agency for Academic Exchange (NAWA) for support of his research. Special thanks are extended to Yu. Maksymiv for technical assistance during all the experiments.

Conflicts of Interest: The authors declare no conflict of interest.

References

1. Ozarska, B. *A Manual for Decorative Wood Veneering Technology*; Project No. PN01.1600; Australian Goverment, Forest & Wood Products Research and Development Corporation: New South Wales, Australia, 2003.
2. Bekhta, P.; Krystofiak, T.; Proszyk, S.; Lis, B. Surface gloss of lacquered medium density fibre-board panels veneered with thermally compressed birch wood. *Prog. Org. Coat.* **2018**, *117*, 10–19. [CrossRef]
3. Hernández, R.E.; Cool, J. Evaluation of three surfacing methods on paper birch wood in relation to water- and solvent-borne coating performance. *Wood Fiber Sci.* **2008**, *40*, 459–469.
4. Richter, K.; Feist, W.C.; Knabe, M.T. The effect of surface roughness on the performance of finishes. Part 1: Roughness characterization and strain performance. *Prod. J.* **1995**, *45*, 91–97.
5. Williams, R.S. Finishing of Wood. In *Wood Handbook—Wood as an Engineering Material*; General Technical Report FPL-GTR-190; U.S. Department of Agriculture, Forest Service, Forest Products Laboratory: Madison, WI, USA, 2010; Chapter 16; pp. 16-1–16-39.
6. Taylor, J.B.; Carrano, A.L.; Lemaster, R.L. Quantification of process parameters in a wood sanding operation. *Prod. J.* **1999**, *49*, 41–46.
7. Bekhta, P.; Proszyk, S.; Krystofiak, T. Colour in short-term thermo-mechanically densified veneer of various wood species. *Eur. J. Wood Prod.* **2014**, *72*, 785–797. [CrossRef]
8. Bekhta, P.; Proszyk, S.; Lis, B.; Krystofiak, T. Gloss of thermally densified alder (*Alnus glutinosa* Goertn.), beech (*Fagus sylvatica* L.), birch (*Betula verrucosa* Ehrh.), and pine (*Pinus sylvestris* L.) wood veneers. *Eur. J. Wood Prod.* **2014**, *72*, 799–808. [CrossRef]
9. Bekhta, P.; Proszyk, S.; Krystofiak, T.; Mamonova, M.; Pinkowski, G.; Lis, B. Effect of thermomechanical densification on surface roughness of wood veneers. *Wood Mater. Sci. Eng.* **2014**, *9*, 233–245. [CrossRef]
10. Bekhta, P.; Proszyk, S.; Krystofiak, T.; Lis, B. Surface wettability of short-term thermo-mechanically densified wood veneers. *Eur. J. Wood Prod.* **2015**, *73*, 415–417. [CrossRef]
11. Bekhta, P.; Proszyk, S.; Krystofiak, T.; Sedliačik, J.; Novák, I.; Mamonova, M. Effects of short-term thermomechanical densification on the structure and properties of wood veneers. *Wood Mater. Sci. Eng.* **2017**, *12*, 40–54. [CrossRef]
12. Bekhta, P.; Krystofiak, T.; Proszyk, S.; Lis, B. Evaluation of dynamic contact angle of loose and tight sides of thermally compressed birch veneer. *Drv. Ind.* **2018**, *69*, 387–394. [CrossRef]
13. Navi, P.; Heger, F. Combined Densification and Thermo-Hydro-Mechanical Processing of Wood. *MRS Bull.* **2004**, *29*, 332–336. [CrossRef]

14. Bekhta, P.; Hiziroglu, S.; Shepelyuk, O. Properties of plywood manufactured from compressed veneer as building material. *Mater. Des.* **2009**, *30*, 947–953. [CrossRef]
15. Bekhta, P.; Krystofiak, T.; Proszyk, S.; Lis, B. Adhesion strength of thermally compressed and varnished wood (TCW) substrate. *Prog. Org. Coat.* **2018**, *125*, 331–338. [CrossRef]
16. Bekhta, P.A.; Sedliacik, J. Effect of Thermo-Mechanical Densification on the Surface Characteristics of Wood Veneers and Glue Consumption. In *Current and Future Trends of Thermo-Hydro-Mechanical Modification of Wood Opportunities for New Markets?* Book of Abstract; COST Action FP0904 Workshop: Nancy, France, 2012; pp. 136–137.
17. Slabejová, G.; Šmidriaková, M.; Fekiač, J. Gloss of transparent coating on beech wood surface. *Acta Fac. Xylologiae Zvolen* **2016**, *58*, 37–44. [CrossRef]
18. Pelit, H.; Budakçı, M.; Sönmez, A.; Burdurlu, E. Surface roughness and brightness of scots pine (*Pinus sylvestris*) applied with water-based varnish after densification and heat treatment. *J. Wood Sci.* **2015**, *61*, 586–594. [CrossRef]
19. Kesik, H.I.; Akyildiz, M.H. Effect of heat treatment on the adhesion strength of water based wood varnishes. *Wood Res.* **2015**, *60*, 987–994.
20. Altgen, M.; Militz, H. Thermally modified Scots pine and Norway spruce wood as substrate for coating systems. *J. Coat. Technol. Res.* **2017**, *14*, 531–541. [CrossRef]
21. Hrčková, M.; Koleda, P.; Koleda, P.; Barcík, Š.; Štefková, J. Color change of selected wood species affected by thermal treatment and sanding. *BioResources* **2018**, *13*, 8956–8975. [CrossRef]
22. Slabejová, G.; Vidholdová, Z.; Šmidriaková, M. Surface finishes for thermally modified beech wood. *Acta Fac. Xylologiae Zvolen* **2019**, *61*, 41–50. [CrossRef]
23. Vidholdová, Z.; Slabejová, G.; Šmidriaková, M. Quality of Oil- and Wax-Based Surface Finishes on Thermally Modified Oak Wood. *Coatings* **2021**, *11*, 143. [CrossRef]
24. Sedliačiková, M.; Aláč, P.; Moresová, M.; Sedliačik, I. Mapping the wood colour preferences among potential customers. *Acta Fac. Xylologiae Zvolen* **2021**, *63*, 163–173. [CrossRef]
25. de Paula, M.H.; de Mesquita, R.R.S.; Costa, M.D.A.; Gonçalez, J.C.; Ananías, R.A.; Janin, G. Effect of applying finishing products and sanding on the surface of marupa wood. *Maderas Cienc. Tecnol.* **2020**, *22*, 45–54. [CrossRef]
26. Čunderlík, I. *Štruktúra dreva*; Technical University in Zvolen: Zvolen, Slovakia, 2009.
27. Bekhta, P.; Niemz, P. Effect of high temperature on the change in color, dimensional stability and mechanical properties of spruce wood. *Holzforschung* **2003**, *57*, 539–546. [CrossRef]
28. Bekhta, P.; Mamonova, M.; Sedliačik, J.; Novák, I. Anatomical study of short-term thermo-mechanically densified alder wood veneer with low moisture content. *Eur. J. Wood Prod.* **2016**, *74*, 643–652. [CrossRef]
29. Bekhta, P.; Marutzky, R. Reduction of glue consumption in the plywood production by using previously compressed veneer. *Holz als Roh-und Werkstoff* **2007**, *65*, 87–88. [CrossRef]
30. Bekhta, P.; Niemz, P.; Sedliačik, J. Effect of pre-pressing of veneer on the glueability and properties of veneer-based products. *Eur. J. Wood Prod.* **2012**, *70*, 99–106. [CrossRef]
31. Bekhta, P.; Sedliačik, J.; Jones, D. Effect of short-term thermomechanical densification of wood veneers on the properties of birch plywood. *Eur. J. Wood Prod.* **2018**, *76*, 549–562. [CrossRef]
32. Reinprecht, L.; Mamoňová, M.; Pánek, M.; Kačík, F. The impact of natural and artificial weathering on the visual, colour and structural changes of seven tropical woods. *Eur. J. Wood Prod.* **2018**, *76*, 175–190. [CrossRef]
33. Kubovský, I.; Oberhofnerová, E.; Kačík, F.; Pánek, M. Surface changes of selected hardwoods due to weather conditions. *Forests* **2018**, *9*, 557. [CrossRef]
34. Pandey, K.K. A note on the influence of extractives on the photo-discoloration and photo-degradation of wood. *Polym. Degrad. Stab.* **2005**, *87*, 375–379. [CrossRef]
35. Chang, H.T.; Yeh, T.F.; Chang, S.T. Comparisons of chemical characteristic variations for photodegraded softwood and hardwood with/without polyurethane clear coatings. *Polym. Degrad. Stab.* **2002**, *77*, 129–135. [CrossRef]
36. Kubovský, I.; Kačík, F. Changes of the wood surface colour induced by CO_2 laser and its durability after the xenon lamp exposure. *Wood Res.* **2013**, *58*, 581–590.
37. Laskowska, A. The influence of ultraviolet radiation on the colour of thermo-mechanically modified beech and oak wood. *Maderas Cienc. Tecnol.* **2020**, *22*, 55–68. [CrossRef]
38. Salca, E.A.; Krystofiak, T.; Lis, B.; Mazela, B.; Proszyk, S. Some coating properties of black alder wood as a function of varnish type and application method. *BioResources* **2016**, *11*, 7580–7594. [CrossRef]
39. Salca, E.A.; Krystofiak, T.; Lis, B. Evaluation of selected properties of alder wood as functions of sanding and coating. *Coatings* **2017**, *7*, 176. [CrossRef]
40. Vidholdová, Z.; Slabejová, G. Colour Stabilisation of Surface of Four Thermally Modified Woods with Saturated Water Vapour by Finishes. *Polymers* **2021**, *13*, 3373. [CrossRef] [PubMed]
41. Kubovský, I.; Kačík, F.; Reinprecht, L. The impact of UV radiation on the change of colour and composition of the surface of lime wood treated with a CO_2 laser. *J. Photochem. Photobiol. A* **2016**, *322*, 60–66. [CrossRef]
42. Kúdela, J.; Štrbová, M.; Jaš, F. Influence of accelerated ageing on colour and gloss changes in tree of heaven surface treated with an iruxil coating system. *Acta Fac. Xylologiae Zvolen* **2016**, *58*, 25–34. [CrossRef]

43. Kúdela, J.; Kubovský, I. Accelerated-ageing-induced photo-degradation of beech wood surface treated with selected coating materials. *Acta Fac. Xylologiae Zvolen* **2016**, *58*, 27–36. [CrossRef]
44. Reinprecht, L.; Pánek, M. Effects of wood roughness, light pigments, and water repellent on the color stability of painted spruce subjected to natural and accelerated weathering. *BioResources* **2015**, *10*, 7203–7219. [CrossRef]
45. Salca, E.-A.; Krystofiak, T.; Lis, B.; Hiziroglu, S. Glossiness evaluation of coated wood surfaces as function of varnish type and exposure to different conditions. *Coatings* **2021**, *11*, 558. [CrossRef]
46. Salca, E.-A.; Cismaru, I. Colour changes evaluation of freshly cut alder veneers under the influence of indoor sunlight. *Pro. Ligno* **2011**, *7*, 15–24.
47. Cirule, D.; Kuka, E.; Kevers, M.; Andersone, I.; Andersons, B. Photodegradation of Unmodified and Thermally Modified Wood Due to Indoor Lighting. *Forests* **2021**, *12*, 1060. [CrossRef]
48. Cividini, R.; Travan, L.; Allegretti, O. White beech: A tricky problem in the drying process. In Proceedings of the International Scientific Conference on Hardwood Processing, Québec City, QC, Canada, 24–26 September 2007; pp. 135–140.
49. Diouf, P.N.; Stevanovic, T.; Cloutier, A.; Fang, C.H.; Blanchet, P.; Koubaa, A.; Mariotti, N. Effects of thermo-hygro-mechanical densification on the surface characteristics of trembling aspen and hybrid poplar wood veneers. *Appl. Surf. Sci.* **2011**, *257*, 3558–3564. [CrossRef]
50. Laskowska, A.; Marchwicka, M.; Boruszewski, P.; Wyszyńska, J. Chemical composition and selected physical properties of oak wood (*Quercus robur* L.) modified by cyclic thermo-mechanical treatment. *BioResources* **2018**, *13*, 9005–9019. [CrossRef]
51. Moya, R.; Fallas, R.S.; Bonilla, P.J.; Tenorio, C. Relationship between wood color parameters measured by the CIELab system and extractive and phenol content in Acacia mangium and Vochysia guatemalensis from fast-growth plantations. *Molecules* **2012**, *17*, 3639–3652. [CrossRef]
52. Yazaki, Y.; Collins, P.J.; Mccombe, B. Variations in hot water extractives content and density of commercial wood veneers from blackbutt (*Eucalyptus pilularis*). *Holzforschung* **1994**, *48*, 107–111. [CrossRef]
53. Chang, S.-T.; Hon, D.N.-S.; Feist, W.C. Photodegradation and photoprotection of wood surfaces. *Wood Fiber* **1982**, *14*, 104–117.
54. Pandey, K.K. Study of the effect of photo-irradiation on the surface chemistry of wood. *Polym. Degrad. Stab.* **2005**, *90*, 9–20. [CrossRef]
55. Hill, C.A.S. *Wood Modification: Chemical, Thermal and Other Processes*; John Wiley & Sons, Ltd.: Chichester, UK, 2006.
56. Tolvaj, L.; Nemeth, R.; Pasztory, Z.; Bejo, L.; Takats, P. Colour Stability of Thermally Modified Wood during Short Term Photodegradation. *BioResources* **2014**, *9*, 6644–6651. [CrossRef]
57. Ayadi, N.; Lejeune, F.; Charrier, F.; Charrier, B.; Merlin, A. Color stability of heat-treated wood during artificial weathering. *Holz als Roh-und Werkstoff* **2003**, *61*, 221–226. [CrossRef]
58. Herrera, R.; Sandak, J.; Robles, E.; Krystofiak, T.; Labidi, J. Weathering resistance of thermally modified wood finished with coatings of diverse formulations. *Prog. Org. Coat.* **2018**, *119*, 145–154. [CrossRef]
59. Deka, M.; Petrič, M. Photo-degradation of water borne acrylic coated modified and non-modified wood during artificial light exposure. *BioResources* **2008**, *3*, 346–362.
60. Miklečić, J.; Jirouš-Rajković, V.; Antonović, A.; Španić, N. Discolouration of thermally modified wood during simulated indoor sunlight exposure. *BioResources* **2011**, *6*, 434–446. [CrossRef]
61. Evans, P.D.; Michell, A.J.; Schmalzl, K.J. Studies of the degradation and protection of wood surfaces. *Wood Sci. Technol.* **1992**, *26*, 151–163. [CrossRef]

forests

MDPI

Communication

Some Properties of Wood Plastic Composites Made from Rubberwood, Recycled Plastic and Silica

Aujchariya Chotikhun [1], Jitralada Kittijaruwattana [1], Wa Ode Muliastuty Arsyad [2], Emilia-Adela Salca [3,*], Yusuf Sudo Hadi [4] and Salim Hiziroglu [5]

[1] Faculty of Science and Industrial Technology, Prince of Songkla University, Surat Thani Campus, Mueang, Surat Thani 84000, Thailand; aujchariya.c@psu.ac.th (A.C.); jitralada.k@psu.ac.th (J.K.)
[2] Forest Products Research and Development, Jalan Gunung Batu, Bogor 16610, Indonesia; waodemuliastuty@gmail.com
[3] Faculty of Furniture Design and Wood Engineering, Transilvania University of Brasov, 5000068 Brasov, Romania
[4] Forest Products Department, Faculty of Forestry and Environment, Bogor Agricultural University, Kampus IPB Darmaga, Bogor 16680, Indonesia; yshadi@indo.net.id
[5] Department of Natural Resource Ecology and Management, Oklahoma State University, Stillwater, OK 74078-6013, USA; salim.hiziroglu@okstate.edu
* Correspondence: emilia.salca@unitbv.ro

Abstract: The objective of this work was to evaluate some of the properties of experimental wood plastic composite (WPC) panels manufactured from a low percentage of rubberwood (*Hevea brasiliensis* Muell. Arg), waste polyethylene terephthalate (PET) and silica at three different ratios. It was determined that water absorption values of the samples decreased with the increasing amount of PET in the panels. The lowest absorption value of 0.34% was determined for the samples having 40% PET in their content as a result of 24-h soaking. The highest hardness value of 4492 N was found for the samples made with the combination of rubberwood, PET and silica at 10%, 40% and 50%, respectively. The compressive strength of WPC specimens also followed a similar trend with the hardness characteristics of the panel and improved with increasing PET percentage. Statistical analyses revealed that values of compression strength, hardness, 2-h and 24-h water absorption of the specimens made with 20, 30, and 40% PET content resulted in significant difference from each other ($p \leq 0.0001$). Based on the findings in this study it appears that increasing silica content in the samples adversely influenced their mechanical properties while creating a certain level of enhancement of water absorption of the specimens. It seems that using a combination of waste PET and a limited amount of silica with a low percentage of wood particles could have the potential to produce value-added environmentally friendly composites to be used for different applications.

Keywords: wood plastic composite; rubberwood; polyethylene terephthalate; silica

Citation: Chotikhun, A.; Kittijaruwattana, J.; Arsyad, W.O.M.; Salca, E.-A.; Hadi, Y.S.; Hiziroglu, S. Some Properties of Wood Plastic Composites Made from Rubberwood, Recycled Plastic and Silica. *Forests* **2022**, *13*, 427. https://doi.org/10.3390/f13030427

Academic Editor: Milan Gaff

Received: 9 February 2022
Accepted: 7 March 2022
Published: 9 March 2022

1. Introduction

Wood-plastic composite (WPCs) are panel or lumber products made from recycled plastic and small wood particles or fibers being low-carbon and environmentally friendly value-added material. Wood plastic composites are relatively new as compared to the long history of natural lumber or traditional wood composites such as particleboard or fiberboard. They are manufactured by mixing wood particles as fine flour and recycled plastics to be used for indoor and outdoor applications in the U.S. and many Asian countries. Due to the ncreasing demand for WPC, new products are being developed such as door stiles, rails, and window lineal.A typical manufacturing process of WPC involves a combination of wood and thermoplastic, such as high-density polyethylene (HDPE), low-density polyethylene (LDPE), and polyvinyl chloride (PVC), which are mixed into a dough-like-consistency, called compounding. Mixing can be carried out by either batch or continuous process. In addition to the main ingredient, which is wood with a grain size ranging from 20 to 60 mesh, plastic coupling agents, stabilizers, foaming agents or dyes are also added to enhance the properties of the final product for specific use, including window and stair rails [1–5].

Wood-plastic composites have satisfactory strength properties as well as excellent hydrophobic characteristics, which prevent water and entrapping air onto its surface, so they are ideal products for outdoor applications [6].

Globally, 6–7 billion tons of plastic wastes have been produced per year in the form of different materials such as disposed of polythene bags, single-use face masks, cups, and water bottles [7]. Polyethylene terephthalate (PET) is a widely used synthetic plastic that is polymerized by terephthalic acid (TPA) and ethylene glycol (EG) [8]. It has become an indispensable part of daily life since disposable plastic bottles were initially made in the 20th century [9]. Currently, the main methods used to manage and eliminate PET waste include landfilling, incineration, as well as physical and chemical recycling [10]. Some of the initial studies investigated the panels manufactured using PET as a partial substitute for sand in concrete [11,12]. Additionally, recent studies investigated the potential of PET to be used as a raw material in WPC production [13,14]. Additional works also evaluated the characteristics of WPCs manufactured from different wood species by the flat platen pressing process [15,16].

Rubberwood (*Hevea brasiliensis* Muell. Arg) is grown in tropical forest zones in the form of plantations and plays a significant role in the economy in Southeast Asian counties. Wood is comparable to oak, containing an amount of cellulose, hemicellulose and lignin with approximate values of 38–40%, 28–31% and 21–24%, respectively [17]. Rubbertree represents a renewable and environmentally friendly material and, typically, any tree that is over 7 years old can produce latex until it reaches 30 years. Nonproductive trees are mainly used for furniture production and wood-based panels including particleboard and fiberboard in Thailand [18,19].

In a previous study carried out by Ramesh et al. different aspects of wood based polymer composites were reviewed and the use of various additives to improve the overall properties, including mositure resistance and bonding strength of the experimental samples was emphasized [20]. The characteristics of wood plastic composites manufactured from pecan orchard waste were also investigated in a study, determining that increasing the amount of pecan flour in the panels increased their tensile properties [21]. Along this line, the dimensional stability and mechanical properties of WPC panels have been improved with different approaches including acetylation, silane treatment, and thermal treatment of wood particles [22–25]. Application of different materials as fillers, such as silica, would also be considered an alternative method within the perspective of these approaches [25]. It is a well-known fact that silica is one of the most abundant available materials with a low cost. It has been used as fillers in the manufacturing of WPCs in past studies [24–26].

It is a fact that waste plastic material is a major problem, creating an adverse influence on the environment. Therefore, it is vital to reduce, reuse, and recycle such waste resources so that they can be managed effectively and efficiently. Scientists have been searching for innovative and sustainable approaches to reuse and recycle plastic wastes to reduce their negative impact on the environment [27]. It is a well accepted fact that manufacturing value-added composite from waste plastic would be considered as one of these approaches. There are past studies that have investigated the properties of WPCs manufactured from different wood species, including rubberwood [28–32]. However, there is almost none or very limited information on the properties of experimental panels with a combination of rubberwood particles, recycled plastic and silica.

Therefore, the main objective of this work is to manufacture experimental panels from such combinations and determine their basic properties to understand how this material can be used with better effectiveness for different applications.

2. Materials and Methods

Commercially produced rubberwood (*Hevea brasiliensis* Muell. Arg) sawdust supplied by BNS Wood Industry Co., Ltd. In Surat Thani, Thailand, was used to produce the samples. Sawdust was classified into particle size on 18–40 mesh screen. Polyethylene terephthalate (PET) from plastic bottom waste was provided and shredded, employing a hammermill manufactured by Wagner Inc, in Austria into small particles. Silica with 18–40 mesh size was bought from Huatanon Co., Ltd., Kanchanadit, Surat Thani province, Thailand. Initially, both rubberwood particles and silica were dried at a temperature of 102 ± 3 °C for 24 h before the mixing process was carried out.

2.1. Manufacturing of the WPC Samples

The WPC samples were manufactured based on three different composition ratios of rubberwood, PET, and silica by weight, as displayed in Table 1. All materials were mixed in a laboratory-type reactor (PSU, Songkhla, Thailand) heated at a temperature of 180 °C and manually stirred for 5–10 min until becoming a homogeneous compound. In the next step it was transfored into a square frame of 300 mm by 300 mm in 5 mm thickness. Each mat was compressed using a pressure of 5.5 MPa at a temperature of 180 °C for 10 min in a computer-controlled press, Chareon Tut Co., Ltd., Bang Phli, Thailand. Afterward, the panels were cooled off for 20 min until they were formed and cured completely. Later, the panels were conditioned in a controlled room having a temperature of 25 ± 2 °C and relative humidity of $65 \pm 2\%$ for a week before the samples were cut for different tests. An average target panel density was 1.46 g/cm^3.

Table 1. The composition of WPC samples.

Sample Type	Rubberwood (%*w*/*w*)	PET (%*w*/*w*)	Silica (%*w*/*w*)
WPC-1	10	20	70
WPC-2	10	30	60
WPC-3	10	40	50

2.2. Water Absorption Test of the Samples

The water absorption test (WA) was carried out according to ASTM D1037-12 standard [33]. A total of nine specimens with dimensions of 50 mm by 50 mm by 15 mm from each type of panel were cut using a bandsaw for the tests. At the end of the 2-h and 24-h tests, the specimens were taken out from the water and all surface water was removed by wiping before they were weighed at an accuracy of 0.01 g.

2.3. Janka Hardness Test of the Samples

The Universal Testing Machine, Tinius Olsen, Series,100KU (Redhill, UK) was employed for the Janka hardness test. The ASTM standard [33] was applied. A total of 10 samples with dimensions of 50 mm by 50 mm by 15 mm from each panel type were used. The samples were embedded by a hemisphere steel having 11.2 mm diameter on their surface as depicted in Figure 1.

2.4. Compressive Strength Test of the Samples

The compressive strength test of the samples was carried out on the Automatic Compression Testing Machine, TTR-D 080G Series: KC-2000 based on ASTM C109/C109M-02 standard [34].

A total of 10 samples with the dimensions of 50 mm by 50 mm by 50 mm from each panel type were considered for the compression test. The compressive strength of each specimen was determined based on the equation below:

$$CS = \frac{F}{A}, \quad (1)$$

where CS is the compressive strength (N/mm^2), F is the maximum force or load (N) at the point of failure and A is the initial cross-section surface area (mm^2).

Figure 1. Hardness test of the samples.

2.5. Micrographs by SEM

Three samples of 5 mm by 10 mm by 6 mm were cut from each type of WPC panel for microscopic evaluation. The samples were coated with a thin gold layer before micrographs were taken on a scanning electron microscope (SEM), FEI Quanta 250.

2.6. Processing of Data

Analysis of variance (ANOVA) was used to evaluate the significant differences among the three types of WPC specimens by using XLSAT in Microsoft Excel 365® (Microsoft, Redmond, WA, USA). A confidence level of the p-value = 0.05 was considered as displayed in Table 2.

Table 2. Statistical analysis of the tests.

Analysis of Variance (Compressive Strength)					
Source	DF	Sum of squares	Mean squares	F	Pr > F
Model	2	113.740	56.870	159.948	**<0.0001**
Error	6	2.133	0.356		
Corrected Total	8	115.874			
Analysis of Variance (Hardness)					
Source	DF	Sum of squares	Mean squares	F	Pr > F
Model	2	11,907,613.333	5,953,806.667	7.861	**0.007**
Error	12	9,088,480.000	757,373.333		
Corrected Total	14	20,996,093.333			
Analysis of Variance (2 h WA):					
Source	DF	Sum of squares	Mean squares	F	Pr > F
Model	2	11.074	5.537	705.830	**<0.0001**
Error	6	0.047	0.008		
Corrected Total	8	11.121			
Analysis of Variance (24 h WA):					
Source	DF	Sum of squares	Mean squares	F	Pr > F
Model	2	24.150	12.075	158.278	**<0.0001**
Error	6	0.458	0.076		
Corrected Total	8	24.607			

3. Results and Discussion

3.1. Water Absorption of the Samples

Table 3 displays the results of water absorption characteristics of the samples.

Table 3. Water absorption values of the samples.

Samples Type	Water Absorption (%)	
	2-h Soaking	24-h Soaking
WPC-1	2.76 (0.08) *	4.28 (0.42)
WPC-2	1.28 (0.13)	1.68 (0.19
WPC-3	0.05 (0.02)	0.34 (0.14)

* Numbers in parentheses are standard deviation values.

Figure 2 also illustrates the images taken of surface of the water-soaked samples for 24 h by a single lens reflex digital camera. Panels with 40% PET and 50% silica along 10% rubberwood particles had the lowest water absorption values of 0.05% and 0.34% for 2-h and 24-h exposure, respectively. The corresponding findings were 1.28% and 1.68% for WPC-2 type panels manufactured with a 30%/60% PET and silica combination. The ANOVA of samples showed that the values are significantly different from each other ($p \leq 0.0001$). It appears that having higher silica content in the panels significantly enhanced their water absorption resistance. In addition, excellent water resistance of any kind of plastic-based materials is a well-known fact. In a previous study, experimental WPCs manufactured from eastern red cedar and polypropylene had improved dimensional stability when polypropylene content was increased in the samples [35]. Silica is widely used, where special applications for moisture resistance are desired. In the case of sample type WPC-3 with 40% PET and 50% silica increased the water resistance of the samples. Of course, using a relatively low amount of only 10% rubberwood with poor dimensional stability as compared to that of both PET and silica did not contribute to any substantial adverse effect in water absorption of the samples. Panel type WPC-1 manufactured with 20% PET and 70% silica had poor water resistance values. Such finding could be related to increased porosity due to gaps created using a large amount of silica content resulting in more void volume to attract water.

Figure 2. Surface of the water-soaked samples for 24-h: (**a**) WPC-1, (**b**) WPC-2, (**c**) WPC-3.

One should note that all three types of samples still had far better water absorption values that any other types of traditional wood composites such as particleboard or fiberboard.

In two previous studies, typical water absorption values for 24-h water soaking were found, ranging from 24.9% to 55.2% for particleboard and 25% to 30% for medium density fiberboard [36,37]. Figure 3 also illustrates the water absorption values of the three types of panels.

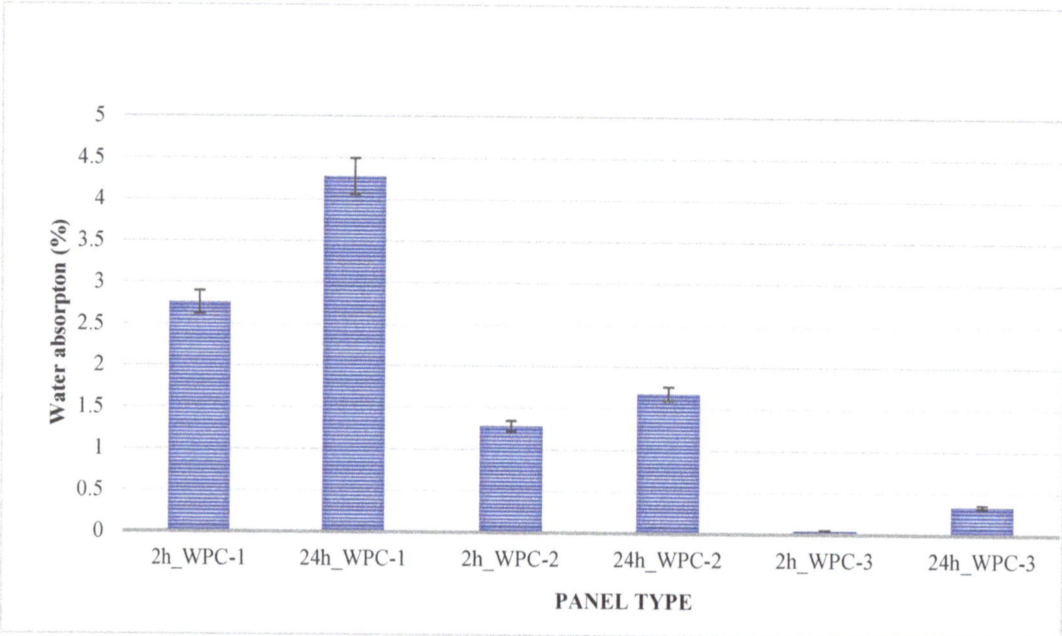

Figure 3. Water absorption of the samples.

3.2. *Janka Hardness of the Samples*

Mechanical properties of the samples are displayed in Table 4. The hardness of any kind of composites is an important characteristic, especially when they are targeted to be used for construction purposes. The hardness of WPCs is also a function of the polymer type, porosity of wood species, as well as its density. In general, a higher polymer load would result in a harder panel [38]. Having 50% silica and 40% PET in the samples showed the highest hardness value of 4922 N. Overall hardness of the samples reduced with decreasing PET content in the panels, as illustrated in Figure 4. An increasing percentage of silica in the samples also did not improve their hardness as in the case of panel type WPC-1 having 70% silica and 20% PET. In a previous study, it was also found that the initial addition of different types of fillers improved the hardness of WPC panels, however, their hardness was adversely influenced by the increase of those additives at the expense of polymer loading [38]. The means of groups are significantly different from each other, as shown in Table 2.

Table 4. Mechanical properties of the WPC samples.

Sample Type	Density (g/cm^3)	Hardness ** (N)	Compressive Strength ** (N/mm^2)
WPC-1	1.41 (0.07) *	2854 (292.80)	10.08(0.51)
WPC-2	1.41 (0.06)	4492 (1152.57)	15.55(0.31)
WPC-3	1.57 (0.02)	4922 (926.27)	18.69(0.84)

* Numbers in parentheses are standard deviations values. ** Highly significantly different ($p \leq 0.01$).

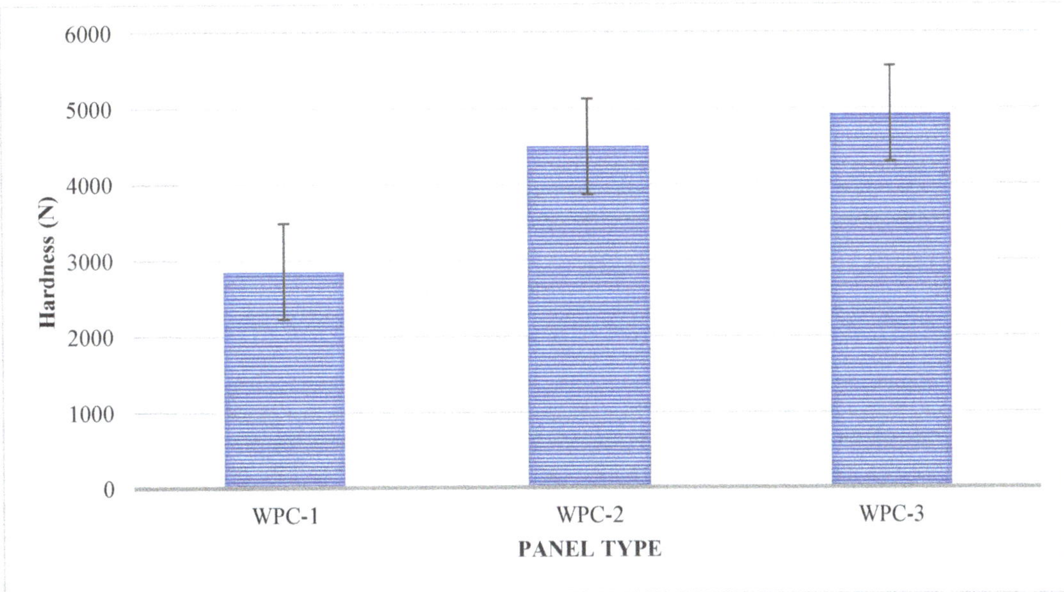

Figure 4. Hardness values of the samples.

3.3. Compression Strength of the Samples

The lowest compression strength value of 10.08 N/mm^2 was determined for the panel type WPC-1, having 20% PET and 70% silica, as shown in Figure 5. It seems that a high percentage of silica in the panels did not mix uniformly, creating a certain amount of gaps between the two major materials, resulting in lower strength values. This finding can be observed from the micrographs taken by SEM, as shown in Figure 6a. Similar to the hardness of the samples, the amount of polymer loading is also a main parameter influencing the overall mechanical properties of WPC. When the percentage of PET in the panels is increased sequentially from 30% to 40% with decreasing silica content, compression strength values of the samples was enhanced to a certain extent. An increased amount of PET in the samples had a more uniform mixture, resulting in higher compression strength values of the samples. The SEM micrographs taken from the panels surface of types WPC 2 and WPC 3 also supported such findings. The WPC-3 sample presented an increased value of the compression strength of 18.69 N/mm^2, which is almost 1.8 times higher than that of WPC-1 panels type.

It appears that both types of panels had a more uniform and homogeneous mixture of each material in the panels, so that their compression strength properties were improved, in WPC-2 and WPC-3, as shown in Figure 5. The compressive strength of WPC samples was also confirmed to be significantly different from each other based on the statistical analysis, as displayed in Table 4. The result showed statistical differences between the means of three independent groups ($p \leq 0.0001$).

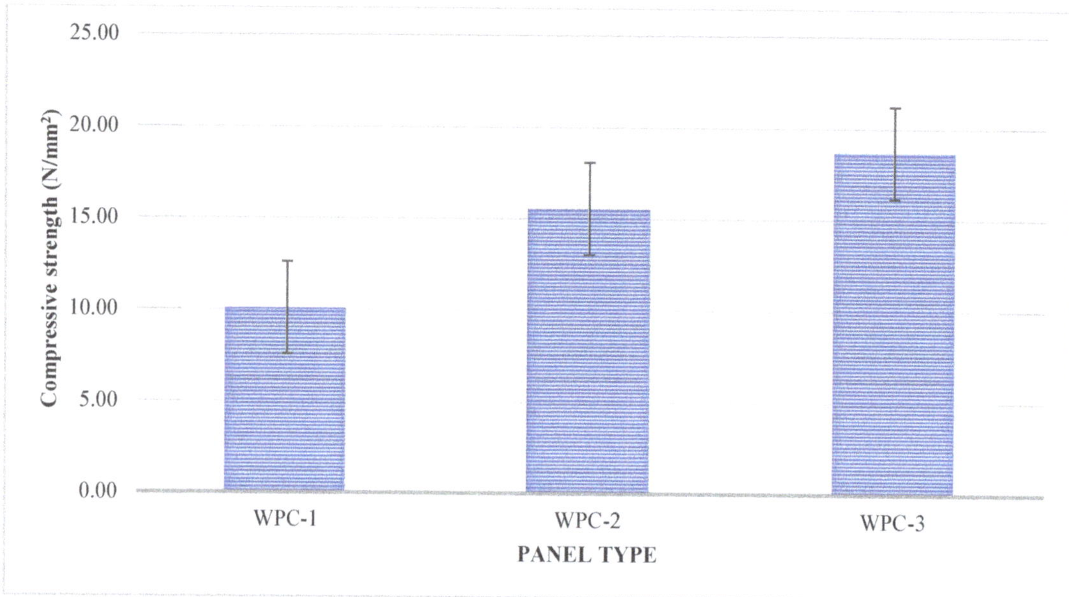

Figure 5. Compressive strength of the samples.

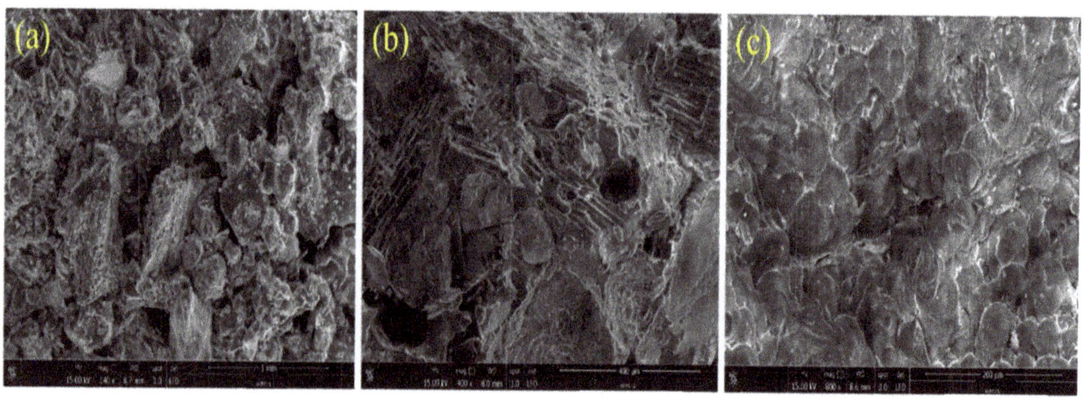

Figure 6. SEM micrographs of WPC samples: (**a**) WPC-1, (**b**) WPC-2, and (**c**) WPC-3.

4. Conclusions

In this work, some properties of experimental wood plastic composite (WPC) samples manufactured from a low percentage of rubberwood, waste polyethylene terephthalate (PET), and silica at three different ratios have been evaluated. Both hardness and compression strength values of the specimens were adversely influenced by increased silica content in the panels, while their water absorption properties were improved at a certain extent. It appears that using a higher amount of silica in the samples, creating a larger void volume for water to be located in, resulted in high water absorption characteristics. Overall, it seems that using a combination of waste PET and silica with a low percentage of wood particles could have the potential to produce value-added environmentally friendly composites to be used for different applications such as window and stairs rails.

Author Contributions: Conceptualization, A.C., Y.S.H., J.K. and S.H.; methodology, A.C., W.O.M.A., Y.S.H.; software, A.C. and Y.S.H.; validation, A.C., Y.S.H. and S.H.; formal analysis, A.C.; investigation, A.C.; resources, A.C. and J.K.; data curation, Y.S.H.; writing—original draft preparation, A.C., S.H. and E.-A.S.; writing—review and editing, S.H. and E.-A.S.; visualization, S.H. and E.-A.S.; supervision, S.H and Y.S.H.; project administration, A.C.; funding acquisition, A.C. All authors have read and agreed to the published version of the manuscript.

Funding: This research received no external funding.

Data Availability Statement: Not applicable.

Acknowledgments: This work was supported by the government budget allocated to Prince of Songkla University.

Conflicts of Interest: The authors declare no conflict of interest.

References

1. Özturk, S. Effect of Fiber Loading on the Mechanical Properties of Kenaf and Fiberfrax Fiber-reinforced Phenol-Formaldehyde Composites. *J. Compos. Mater.* **2010**, *44*, 2265–2288. [CrossRef]
2. Hiziroglu, S. What Is Wood Plastic Composite. Fact Sheet, FAPC 170. 2013. Available online: https://extension.okstate.edu/fact-sheets/print-publications/fapc-food-and-agricultural-products-center/what-is-wood-plastic-composite-fapc-170.pdf (accessed on 8 February 2022).
3. Ashori, A.; Behzad, H.M.; Tarmian, A. Effects of chemical preservative treatments on durability of wood flour/HDPE composites. *Compos. B Eng.* **2013**, *47*, 308–313. [CrossRef]
4. Taufiq, M.; Mansor, M.R.; Mustafa, Z. Characterisation of wood plastic composite manufactured from kenaf fibre reinforced recycled-unused plastic blend. *Compos. Struct.* **2018**, *189*, 510–515. [CrossRef]
5. Delviawan, A.; Suzuki, S.; Kojima, Y.; Kobori, H. The Influence of Filler Characteristics on the Physical and Mechanical Properties of Wood Plastic Composite(s). *Rev. Agric. Sci.* **2019**, *7*, 1–9. [CrossRef]
6. Li, X.; Ling, T.; Hung Mo, K. Functions and impacts of plastic/rubber wastes as eco-friendly aggregate in concrete—A review. *Constr. Build. Mater.* **2020**, *240*, 117869. [CrossRef]
7. Shahani, S.; Gao, Z.; Qaisrani, M.A.; Ahmed, N.; Yaqoob, H.; Khoshnaw, F.; Sher, F. Preparation and Characterisation of Sustainable Wood Plastic Composites Extracted from Municipal Solid Waste. *Polymers* **2021**, *13*, 3670. [CrossRef]
8. Qi, X.; Yan, W.; Cao, Z.; Ding, M.; Yuan, Y. Current Advances in the Biodegradation and Bioconversion of Polyethylene Terephthalate. *Microorganisms* **2022**, *10*, 39. [CrossRef]
9. Koshti, R.; Mehta, L.B.; Samarth, N. Biological Recycling of Polyethylene Terephthalate: A Mini-Review. *J. Polym. Environ.* **2018**, *26*, 3520–3529. [CrossRef]
10. Peng, R.; Xia, M.; Ru, J.; Huo, Y.; Yang, Y. Microbial degradation of polyurethane plastics. *Sheng Wu Gong Cheng Xue Bao (Chin. J. Biotechnol.)* **2018**, *34*, 1398–1409. [CrossRef]
11. Sambhaji, P.P. Use of Waste Plastic in Concrete Mixture as Aggregate Replacement. *Int. J. Adv. Eng. Res. Sci.* **2016**, *3*, 236956. [CrossRef]
12. Almeshal, I.; Tayeh, B.A.; Alyousef, R.; Alabduljabbar, H.; Mohamed, A.M. Eco-friendly concrete containing recycled plastic as partial replacement for sand. *J. Mater. Res. Technol.* **2020**, *9*, 4631–4643. [CrossRef]
13. Rahman, K.; Islam, M.N.; Rahman, M.M.; Hannan, M.O.; Dungani, R.; Khalil, H.A. Flat-pressed wood plastic composites from sawdust and recycled polyethylene terephthalate (PET): Physical and mechanical properties. *SpringerPlus* **2013**, *2*, 629. [CrossRef]
14. Chiang, C.L.; Li, Y.L.; Shen, M.Y. Effects of Environmental Aging on the Durability of Wood-Flour Filled Recycled PET/PA6 Wood Plastic Composites. *J. Polym. Environ.* **2021**. [CrossRef]
15. Fabiyi, J.S.; McDonald, A.G. Effect of wood species on property and weathering performance of wood plastic composites. *Compos. Part A Appl. Sci. Manuf.* **2010**, *41*, 1434–1440. [CrossRef]
16. Hung, K.; Yeh, H.; Yang, T.; Wu, T.; Xu, J.; Wu, J. Characterization of Wood-Plastic Composites Made with Different Lignocellulosic Materials that Vary in Their Morphology, Chemical Composition and Thermal Stability. *Polymers* **2017**, *9*, 726. [CrossRef]
17. Riyaphan, J.; Phumichai, T.; Neimsuwan, T.; Witayakran, S.; Sungsing, K.; Kaveata, R.; Phumichai, C. Variabilityin chemical and mechanical properties of Para rubber (*Hevea brasiliensis*) trees. *Sci. Asia* **2015**, *41*, 251–258. [CrossRef]
18. Verheye, W. Growth and production of rubber. In *Land Use, Land Cover and Soil Sciences*; Verheye, W.H., Ed.; UNESCO-EOLSS Publishers: Oxford, UK, 2010; pp. 295–300.
19. Chotikhun, A.; Kittijaruwattana, J.; Salca, E.A.; Hiziroglu, S. Selected Physical and Mechanical Properties of Microwave Heat Treated Rubberwood (*Hevea brasiliensis*). *Appl. Sci.* **2020**, *10*, 6273. [CrossRef]
20. Ramesh, M.; Rajeshkumar, L.; Sasikala, G.; Balaji, D.; Saravanakumar, A.; Bhuvaneswari, V.; Bhoopathi, R. A Critical Review on Wood-Based Polymer Composites: Processing, Properties, and Prospects. *Polymers* **2022**, *14*, 589. [CrossRef]
21. Díaz-Mendoza, J.M.; Valles-Rosales, D.J.; Park, Y.H.; Sabo, R.C. Micromechanical Modeling for Tensile Properties of Wood Plastic Composites: Use of Pruned Waste from Pecan Orchards as Sustainable Material for Reinforcement of Thermoplastic Composite. *Polymers* **2022**, *14*, 504. [CrossRef]

22. Ichazo, M.N.; Albano, C.; Gonzalez, J.; Perera, R.; Candal, M.V. Polypropylene wood flour composites: Treatments and properties. *Compos. Struct.* **2001**, *54*, 207–214. [CrossRef]
23. Lee, S.H.; Wang, S. Biodegradable polymers/bamboo fiber biocomposite with bio-based coupling agent. *Compos. A Appl. Sci. Manuf.* **2006**, *37*, 80–91. [CrossRef]
24. Ayrilmis, N.; Jarusombuti, S.; Fueangvivat, V.; Bauchongkol, P. Effect of thermal-treatment of wood fibers on properties of flat-pressed wood plastic composites. *Polym. Degrad. Stab.* **2011**, *96*, 818–822. [CrossRef]
25. Ayrilmis, N.; Kwon, J.H.; Han, T.H.; Durmus, A. Effect of Wood-derived Charcoal Content on Properties of Wood Plastic Composites. *Mater. Res.* **2015**, *18*, 654–659. [CrossRef]
26. Deka, B.K.; Maji, T.K. Effect of silica nanoflour on the properties of wood flour/polymer composite. *Polym. Eng. Sci.* **2012**, *52*, 1516–1523. [CrossRef]
27. Lamba, P.; Kaur, D.P.; Raj, S.; Sorout, J. Recycling/reuse of plastic waste as construction material for sustainable development: A review. *Environ. Sci. Pollut. Res.* **2021**, *16*, 1–24. [CrossRef]
28. Ayrilmis, N.; Jarusombuti, S. Flat-pressed Wood Plastic Composite as an Alternative to Conventional Wood-based Panels. *J. Compos. Mater.* **2011**, *45*, 103–112. [CrossRef]
29. Ratanawilai, T.; Lekanukit, P.; Urapantamas, S. Effect of rubberwood and palm oil content on the properties of wood-polyvinyl chloride composites. *J. Thermoplast. Compos. Mater.* **2014**, *27*, 719–730. [CrossRef]
30. Acosta, A.P.; Labidi, J.; Schulz, H.R.; Gallio, E.; Barbosa, K.T.; Beltrame, R.; Delucis, R.A.; Gatto, D.A. Thermochemical and Mechanical Properties of Pine Wood Treated by In Situ Polymerization of Methyl Methacrylate (MMA). *Forests* **2020**, *11*, 768. [CrossRef]
31. Feng, L.; Xie, W. Analysis of Factors Affecting Creep of Wood–Plastic Composites. *Forests* **2021**, *12*, 1146. [CrossRef]
32. Lv, X.; Hao, X.; Ou, R.; Liu, T.; Guo, C.; Wang, Q.; Yi, X.; Sun, L. Rheological Properties of Wood–Plastic Composites by 3D Numerical Simulations: Different Components. *Forests* **2021**, *12*, 417. [CrossRef]
33. *ASTM D 1037-12*; Standard Test Methods for Evaluating Properties of Wood-Base Fiber Panel Materials. ASTM International: West Conshohocken, PA, USA, 2020.
34. *ASTM C109/C109-M-02*; Standard Test Method for Compressive Strength of Hydraulic Cement Mortars. ASTM International: West Conshohocken, PA, USA, 2017.
35. Karumuri, S.; Hiziroglu, S.; Kalkan, A.K. The distribution and role of nanoclay in lignocellulose–polymer blends. *RSC Adv.* **2017**, *7*, 19406–19416. [CrossRef]
36. Copur, Y.; Guler, C.; Akyol, M.; Tascioglu, C. Some chemical properties of hazelnut husk and its suitablility for particleboard production. *Build Environ.* **2007**, *42*, 2568–2572. [CrossRef]
37. Ates, S.; Kara, H.; Olgun, C.; Ozkan, O. Effect of heat treatment on some propeteis of MDF. *Wood Mater. Sci. Eng.* **2017**, *12*, 158–164. [CrossRef]
38. Huuhilo, T.; Martikka, O.; Butylina, S.; Kärki, T. Mineral fillers for wood–plastic composites. *Wood Mater. Sci. Eng.* **2010**, *5*, 34–40. [CrossRef]

forests

MDPI

Article

Creep Properties of Densified Wood in Bending

Lei Han [1,2,*], Andreja Kutnar [1,2], José Couceiro [3] and Dick Sandberg [3]

1 Faculty of Mathematics, Natural Sciences and Information Technologies, University of Primorska, Glagoljaška 8, 6000 Koper, Slovenia; andreja.kutnar@innorenew.eu
2 InnoRenew CoE, Livade 6, 6310 Izola, Slovenia
3 Wood Science and Engineering, Luleå University of Technology, 93187 Skellefteå, Sweden; jose.couceiro@ltu.se (J.C.); dick.sandberg@ltu.se (D.S.)
* Correspondence: lei.han@innorenew.eu; Tel.: +386-069751131

Abstract: Thermo-hydro-mechanical (THM)-densified timber is rarely used in construction, although its mechanical properties are in many cases excellent. The main reason for its rare use is set-recovery, which reduces the degree of densification over time so that the mechanical properties deteriorate. Our knowledge of the long-term creep of densified timber is insufficient and a full understanding of its long-term behaviour is still lacking. The purpose of this study was to examine the behaviour under long-term loading of Scots pine sapwood densified in an open system at 170–200 °C. The influence of the THM densification process on the creep properties was studied on (1) unmodified specimens, (2) THM-densified specimens, (3) THM-densified specimens that had been further thermally treated, and (4) low-molecular-weight phenol-formaldehyde resin-impregnated and THM-densified specimens. All specimens were loaded at 20 ± 2 °C and 65 ± 5% relative humidity for 14 days under 3-point bending at 35% of the short-term ultimate load, and the bending deformation was registered. The THM densification doubled the density, causing a significant increase in the modulus of rupture but no change in the modulus of elasticity, and reduced the equilibrium moisture content and creep compliance. Post-thermal modification and resin impregnation improved the dimensional stability and further reduced the creep compliance in bending. The results demonstrate that THM densification combined with resin-impregnation or thermal modification reduces the creep of Scots pine timber under a long-term bending load in a constant climate.

Keywords: thermo-hydro-mechanical densification; Scots pine; creep compliance; thermal modification; resin impregnation

Citation: Han, L.; Kutnar, A.; Couceiro, J.; Sandberg, D. Creep Properties of Densified Wood in Bending. *Forests* **2022**, *13*, 757. https://doi.org/10.3390/f13050757

Academic Editors: Tomasz Krystofiak and Pavlo Bekhta

Received: 4 May 2022
Accepted: 12 May 2022
Published: 14 May 2022

1. Introduction

Timber has become an increasingly popular building material for commercial, medium and high-rise buildings worldwide due to the general concern regarding the environment and climate change. For timber to be competitive with other construction materials, wood material must be attractive from both a technical and an economic and environmental perspective. Most engineered timber construction elements such as cross-laminated timber (CLT) and glued-laminated timber (GLT) are manufactured from low-density softwoods (typically 300–600 kg m^{-3}) like Norway spruce and Scots pine. These low-density woods have, however, a limited capacity to support heavy loads, and the load-bearing span is limited due to the large deformations under load. Densification is a process to considerably improve the mechanical properties of low-density woods and make them more attractive as a raw material in timber construction elements, for example.

It is well known that there is a strong correlation between many important timber properties and the density, and densification, i.e., the transversal compression of wood, is a method of increasing the density and improving the properties primarily of low-density timber species [1]. In most cases, densification involves a combination of heat, moisture and pressure, i.e., thermo-hydro-mechanical (THM) densification, to deform the wood cells without fracturing the cell walls, leading to a reduction in the volume of voids within the cells. This means that the maximum bulk density achievable through densification is close to the density of the wood's cell wall, which is about 1500 kg m^{-3} [2]. The increase in hardness, modulus of elasticity (MOE) and modulus of rupture (MOR) in bending usually matches the increase in density achieved during densification [3–5].

During THM densification, the main wood components, except cellulose, become plastic and deform when the wood is under the combined influence of moisture, high temperature and mechanical compression. The semicrystalline cellulose macromolecules are, however, elastically deformed, and elastic strain energy is then stored in the helical semicrystalline microfibrils. The deformation is temporarily fixed by two phenomena: (1) the transfer of lignin from a rubbery state to a glassy state when the wood is cooled to below the glass-transition temperature, and (2) the formation of hydrogen bonds between cellulose and hemicellulose during drying. When the densified wood is again exposed to moisture and heat, the lignin is softened and hydrogen bonds between cellulose and hemicellulose are broken and combined with water molecules. As a result, densified wood cells tend to recover to their original shape and the improvement in properties resulting from densification is lost—a phenomenon known as set-recovery [6,7].

There are three basic ways to avoid set-recovery: (1) to make the cell wall hydrophobic to prevent the wood from being re-softened by the absorption of moisture, (2) to create covalent crosslinks between the wood constituents during densification, and (3) to release the elastic stresses stored inside the microfibrils [8,9]. In practice, steaming wood before or after densification is widely used to prevent the densified wood from recovering to its original shape due to stress relaxation, the degradation of hygroscopic hemicellulose or the breaking of crosslinks between adjacent cellulose molecules [7,10,11]. Other ways to reduce set-recovery are to impregnate the wood with a low molecular-weight resin that acts as a plasticizer before densification, and after curing, locks the wood in its densified shape [12] by modification with ionic liquids [13] or by mechanical locking [14].

With the successful elimination of set-recovery, densification creates the potential for the use of low-density wood in load-bearing structures. Timber used in construction (whether it is in its natural state or not), in engineered wood products or that is modified in some way, will be exposed to long-term loading and will thereby exhibit creep deformation and, in extreme circumstances, lead to early failure of the structure or at least give a negative impact on the serviceability during its service life. With insufficient knowledge of the long-term loading behaviour limits, however, the use of densified wood in timber construction may involve a potential safety risk [15,16].

Wood is a viscoelastic material, i.e., its behaviour under load is time-dependent, showing the characteristics of both elastic and viscous materials. The total creep behaviour can be divided into three main stages, as shown in Figure 1: (1) primary creep, which increases rapidly and then slows down progressively with time; (2) secondary creep, which is almost zero; and (3) tertiary creep where the deformation increases rapidly to failure [17]. Whether or not the tertiary stage occurs in a loaded timber element depends on the stress level (σ). In order to guarantee that the deformation never reaches the tertiary stage resulting in wood failure, Eurocode 5 (CEN 2004) specifies a stress limit (σ_{lim}), which depends on the climate and on the duration of the load [18].

Like all polymeric materials, wood exhibits a complex viscoelastic behaviour depending on the structure, temperature and moisture content (MC), and it is extremely anisotropic [20,21]. In general, the creep behaviour in wood can be divided into two categories that are dependent on the ambient temperature (T) and relative humidity (RH): viscoelastic creep and mechano-sorptive creep [22].

Viscoelastic creep is a deformation beyond the elastic recoverable deformation and arises when wood that is kept at a constant MC in the green state or at an equilibrium MC (EMC) below fibre saturation is instantly loaded to a constant stress level. Viscoelastic creep may occur as an effect of a change in temperature during loading while the MC is kept constant. Mechano-sorptive creep occurs as a consequence of a change in the MC of the material. When wood is under constant load and the RH cycles from wet to dry, the creep deformation also follows a cyclic pattern, a phenomenon called mechano-sportive creep. Unlike viscoelastic creep, which is time-dependent, the mechano-sportive component of the total creep is directly related to the rate of change and to the range of the change in MC. The accumulative deformation as a result of RH cycles is considerably greater than that of a constant RH, at a given stress level, and mechano-sportive creep can result in premature failure at a low stress level [23].

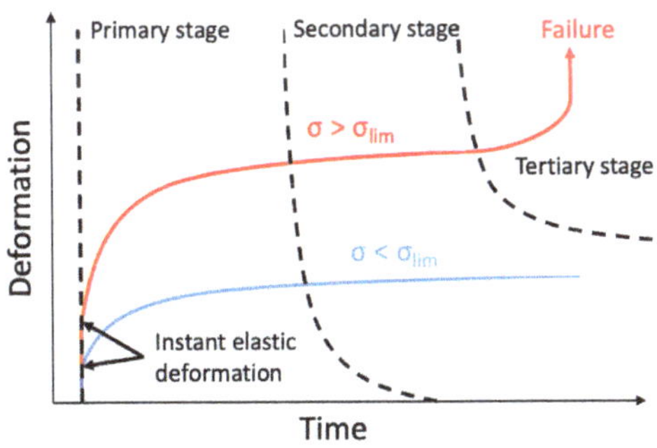

Figure 1. Creep deformation of timber under load. σ = applied load stress, σ_{lim} = ultimate stress limit [19].

Under conditions of moderate loading, constant temperature and constant MC, the time-dependent characteristics of wood can be measured in static (creep, stress relaxation) and dynamic tests. To characterise the linear viscoelastic behaviour of wood, creep and stress relaxation tests are generally preferred to dynamic tests. In static tests, the creep and relaxation functions are obtained explicitly on a time scale, whereas dynamic tests give a complex function related to the frequency. It is not always possible to convert complex dynamic functions to static creep or relaxation functions, but within the limits of linear viscoelasticity, the results obtained from these three methods are mathematically related [24].

Water acts as a natural plasticizer for wood, where the replacement of hydrogen bonds within the amorphous cellulose and hemicellulose increases the movability of the cell-wall constituents. Several studies have reported that viscoelastic creep compliance increases when the MC of wood increases [25–27]. Hering and Niemz [26] studied the viscoelastic creep of European beech timber loaded in bending to 25% of its ultimate strength at different MCs (8, 15 and 23%), and showed that the viscoelastic creep compliance increased by about 8 times when the MC increased from the lowest to the highest tested MC.

The temperature of the wood material affects the viscoelastic creep behaviour both directly and indirectly. With increasing temperature, the thermal action weakens the inter- and intra-molecular interactions such as hydrogen and Van der Waals forces between the main wood components and makes the macromolecular more flexible [28]. Engelund and Salmén [29] showed that the creep deformation of Norway spruce at constant MC increased when the temperature was raised from 10–45 °C. The influence of temperature on the creep becomes more apparent when the glass-transition temperature (Tg) of wood is passed [30]. The Tg of wood is related to the proportions of the main wood components and their MC. The softening of cellulose is limited due to its crystalline and hydrophobic nature, however, and the strong secondary forces between the molecules of crystalline cellulose raise the Tg above the temperature where wood begins to show severe thermal degradation. Because of the branched, highly crosslinked and hydrophobic molecular structure of lignin, its Tg is still above 80 °C when wood reaches the fibre-saturation point. In contrast, the amorphous cellulose and hemicelluloses are hydrophilic, and their Tg can be as low as 20 °C when the MC is high [31].

Because of the natural orthotropic structure of wood, the viscoelastic creep behaviour is also highly dependent on the loading direction, the loading mode and the stress level. The creep of timber when it is loaded perpendicular to the fibre direction can be several times greater than when it is loaded parallel to the fibres, but due to the existence of wood rays, the creep in the radial direction is less than that in the tangential direction. Shear loading results in greater creep than the creep in compression, tension or bending [32]. The ultimate stress level also has a significant effect on the creep level. Roszyk and Kaboorani [27,33] found that wood showed a linear viscoelastic behaviour in bending at stress levels up to 45% of its ultimate strength. Because the creep behaviour of wood is strongly dependent on the MC, Eurocode 5 (CEN 2004) takes into consideration the ambient climate by using the concept of "service class". The standard provides different strength reduction factors (k_{mod}) and different creep coefficients (k_{def}) at different service classes to estimate the load-bearing capacity and total deformation during the structural life. The service class is defined as the temperature and RH of the material in service [18]. For modified timber, the safety factors may differ from those of unmodified timber, but this is not yet well understood.

The objective of the present study was to examine the behaviour of THM densified Scots pine sapwood under long-term loading at a constant climate (Service Class 1 of Eurocode 5). The effects of resin impregnation prior to the THM treatment and thermal modification after the THM densification have also been examined.

2. Materials and Methods

2.1. Specimen Description

Kiln-dried Scots pine (*Pinus sylvestris* L.) sapwood with an MC of 9.4% and an average density of 480 kg m^{-3} (at ≈12% MC) was used. Sixty straight-grained, knot-and defect-free specimens 200 mm (longitudinal) × 20 mm (radial) × 20 mm (tangential) in size were prepared from a single piece of sawn timber. The specimens were randomly distributed into four different groups: (1) reference (R), (2) densified (D), (3) resin-impregnated and densified (RI-D) and, (4) densified and thereafter thermally modified (D-TM), as shown in Figure 2. Each group was divided into sub-groups of five replicates for set-recovery, three-point bending, and creep tests (Table 1).

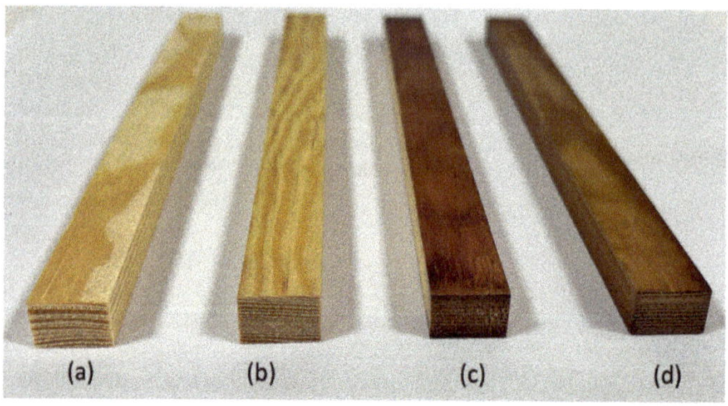

Figure 2. Specimens in the four groups: (**a**) reference without any treatment, (**b**) densified, (**c**) resin-impregnated and densified, and (**d**) densified and thereafter thermally modified.

Table 1. Test groups and numbers of specimens in each group.

Title 1	Set-Recovery	Three-Point Bending Short-Term Test	Three-Point Bending Creep Test
Reference (R)	5	5	5
Densified (D)	5	5	5
Resin-impregnated and densified (RI-D)	5	5	5
Densified and thereafter thermally modified (D-TM)	5	5	5

2.2. Wood Modification

An open-system hydraulic hot-press (Langzauner "Perfect" LZT-UK-30-L, Lambrechten, Austria) equipped with a water-cooling system was used for the THM densification of groups D, RI-D and D-TM. The specimens were compressed in the radial direction from 20 to 10 mm in thickness (50% target compression ratio). A metal stop with a height of 10 mm determined the endpoint. The THM densification included the following stages: (a) the upper and lower platens of the hot press were pre-heated to ≈170 °C, (b) the specimens were placed in the press, and a pressure of 4 MPa was applied in the radial direction with a closing speed of 3 mm s^{-1}, (c) the pressing conditions were held for 3 min after the press platen reached the metal stop, (d) the temperature of the upper and lower platens was raised to ≈200 °C and kept at this temperature for 2 min, and (e) the upper and lower platens were cooled to ≈60 °C with the specimens remaining under compression. The press was then opened, and the specimen dimensions were measured immediately (Figure 3).

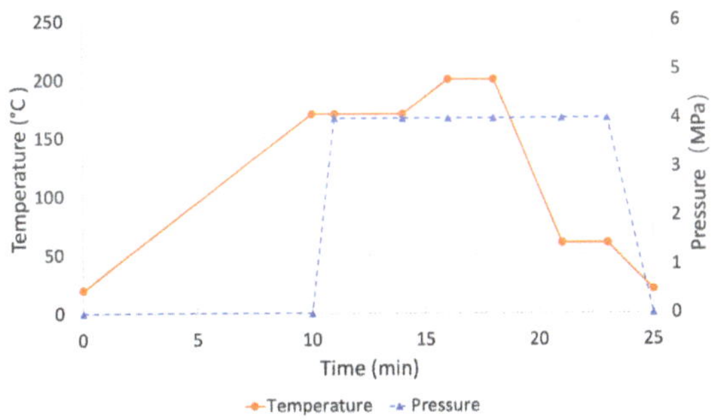

Figure 3. Schedule of thermo-hydro-mechanical (THM) treatment. Closing speed 3 mm s^{-1}.

An aqueous, low-molecular-weight phenol-formaldehyde (PF) resin was used for the impregnation process of group RI-D, supplied by Metadynea Austria GmbH (Krems an der Donau, Austria). A 30% solids content solution was prepared with distilled water. During the impregnation, the wood specimens were submerged in the solution and vacuum impregnated at 0.001 bar for 30 min. The specimens were then dried under indoor conditions for 12 h and afterwards at ≈60 °C in an oven for 24 h.

The thermal modification was carried out for group D-TM in a non-pressurized chamber, heated with a 4 kW electric resistance and a 1.4 bar steam generator following the schedule shown in Figure 4. The process was divided into three stages: (a) the temperature was progressively raised from room temperature (≈20 °C) to ≈200 °C (dry-bulb temperature) for 8 h, (b) the temperature was held at 200 °C for 2 h (thermal-modification stage), and (c) the chamber was cooled down to a temperature of ≈100 °C, and the specimens were removed after the chamber had reached room temperature.

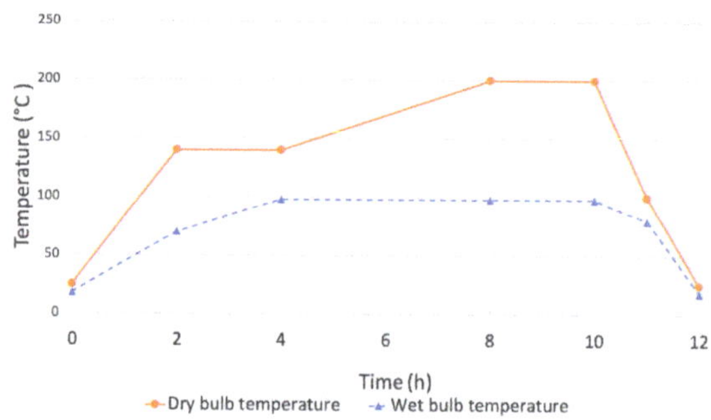

Figure 4. Schedule of thermal modification.

The specimens in all four groups were conditioned at 20 ± 2 °C and 65 ± 5% RH for 2 weeks, and then sawn to final dimensions of 200 mm (longitudinal) × 10 mm (radial) × 15 mm (tangential) before the physical tests were carried out.

2.3. Set-Recovery Test

Before densification, after press opening, and after conditioning at 20 ± 2 °C and 65 ± 5% RH for 2 weeks, the dimensions of the specimens were measured by a digital caliper with a precision of ±0.03 mm, and volume (*V*) was calculated. The mass (*M*) of each specimen was also measured on a balance at the same time. The density (ρ) was calculated as Equation (1).

$$\rho = \frac{M}{V} \tag{1}$$

The specimens with a dimension of 200 mm (longitudinal) × 10 mm (radial) × 15 mm (tangential) were kept in a convection oven at a constant temperature of ≈103 °C for 24 h, and the dimension in the densification direction (the radial direction) was measured. The specimens were then immersed in water (≈20 °C) for 24 h followed by oven-drying at ≈103 °C for 24 h. This wet–dry cycle was repeated twice. After each cycle, the specimens were weighed and the dimension in the radial direction was measured at three locations (5, 150 and 195 mm) in the length direction of the specimens with a digital calliper with a precision of ±0.03 mm. The water uptake (*W*) and the set-recovery (*SR*) were calculated as Equations (2) and (3).

$$W = \frac{M_1 - M_0}{M_0} \tag{2}$$

where M_1 is the mass after being immersed in water for 24 h, and M_0 is the oven-dry mass, and

$$SR = \frac{T'_0 - T_d}{T_0 - T_d} \tag{3}$$

where T'_0 is the oven-dry thickness after the wet-dry cycles, T_d is the actual thickness after densification and T_0 is the initial oven-dry thickness before densification.

2.4. Three-Point Bending Short-Term Test

The three-point bending test based on EN 408 (CEN 2012) [34] was carried out using a universal machine Zwick/Roell UTM Z100 (Zwick GmbH & Co. KG, Ulm, Germany) with a 100 kN load cell to determine the modulus of elasticity (*MOE*) and modulus of rupture (*MOR*). The loading span was set to 160 mm, and the speed of loading was controlled by a displacement rate of 5 mm min^{-1}. The specimens were loaded in the radial direction. *MOE* and *MOR* values were calculated as Equations (4) and (5).

$$MOE = \frac{Pl^3}{4bd^3\,\Delta} \tag{4}$$

$$MOR = \frac{3F_{max}l}{2bd^2} \quad (5)$$

where P is the load difference (N) in the elasticity zone, l is the supporting span (mm), b is the width (mm) of the specimens, d is the thickness (mm) of the specimens, Δ is the deflection (mm) at mid-length below the proportion of deflection limit, and F_{max} is the maximum load (N) when the specimen breaks.

2.5. Three-Point Bending Creep Test

The creep test was carried out in a custom KAPPA Multistation testing machine (Zwick GmbH & Co. KG, Ulm, Germany), able to accommodate five specimens simultaneously. The specimens were loaded as in the three-point bending with an applied stress level at 35% of the mean MOR value for each group given in Table 2. Each test was run for 14 days at 20 ± 2 °C and 65 ± 5% RH, in accordance with Service Class 1 as defined in Eurocode 5 (CEN 2004). The mid-span vertical deflection was recorded by an extensometer.

Table 2. Physical and flexural properties of untreated and treated specimens.

	After Press Opening	After 2 Weeks Conditioning at 20 ± 2 °C and 65 ± 5% RH						
Group	Thickness (mm)	Thickness (mm)	Density (kg m^{-3})	EMC (%)	MOE (GPa)	MOR (MPa)	Water Uptake (%)	Set-Recovery (%)
R	/	/	480 (±0.04) C	9.4 (±1.6) C	12.4 (±1.9) B	101.5 (±10.4) C	47.5 (±11.7) C	/
D	9.85 (±0.05) A	10.67 (±0.18) A	940 (±0.04) B	7.0 (±0.7) B	14.0 (±3.4) AB	166.8 (±25.9) AB	81.1 (±21.3) A	72.3 (±4.1) C
RI-D	9.83 (±0.05) A	10.09 (±0.06) B	980 (±0.04) A	5.2 (±0.7) A	19.4 (±4.1) A	212.7 (±38.5) A	23.8 (±14.5) B	24.7 (±11.3) A
D-TM	9.83 (±0.03) A	10.12 (±0.12) B	910 (±0.03) B	6.2 (±0.6) AB	17.0 (±4.3) AB	157.4 (±40.4) B	35.3 (±14.0) B	6.5 (±1.2) B

R is the reference group without any treatment; D is the densified group; RI-D is the resin-impregnated and densified group; D-TM is the densified and thereafter thermally modified group. Values with different letters behind indicate a statistically significant difference (Scheff's post-test, $p < 0.05$).

To facilitate comparison, the creep is expressed as the total creep compliance $D_t(t)$, creep compliance $D_c(t)$, initial compliance D_i, and the relative creep compliance $R_c(t)$, which were calculated according to Equations (6)–(8), respectively.

$$D(t) = \frac{4bd^3}{Fl^3}\delta \quad (6)$$

where b is the width (mm) of the specimens, d is the thickness (mm) of the specimens, F is the applied load (N), l is the span (mm) between support, δ is the deflection (mm) at time t,

$$D_t(t) = D_c(t) + D_i \quad (7)$$

where D_i is the initial compliance after the loading reached the applied stress target, and

$$R_c(t) = \frac{D_c(t)}{D_i} \quad (8)$$

where $D_c(t)$ is the creep compliance at time t and D_i is the initial creep compliance after the load reaches the applied stress target.

The anti-creep efficiency (ACE) is a value to quantify the ability of a specific modification to reduce the creep deformation [35], calculated as Equation (9).

$$ACE = \frac{DJ_u - DJ_t}{DJ_u} \quad (9)$$

2.6. Statistical Analysis

All the results are expressed as mean values with standard deviations. The normality and homogeneity of the data's variance were verified by the Shapiro–Wilk and the Levene's tests, respectively. The results were satisfactory for the application of parametric tests.

Therefore, a comparison between the results was performed by one-way analysis of variance (ANOVA) with Scheff's post-test at a significance level of 5%.

3. Results and Discussion

3.1. Physical and Flexural Properties

Table 2 shows the physical and flexural properties of the different groups of specimens. After press opening, the dimensions of the densified specimens in the compression direction (thickness direction) were slightly less than the end-stop limit (10 mm) due to shrinkage, which means that the 50% compression ratio was successfully achieved. The density of the RI-D group was slightly higher than that of the D group due to the mass gain during resin impregnation, and the D-TM group had a slightly lower density due to thermal degradation of the wood substance.

As shown in Table 2, after conditioning at 20 ± 2 °C and 65 ± 5% RH for 2 weeks, a small set-recovery was recorded for the modified specimens (D, RI-D, and D-TM). All the modified groups had a statistically significant lower EMC 20 ± 2 °C and 65 ± 5% RH compared with the untreated R group, especially the EMC of the RI-D group, which was only 5.2%. The mean MOE was slightly improved by densification, but there was no significant difference between the D and R groups. Nevertheless, the resin impregnation combined with densification (RI-D) gave a considerable increase in the MOE. The densification leads to an improvement by about 64% on the MOR. When combining resin-impregnation with densification, RI-D showed further increases on the MOR due to the impregnation induced higher density and permanent swelling of the cell wall [36,37]. The thermal modification, however, slightly reduced the densification-induced MOR improvement, presumably because the non-crystalline cellulose, hemicellulose and extractives were thermally degraded [7,10].

The resin impregnation and thermal modification reduced the water uptake and set-recovery considerably compared to the D group. The high plasticizing effect of the PF resin on the wood cell wall [12] can effectively decrease the inner stresses generated during densification and therefore prevent set-recovery if specimens are exposed to moisture, and the cured resin in the cell wall can provide excellent anti-swelling efficiency when exposed to a varying climate [36]. The post-thermal modification weakens the connection between the microfibrils and lignin and leads to hydrolysis of the hemicelluloses and, therefore, allows internal stresses to relax [38].

3.2. Three-Point Bending Creep

The creep compliance of Scots pine under constant climate was significantly changed by the densification (Table 3). The initial creep compliance of the modified groups (D, RI-D, and D-TM) was about half that of the untreated R-group, which can be attributed to the EMC reduction. There was no significant difference in the relative creep between the untreated R groups and the modified groups (D, RI-D, and D-TM) after 14 days.

Table 3. Result of bending creep test.

Groups	D_i (GPa^{-1})	D_c (14d) (GPa^{-1})	R_c (14d) (%)	ACE (14d) (%)
R	1.02 (±0.11) A	0.36 (±0.07) A	34.88 (±3.69) AB	/
D	0.57 (±0.11) B	0.21 (±0.04) B	36.88 (±3.71) A	41 A
RI-D	0.50 (±0.11) B	0.16 (±0.02) B	33.45 (±4.17) AB	54 A
D-TM	0.65 (±0.08) B	0.16 (±0.05) B	25.43 (±8.48) A	54 A

R is the reference group without any treatment; D is the densified group; RI-D is the resin-impregnated and densified group; D-TM is the densified and thereafter thermally modified group. D_i is the creep compliance after loading achieved target applied stress and called initial compliance; D_c (14d) is the creep compliance except for initial compliance at 14 days; R_c (14d) is relative creep at day 14; ACE (14d) is Anti-creep efficiency at day 14. Values with different letters behind indicate a statistically significant difference (Scheff's post-test, $p < 0.05$).

The creep compliance of all the groups showed a rapid increase in the first 2 days and followed by a gradual slowing down with time (Figure 5). At the end of the test, the creep compliance of the untreated R group was 0.36 GPa^{-1}, whereas that of the modified groups (D, RI-D, and D-TM) was only about half this value. Compared to the untreated R group, the creep compliance of all the modified groups (D, RI-D, and D-TM) was reduced by 40–55% after 14 days, which means that the densification reduced the creep deformation and that both resin impregnation and thermal modification enhanced this effect. This result agrees with a previous study of the viscoelastic properties of densified wood using dynamic mechanical analysis, which showed a lower creep compliance of densified wood under controlled RH [39]. None of the specimens reached the tertiary creep stage, where the deformation rapidly increases and leads to failure of the material.

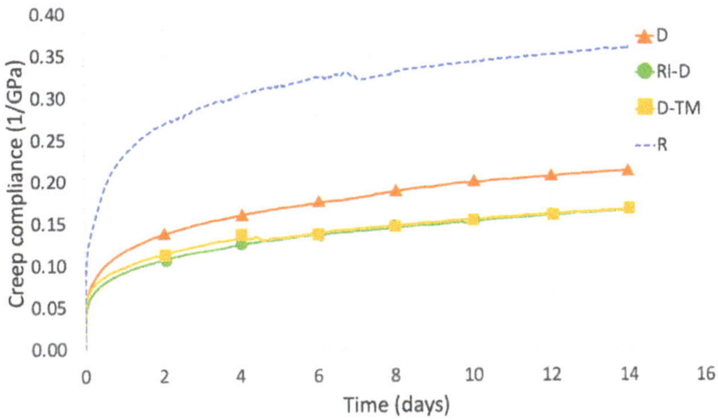

Figure 5. Average creep compliance of untreated and treated wood for 14 days. R is the reference group without any treatment; D is the densified group; RI-D is the resin-impregnated and densified group; D-TM is the densified and thereafter thermally modified group.

It was assumed that the modification-induced change in the chemical components and the change in hygroscopicity contribute to the creep compliance reduction. It has been reported that the crystallinity increases during densification due to the hydrolysis and depolymerization of amorphous wood components [40–42]. Since the change in cell wall crystallinity contributes to about 60% variation in the total creep deformation in a constant environment [43], the densification-induced crystallinity increase may also reduce the creep deformation. The lower viscoelastic creep at low EMC is because water is a natural plasticizer for wood. Compared with wood at a low EMC, there are more water molecules in wood with a high EMC and these break the hydrogen bonds within the amorphous components. As a result, there is more free space for the movement of polymer molecules, the intermolecular bonding energy is weakened and finally the slipping between molecular chains and the loosening of cell wall structure increases the creep compliance [27,44].

Impregnation with a thermosetting resin enhances the bonding between wood microfibres and therefore reduces the creep compliance [45,46]. Xiang [47] reported that thermal modification can also significantly increase the crystallinity, crystalline thickness, and crystalline length of wood, which reinforces the arrangement of cellulose molecular chains and therefore reduces creep compliance. In addition, the thermal modification induced condensation polymerization of lignin also increases the creep resistance [48].

4. Conclusions

THM densification leads to a significant increase in density and therefore increases the bending strength but not the MOE of Scots pine sapwood. The THM densified wood showed a reduced time-dependent creep under long-term constant load in a stable environment compared to that of untreated wood. The initial compliance and creep compliance of THM densified wood after 14 days of loading was reduced by 40% compared to that of untreated wood. In summary, THM densification shows the potential to increase the use of low-density wood in construction for long-term loading.

Both resin impregnation and thermal modification considerably reduced the set-recovery of the THM densified wood. In addition, resin impregnation and thermal modification combined with densification can maintain and enhance the densification-induced property improvement. The EMC is an essential factor influencing the creep behaviour of wood, and since the treatments applied in this study lead to a lower EMC of modified wood (group D, group RI-D, and group D-TM), this could be the primary reason for the reduced creep compliance. A future study should therefore focus on how the modification-induced change in EMC affects the bending creep behaviour.

Author Contributions: Conceptualization, L.H. and A.K.; methodology, L.H., J.C. and A.K.; software and analysis, L.H.; writing—original draft preparation, L.H.; writing—review and editing, L.H., A.K. and D.S.; supervision, A.K. and D.S.; funding acquisition, A.K. and D.S. All authors have read and agreed to the published version of the manuscript.

Funding: The authors acknowledge the European Commission for funding the InnoRenew project (Grant Agreement #739574) under the Horizon 2020 Widespread-Teaming program and the Republic of Slovenia (investment funding of the Republic of Slovenia and the European Regional Development Fund) and the Slovenian Research Agency ARRS for funding project J4-3087. The support of CT WOOD—a centre of excellence at Luleå University of Technology supported by the Swedish wood industry—is also gratefully acknowledged.

Conflicts of Interest: The authors declare no conflict of interest. The funders had no role in the design of the study; in the collection, analyses, or interpretation of data; in the writing of the manuscript, or in the decision to publish the results.

References

1. Seborg, R.M.; Millett, M.A.; Stamm, A.J.H. Heat-Stabilized Compressed Wood (Staypak). 1945. Available online: https://ir.library.oregonstate.edu/concern/defaults/vd66w413h (accessed on 10 April 2006).
2. Kutnar, A.; Sandberg, D.; Haller, P. Compressed and moulded wood from processing to products. *Holzforschung* **2015**, *69*, 885–897. [CrossRef]
3. Pelit, H.; Budakçı, M.; Sönmez, A. Density and some mechanical properties of densified and heat post-treated uludağ fir, linden and black poplar woods. *Eur. J. Wood Wood Prod.* **2018**, *76*, 79–87. [CrossRef]
4. Li, L.; Gong, M.; Yuan, N.; Li, D. An optimal Thermo-Hydro-Mechanical Densification (THM) process for densifying balsam fir wood. *BioResources* **2014**, *8*, 3967–3981. [CrossRef]
5. Pertuzzatti, A.; Missio, A.L.; de Cademartori, P.H.G.; Santini, E.J.; Haselein, C.R.; Berger, C.; Gatto, D.A.; Tondi, G. Effect of process parameters in the thermomechanical densification of *Pinus elliottii* and *Eucalyptus grandis* fast-growing wood. *BioResources* **2018**, *13*, 1576–1590. [CrossRef]
6. Navi, P.; Heger, F. Combined densification and thermo-hydro-mechanical processing of wood. *MRS Bull.* **2004**, *29*, 332–336. [CrossRef]
7. Navi, P.; Pizzi, A. Property changes in thermo-hydro-mechanical processing: COST action FP0904 2010-2014: Thermo-hydro-mechanical wood behavior and processing. *Holzforschung* **2015**, *69*, 863–873. [CrossRef]
8. Makinaga, M.; Norimoto, M.; Inoue, M. Permanent fixation of bending deformation of wood by steam treatment. *Wood Res. Bull. Wood Res. Inst. Kyoto Univ.* **1997**, *84*, 39–41.
9. Morsing, N. *Densification of Wood: The Influence of Hygrothermal Treatment on Compression of Beech Perpendicular to the Grain*; Series R; Department of Structural Engineering and Materials Technical, University of Denmark: Kgs. Lyngby, Denmark, 2000; Volume 79, p. 138.
10. Inoue, M.; Normotoi, M.; Tanahashi, M.; Rowell, R. Steam or heat fixation of compressed wood. *Wood Fiber Sci.* **1993**, *25*, 224–235.
11. Pelit, H.; Budakçi, M.; Sönmez, A. Effects of heat post-treatment on dimensional stability and water absorption behaviours of mechanically densified uludağ fir and black poplar woods. *BioResources* **2016**, *11*, 3215–3229. [CrossRef]
12. Shams, M.I.; Yano, H.; Endou, K. Compressive deformation of wood impregnated with low molecular weight Phenol Formaldehyde (PF) resin I: Effects of pressing pressure and pressure holding. *J. Wood Sci.* **2004**, *50*, 337–342. [CrossRef]
13. Neyses, B.; Karlsson, O.; Sandberg, D. The effect of ionic liquid and superbase pre-treatment on the spring-back, set-recovery and brinell hardness of surface-densified scots pine. *Holzforschung* **2020**, *74*, 303–312. [CrossRef]
14. Nilsson, J.; Johansson, J.; Kifetew, G.; Sandberg, D. Shape stability of modified engineering wood product subjected to moisture variation. *Wood Mater. Sci. Eng.* **2011**, *6*, 132–139. [CrossRef]
15. *ASTM D6815-09*; Standard Specification for Evaluation of Duration of Load and Creep Effects of Wood and Wood-Based Products. 2015. Available online: https://www.astm.org/Standards/D6815.htm (accessed on 27 December 2016).
16. Huang, Y. Creep behavior of wood under cyclic moisture changes: Interaction between load effect and moisture effect. *J. Wood Sci.* **2016**, *62*, 392–399. [CrossRef]

17. Findley, W.N.; Lai, J.S.; Onaran, K.; Christensen, R.M. Creep and relaxation of nonlinear viscoelastic materials with an introduction to linear viscoelasticity. *J. Appl. Mech.* **1977**, *44*, 364. [CrossRef]
18. *EN 1995-1-1*; Eurocode 5: Design of Timber Structures—Part 1.1: General Rules and Rules for Buildings. European Committee for Standardization: Brussels, Belgium, 2004.
19. Granello, G.; Palermo, A. Creep in timber: Research overview and comparison between code provisions. *N. Z. Timber Des. J.* **2019**, *27*, 6–22.
20. Hermawan, A.; Sakagami, H.; Fujimoto, N. Creep behaviour of Japanese cypress timber under various hygrothermal conditions. *Wood Mater. Sci. Eng.* **2021**, *16*, 1–9. [CrossRef]
21. Li, Z.; Jiang, J.; Lyu, J. Temperature and time dependence of orthotropic viscoelastic properties of moist wood determined in tensile mode. *Wood Mater. Sci. Eng.* **2022**, *17*, 1–7. [CrossRef]
22. Holzer, S.; Loferski, J.; Dillard, D. A review of creep in wood: Concepts relevant to develop long-term behavior predictions for wood structures. *Wood Fiber Sci.* **1989**, *21*, 376–392.
23. Armstrong, L.D.; Kingston, R.S.T. Effect of moisture changes on creep in wood. *Nature* **1960**, *185*, 862–863. [CrossRef]
24. Navi, P.; Sandberg, D. *Thermo-Hydro-Mechanical Processing of Wood*; EPFL Press: Lausanne, Switzerland, 2012; ISBN 9782940222411.
25. Moutee, M.; Fortin, Y.; Laghdir, A.; Fafard, M. Cantilever experimental setup for rheological parameter identification in relation to wood drying. *Wood Sci. Technol.* **2010**, *44*, 31–49. [CrossRef]
26. Hering, S.; Niemz, P. Moisture-dependent, viscoelastic creep of european beech wood in longitudinal direction. *Eur. J. Wood Wood Prod.* **2012**, *70*, 667–670. [CrossRef]
27. Kaboorani, A.; Blanchet, P.; Laghdir, A. A rapid method to assess viscoelastic and mechanosorptive creep in wood. *Wood Fiber Sci.* **2013**, *45*, 370–382.
28. Placet, V.; Passard, J.; Perré, P. Viscoelastic properties of wood across the grain measured under water-saturated conditions up to 135 °C: Evidence of thermal degradation. *J. Mater. Sci.* **2008**, *43*, 3210–3217. [CrossRef]
29. Engelund, E.T.; Salmén, L. Tensile creep and recovery of norway spruce influenced by temperature and moisture. *Holzforschung* **2012**, *66*, 959–965. [CrossRef]
30. Wang, J.; Wang, X.; He, Q.; Zhang, Y.; Zhan, T. Time-temperature-stress equivalence in compressive creep response of chinese fir at high-temperature range. *Constr. Build. Mater.* **2020**, *235*, 117809. [CrossRef]
31. Fengel, D.; Wegener, G. *Wood: Chemistry, Ultrastructure, Reactions*; Walter de Gruyter: Berlin, Germany, 1989; ISBN 3110839652.
32. Morlier, P. *Creep in Timber Structures*; CRC Press: Boca Raton, FL, USA, 1994; ISBN 9780429078767.
33. Roszyk, E.; Mania, P.; Moliński, W. The influence of microfibril angle on creep scotch pine wood under tensile stress along the grains. *Wood Res.* **2012**, *57*, 347–358.
34. *EN 408*; Timber Structures, Structural Timber and Glued Laminated Timber, Determination of Some Physical and Mechanical Properties. European Committee for Standardization: Brussels, Belgium, 2010.
35. Norimoto, M.; Gril, J.; Rowell, R.M. Rheological properties of chemically modified wood: Relationship between dimensional and creep stability. *Wood Fiber Sci.* **1992**, *24*, 25–35.
36. Gabrielli, C.P.; Frederick, A.E.; Kamke, A.; Gabrielli, C.P.; Kamke, F.A. Phenol–formaldehyde impregnation of densified wood for improved dimensional stability. *Wood Sci. Technol.* **2010**, *44*, 95–104. [CrossRef]
37. Schwarzkopf, M. Densified wood impregnated with phenol resin for reduced set-recovery. *Wood Mater. Sci. Eng.* **2021**, *16*, 35–41. [CrossRef]
38. Heger, F.; Groux, M.; Girardet, F.; Welzbacher, C.; Rapp, A.O.; Navi, P. Mechanical and durability performance of THM-densified wood. In Proceedings of the Final Workshop COST Action E22' Environmental Optimisation of Wood Protection, Lisboa, Portugal, 22–23 March 2004; pp. 1–10.
39. Kutnar, A.; O'Dell, J.; Hunt, C.; Frihart, C.; Kamke, F.; Schwarzkopf, M. Viscoelastic properties of thermo-hydro-mechanically treated beech (*Fagus Sylvatica*, L.) determined using dynamic mechanical analysis. *Eur. J. Wood Wood Prod.* **2021**, *79*, 263–271. [CrossRef]
40. Tanahashi, M.; Goto, T.; Horii, F.; Hirai, A.; Higuchi, T. Characterization of steam-exploded wood III. transformation of cellulose crystals and changes of crystallinity. *Mokuzai Gakkaishi* **1989**, *35*, 654–662.
41. Reiniati, I.; Osman, N.B.; Mc Donald, A.G.; Laborie, M.P. Linear viscoelasticity of hot-pressed hybrid poplar relates to densification and to the in situ molecular parameters of cellulose. *Ann. For. Sci.* **2015**, *72*, 693–703. [CrossRef]
42. Shi, J.; Peng, J.; Huang, Q.; Cai, L.; Shi, S.Q. Fabrication of densified wood via synergy of chemical pretreatment, hot-pressing and post mechanical fixation. *J. Wood Sci.* **2020**, *66*, 5. [CrossRef]
43. Lofty, M.; El-osta, M.; Wellwood, R.W. Short-term creep as related to cell-wall crystallinity. *Wood Fiber Sci.* **1972**, *4*, 204–211.
44. Navi, P.; Pittet, V.; Plummer, C.J.G. Transient moisture effects on wood creep. *Wood Sci. Technol.* **2002**, *36*, 447–462. [CrossRef]
45. Ranta-Maunus, A.; Kortesmaa, M. Creep of timber during eight years in natural environments. *World Conf. Timber Eng. Br. Columbia* **2000**, 2–7. Available online: https://www.semanticscholar.org/paper/Creep-of-Timber-during-Eight-Years-in-Natural-Ranta-Maunus-Kortesmaa/b251c80424177158a099e73f399858e49a4dbc20 (accessed on 3 May 2022).
46. Yue, K.; Liu, W.; Chen, Z.; Lu, X.; Lu, W. Investigation of the creep property of fast-growing poplar wood modified with low molecular weight resins. *BioResources* **2016**, *11*, 1620–1633. [CrossRef]
47. Xiang, E.; Huang, R.; Yang, S. Change in micromechanical behavior of surface densified wood cell walls in response to superheated steam treatment. *Forests* **2021**, *12*, 693. [CrossRef]

48. Hoseinzadeh, F.; Zabihzadeh, S.M.; Dastoorian, F. Creep behavior of heat treated beech wood and the relation to its chemical structure. *Constr. Build. Mater.* **2019**, *226*, 220–226. [CrossRef]

forests

Article

Evaluation of the Effect of a Combined Chemical and Thermal Modification of Wood though the Use of Bicine and Tricine

Dennis Jones [1,2,*], Davor Kržišnik [3], Miha Hočevar [3], Andreja Zagar [3], Miha Humar [3], Carmen-Mihaela Popescu [4,5], Maria-Cristina Popescu [4], Christian Brischke [6], Lina Nunes [7], Simon F. Curling [8], Graham Ormondroyd [8] and Dick Sandberg [1,2]

[1] Wood Science and Engineering, Luleå University of Technology, Forskargatan 1, SE-931 87 Skellefteå, Sweden; dick.sandberg@ltu.se
[2] Department of Forestry and Biomaterials, Czech University of Life Sciences Prague, 16500 Prague, Czech Republic
[3] Biotechnical Faculty, University of Ljubljana, Jamnikarjeva 101, 1000 Ljubljana, Slovenia; davor.krzisnik@bf.uni-lj.si (D.K.); miha.hoc@gmail.com (M.H.); andreja.zagar@bf.uni-lj.si (A.Z.); miha.humar@bf.uni-lj.si (M.H.)
[4] Petru Poni Institute of Macromolecular Chemistry of the Romanian Academy, 41A Grigore Ghica Voda Alley, 700487 Iasi, Romania; mihapop@icmpp.ro (C.-M.P.); cpopescu@icmpp.ro (M.-C.P.)
[5] Centre of Wood Science and Technology, Edinburgh Napier University, Edinburgh EH11 4EP, UK
[6] Wood Biology and Wood Products, University of Goettingen, Buesgenweg 4, D-37077 Goettingen, Germany; christian.brischke@uni-goettingen.de
[7] Structures Department, LNEC, National Laboratory for Civil Engineering, Av. do Brasil, 101, 1700-066 Lisbon, Portugal; linanunes@lnec.pt
[8] The Biocomposites Centre, Bangor University, Deiniol Road, Bangor LL57 2UW, UK; s.curling@bangor.ac.uk (S.F.C.); g.ormondroyd@bangor.ac.uk (G.O.)
* Correspondence: dennis.jones@ltu.se

Citation: Jones, D.; Kržišnik, D.; Hočevar, M.; Zagar, A.; Humar, M.; Popescu, C.-M.; Popescu, M.-C.; Brischke, C.; Nunes, L.; Curling, S.F.; et al. Evaluation of the Effect of a Combined Chemical and Thermal Modification of Wood though the Use of Bicine and Tricine. *Forests* **2022**, *13*, 834. https://doi.org/10.3390/f13060834

Academic Editors: Tomasz Krystofiak, Pavlo Bekhta and Angela Lo Monaco

Received: 5 April 2022
Accepted: 25 May 2022
Published: 27 May 2022

Abstract: The effects of thermal modification of wood have been well established, particularly in terms of reductions in mechanical performance. In recent years, there has been an increase in studies related to the Maillard reaction. More commonly associated with food chemistry, it involves the reaction of amines and reducing sugars during cooking procedures. This study has attempted to combine the use of amines and thermal modification, with subsequent properties investigated for the treatment of spruce (*Picea abies* (L.) H. Karst) and beech (*Fagus sylvatica* L.). In this initial study, the combined effects of chemical treatments by tricine and bicine were investigated with thermal modification. Along with some preliminary data on mechanical properties, the modifications which appeared in the wood structure were evaluated by infrared spectroscopy and biological studies according to EN113 and EN117 methodologies. The hierarchal study interpretation of FTIR suggested interactions between the bicine or tricine and the wood, which was partly supported by the analysis of volatile organic compounds (VOC), though other tests were not as conclusive. The potential of the method warrants further consideration, which will be described.

Keywords: wood; Maillard reaction; thermal/chemical treatment; mechanical properties; infrared spectroscopy; biological properties

1. Introduction

Wood has long been associated as an important natural resource, given its use by humanity to produce such diverse items as art and furniture and its use in construction. The use of wood is seen as a necessity for sustainable construction in our modern society [1]. As greater importance is placed upon the performance of wood as a material, it has become necessary to alter some of the inherently undesirable properties limiting the longevity of service performance. Of particular importance are issues relating to limitations related to stability in service, susceptibility to fungal decay or weathering, etc., [2,3]. Thus, ways of improving factors such as moisture sensitivity, low dimensional stability, low hardness and wear resistance, low resistance to bio-deterioration against fungi, termites, marine borers and low resistance to UV radiation have become important parameters in wood treatment technologies and in particular wood modification treatments.

Nowadays, wood modification is defined as a process adopted to improve the physical, mechanical or aesthetic properties of sawn timber, veneer or wood particles used in the production of wood composites. This process produces a material that can be disposed of at the end of a product's life cycle without presenting any environmental hazards greater than those that are associated with the disposal or combustion of unmodified wood. Recent publications [4,5] indicated how wood modification is becoming more established across Europe at both the commercial large and the niche manufacturing scale. A range of methods have become available, with the focus on thermal [6–9] and chemical treatments [10–13] being the most studied. The subject has attracted several reviews, e.g., [14–16], and is hereby introduced in more detail.

The thermal modification of wood is now an accepted means of treating wood, with several commercialised processes. Depending on the process, humidity, temperature and time [17], a range of reactions may occur, such as hydrolysis, oxidation and decarboxylation reactions, along with physical changes. The results of such treatments are end products presenting improved physical characteristics, such as lower hygroscopicity, better dimensional stability and durability, though mechanical properties are often significantly reduced, with dynamic mechanical properties (such as impact bending, fatigue) typically being more affected than static properties. Improvements in stability and durability are of importance where wood is exposed to different chemicals or biological agents such as fungi and bacteria, or to frequent use under natural environmental conditions.

Another possible type of wood modification is the reaction with different chemicals—chemical modification—where chemical moieties are covalently bonded to the wood cell wall polymers. Different chemical modifications [11–13,18–20] may be used to alter the properties of wood [20], particularly for improving the dimensional stability, decay resistance and water performance of wood, and to improve mechanical properties compared to those from typical thermal modification processes [21]. Most modification methods involve the hydroxyl groups in the cell wall, which are partially substituted, and the cell wall of the wood is bulked with the bonded chemicals. The substitution of the hydroxyl groups reduces the number of primary sorption sites (typically the OH groups), while the bulking reduces the volume in the wood cell wall which is available to water molecules. The most common chemical modification process involved reaction with anhydrides, with a multitude of examples listed in the literature. However, it is the acetylation process using acetic anhydride that has been the most studied, resulting in its commercialisation [2]. Apart from anhydrides, many other chemicals have been used to try to improve the properties of wooden material. For example, treatments with dimethyloldihydroxyethylene urea (DMDHEU), melamine resin, silane or silicon polymers were found to improve the mechanical strength of wood [3].

More recently, the use of Maillard reactions for wood treatment proved to be a promising modification method to improve some of the wood properties. This type of reaction is well-known in food chemistry, where it is responsible for the browning in many foods during baking [22,23]. The essence of the reaction is that a reducing sugar condenses with a compound possessing a free amino group to give a condensation product [24]. Subsequently, a range of reactions takes place, including cyclisations, dehydrations, retroaldolisations, rearrangements, isomerisations and further condensations, which ultimately lead to the formation of polymers and co-polymers, known as melanoidins [24]. The composition of its chemical structure is relatively unknown due to the complexity of the products that are generated in the reaction [25]. The advantage of this reaction is that it is an aqueous process and initiated by heat only, making it relatively straightforward to apply to wood in a commercial process. In addition, the reaction does not require the use of strong acids or bases, which could degrade the wood structure.

In earlier experiments [26], the influence of the Maillard reaction on European beech and Scots pine was investigated. The wood was impregnated with an amine (glucosamine, lysine or glycine), sugars (glucose or xylose) and an extra reagent to improve the reaction (magnesium chloride, maleic acid or citric acid). The results showed that when lysine, glucose or citric acid were reacted, a high weight percentage gain (WPG) was obtained (18% for beech and 40% for pine). After a leaching procedure, the WPG for beech was 11% and 25% for pine, respectively. This preliminary screening reaction has shown that the Maillard reaction offers potential as a potential new wood modification system, though a reduction in leaching would help maintain the modification benefits.

In this context, a recent work of Hauptmann and co-workers [3] considered the use of tricine for the modification of wood. This is capable of binding with reduced sugars, though studies have been limited to a maximum temperature of 103 °C. They observed increased hardness and the tensile strength of their modified wood species. As indicated earlier, Maillard type reactions are a cascade of reactions (known also as nonenzymatic browning). The initial step consists in a reaction between a reducing end of a saccharide and an amino acid [3,26]. This reaction takes place usually during the heat processing of food with a low relative humidity [27] and it is considered to cause the brown stains during wood kiln drying with temperatures higher than 80 °C due to the presence of natural amino acids and reduced sugars [3]. The most efficient characterisation method to identify the modifications appearing in the wood structure during heating and to identify the possible interaction and bonds formed between the wood components and the reagents used in chemical modification is infrared spectroscopy. This technique proved to be an efficient tool to identify small modifications in the wood structure appearing during different treatments [17,28,29] such as decay, i.e., [30,31] or photodegradation, i.e., [32,33]. Moreover, due to the use of the small amount of sample and little or no processing, this method is considered by many to be non-destructive.

In this study, the combined effects of chemical treatment and the thermal modification are investigated in terms of the use of bicine [2-(Bis (2-hydroxyethyl) amino) acetic acid] and tricine [N-(2-Hydroxy-1,1-bis (hydroxymethyl)ethyl) glycine] (Figure 1). The concept of the Maillard reaction depends upon the availability of an active site on the nitrogen within the amino group (i.e., a N-H group). Whilst this is not present in bicine, it was postulated within this work that the compound may undergo side group displacement, allowing the general concepts within the Maillard reaction to proceed. Some of the resulting properties for the treatment of beech are considered (mechanical, physical and bioresistivity), along with infrared spectroscopic analysis in an attempt to demonstrate actual chemical bonding.

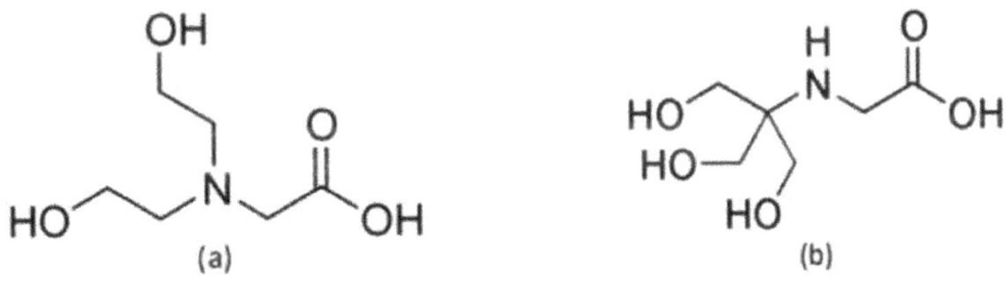

Figure 1. Chemical structure of bicine (**a**) and tricine (**b**).

2. Materials and Methods

In this study, the potential of combining chemical and thermal treatments for European beech (*Fagus sylvatica* L.) and Norway spruce (*Picea abies* (L.) H. Karst) was determined. For both species, material sourced and felled in Slovenia was used. For each treatment set, separate sets without visible growth features and varying ring widths, mainly in semi-radial orientation, were prepared from heartwood sections. Thus, only mature wood was used for these experiments. The density ranges of the samples were within the typical oven-dry ranges commonly noted in the scientific literature. Thus, bicine (Bi, CAS Number 150-25-4, Fisher Scientific, Waltham, MA, USA) and tricine (Tri, CAS Number 5704-04-1,

Fisher Scientific, Waltham, MA, USA) were impregnated into wood specimens, using a 10% solution (weight:volume, *w:v*) of the respective compound per impregnation.

2.1. Specimen Treatment

Beech (hereafter referred to as B) and spruce (hereafter referred to as S) specimens of varying dimensions depending on subsequent testing were prepared and subjected to the following treatment regimes (Table 1).

Table 1. Overview of treatments undertaken.

Species	Code	Treatment Description
Beech	B_HT	Heat treatment only
	B_Bi	Bicine pre-treatment and drying
	B_Bi_HT	Bicine pre-treatment and heat treatment
	B_Tri	Tricine pre-treatment and drying
	B_Tri_HT	Tricine pre-treatment and heat treatment
	B_C	Control
Spruce	S_HT	Heat treatment only
	S_Bi	Bicine pre-treatment and drying
	S_Bi_HT	Bicine pre-treatment and heat treatment
	S_Tri	Tricine pre-treatment and drying
	S_Tri_HT	Tricine pre-treatment and heat treatment
	S-C	Control

For the impregnation procedure, the solutions were introduced using a vacuum-pressure impregnation (VPI) process according to the full-cell process in a laboratory impregnation setup (Kambič, Semič, Slovenia). It consisted of 30 min of vacuum (1.0×10^4 Pa), 40 min of pressure (10×10^5 Pa) and 10 min of vacuum (1.5×10^4 Pa). Reagent uptakes were subsequently determined gravimetrically. The impregnated specimens were conditioned (23 °C; 65% relative humidity (RH)) for 2 weeks prior to thermal modification. Non-impregnated and conditioned specimens served as controls.

The thermal modification (HT) was performed according to a modified Silvapro® commercial procedure [34] limited to 165 °C, due to the risk of thermal degradation of the chemicals if exposed to temperatures above 180 °C. Control specimens were only heated up to 100 °C during the drying procedure in atmospheric conditions. The time of thermal modification at the target temperature was 3 h and mass loss (ML) of the specimens after thermal modification was determined gravimetrically. The HT specimens were stored in the laboratory for 4 weeks (23 °C; 65% RH) before subsequent testing.

2.2. Physical Tests

2.2.1. Colour Analysis

Colour was determined on the semi-radial surfaces of a selection of specimens. The colour measurements were performed according to the CIE L*a*b* system, a method created by the Commission International de l'Eclairage. The CIE L*a*b* system is characterised by three parameters: L*, a* and b*. The L* axis represents the lightness, which varies from a hundred (white) to zero (black), representing the achromatic axis of greys, whereas a* and b* are the chromaticity coordinates. A positive value of a* denotes a redder colour on a green–red scale, whereas a positive value of b* denotes a more yellow colour on a blue–yellow scale. Together, those three components form a three-dimensional colour space. Colour measurements of in-service testing were performed several times a year with a portable Colour Measuring Device (EasyCo 566, Erichsen, Hemer, Germany) and expressed in the CIE L*a*b* system. This device enables contact-free precise colour measurement. The diameter of the measurement spot is 20 mm. However, laboratory test specimens were scanned and processed with Corel Photo-Paint 8 software. Corel Photo-Paint was used as colour analysis as this technique provides the colour of the whole surface and not of individual spots. Total colour difference ΔE* (Equation (1)) from a reference colour (L*0, a*0, b*0) to a target colour (L*1, a*1, b*1) in the CIE Lab space is calculated by determining the Euclidean distance between two colours given by:

$$\Delta E = ((\Delta L^*)^2 = (\Delta a^*)^2 + (\Delta b^*)^2)^{0.5} \quad (1)$$

2.2.2. Contact Angle Measurements

Contact angles were detected using a Theta optic tensiometer (Biolin Scientific Oy, Espoo, Finland) and OneAttesion 2.4 (r4931) software (Biolin Scientific, Espoo, Finland). Five replicates of each treatment were used, to which were added 2 droplets of water (4 µL each). The samples used were 25 × 15 × 50 mm in size, having been thoroughly dried (103 °C, 24 h). Subsequent changes in contact angle were determined using a 7.6-megapixel camera, with software constantly monitoring subsequent wetting over a period of 60 s.

2.2.3. Dynamic Vapour Sorption (DVS) Analysis

Samples for dynamic vapour sorption (DVS) analysis were milled in a Retsch SM 2000 cutting mill (Retsch GmbH, Haan, Germany) with a Conidur® perforation sieve with 1.0 mm perforations. Thus, several samples were milled together to create an average mix of fibres representing each treatment. The milled wood samples were conditioned at 20 ± 0.2 °C and 1 ± 1% RH through blowing with dry air. Analysis of the wood samples was performed using a DVS apparatus (DVS Intrinsic, Surface Measurement Systems Ltd., London, UK). A small amount (approximately 400 mg) of pre-conditioned wood chips was placed on the sample holder, which was suspended in a microbalance within a sealed thermostatically controlled chamber, in which a constant flow of dry compressed air was passed over the sample at a flow rate of 200 cm^3/s and a temperature of 25 ± 0.2 °C. The schedule for DVS had two steps: 0% and 95% RH. The DVS maintained a given RH until the weight change of the sample was less than 0.002%/min for at least 10 min. The running times, target RH, actual RH and sample weights were recorded three times per min throughout the isotherm run. After the DVS analysis, equilibrium moisture content at 95% RH (EMC95% RH) was determined along with the hysteresis values, determined as the difference between desorption and adsorption values at a given RH.

2.3. Mechanical Tests

2.3.1. Mechanical Performance Tests

Modulus of elasticity (MOE) and modulus of rupture (MOR) were determined according to EN 310 [35] with a static three-point bending test on a Zwick Z005 universal testing machine (Zwick-Roell, Ulm, Germany). In total, 60 specimens (10 replicates for each tested material) with dimensions 360 × 20 × 20 mm were prepared and oven dried (103 °C) until a constant mass was achieved. They were dried in order to eliminate the influence of different moisture contents between control and modified wood. The specimens were tested for MOE and MOR immediately after drying.

Compressive strength was determined according to the ASTM D1037-99 standard [36] on a Zwick Z100 universal testing machine (Zwick-Roell, Ulm, Germany). In total, ten specimens for each test group with dimension 50 × 20 × 20 mm were prepared and oven dried until a constant mass was achieved. The specimens were tested for compressive strength immediately after conditioning under standard conditions. After the test, the compressive strength (F_m) was calculated.

The Brinell hardness (HB) was determined by a standard test method according to EN 1534 [37]. The penetration depth (h) of an iron sphere (D = 10 mm) was used for 5 replicates, each undergoing 4 individual measurements, in calculations at load F = 1000 N to determine the Brinell hardness according to:

$$HB = \frac{2 \cdot F}{\pi \cdot D \cdot (D - \sqrt{\left(D^2 - 4 \cdot h \cdot (D - h)\right)}} \quad (2)$$

2.3.2. High Energy Multiple Impact (HEMI) Test

The development and optimisation of the high energy multiple impact (HEMI) test has been previously described by Rapp et al. [38]. For testing modified and non-modified control specimens in a high-energy multiple impact (HEMI) test, specimens of 20 × 20 × 10 mm were split in four specimens of 5 × 20 × 10 mm (radial × tangential × longitudinal). Five times, n = 20 samples of 5 × 20 × 10 mm were submitted to the HEMI tests. HEMI tests were performed in a heavy-impact ball mill (Herzog HSM 100-H; Herzog Maschinenfabrik, Osnabrück, Germany). In short, 20 oven-dried (103 °C) samples were placed in the bowl (140 mm in diameter) of the mill, together with one steel ball of 35 mm in diameter plus three balls of 12 mm and 6 mm in diameter, respectively. For crushing the specimens, the bowl was shaken for 60 s at a rotary frequency of 23.3 s^{-1} and a stroke of 12 mm. The fragments of the 20 specimens were fractionated on a slit sieve according to EN ISO 5223 [39] with a slit width of 1 mm using an orbital shaker at an amplitude of 25 mm and a rotary frequency of 200 min^{-1} for 2 min. The degree of integrity (I), fine percentage (F) and resistance to impact milling (RIM) were calculated following Equations (3)–(5), where m_{20} is the oven-dry mass of the 20 biggest fragments, m_{all} is the oven-dry mass of all fragments and $m_{fragments<1mm}$ is the oven-dry mass of fragments smaller than 1 mm. The RIM was calculated according to the optimised method [38], which contains a three-fold weighting of the fine fraction (3 × F), with the values 300 and 400 guaranteeing to achieve a maximum RIM value of 100%.

$$I = m_{20}/m_{all} \quad (3)$$

$$F = m_{fragments<1mm}/m_{all} \quad (4)$$

$$RIM = ((I - 3 \times F) + 300)/400 \quad (5)$$

2.4. Chemical Tests

2.4.1. Volatile Organic Compound (VOC) Analysis

In order to aid FTIR analysis and attempt to identify Maillard-type reaction products, it was decided to undertake gas chromatography–mass spectrometry (GCMS) analysis of samples. Thus, subsamples of the materials were cut and immediately placed into a minichamber (Markes International, Llantrissant CF72 8XL, UK). A flow of nitrogen at 2 mL/min was passed over the samples and through Tenax columns to collect any volatile organic compounds (VOCs) released. Samples were kept in the chamber for 2, 8 or 16 h. Collected VOCs were eluted form the Tenax columns using methanol and analysed using an electron impact (EI) capillary GCMS (Glarus 680 gas chromatograph, Perkin Elmer, using an Agilent VF5-MS column (30 m × 0.25 mm × 0.25 µm), coupled with a Clarus 600 C mass spectrometer, Perkin Elmer). GC conditions were an initial temperature of 60 °C for 1 min, followed by a temperature increase of 6.0 °C/min to a maximum temperature of 300 °C, with the final temperature being held for a further 10 min. Mass spectra were collected between elution times of 3–51 min from the gas chromatograph over an electron impact mass range of 40.00 to 600.00.

2.4.2. Infrared Spectroscopy

The infrared spectra of the reference and treated wood specimens were recorded in potassium bromide (KBr) pellets on a Bruker ALPHA FT-IR spectrometer with 4 cm^{-1} resolution. The concentration of the sample was a constant of 2 mg/200 mg KBr. Processing of the spectra was performed using the Grams 9.1 program (Thermo Fisher Scientific, Waltham, MA, USA).

2.5. *Effects against Biological Deterioration Tests*

In order to ascertain any biological effects of the treatments, the following aspects were evaluated within this section.

2.5.1. Resistance to Fungal Decay

In order to determine any potential effectiveness of treatment against wood-destroying fungi, tests were carried out according to EN113 [40] once samples had undergone artificial ageing to remove possible extractives and other components that could act like surfactants. A similar process occurs in nature with the first extensive rain period. The European EN 84 standard [41] describes a method for artificial ageing (leaching) of wood before testing the biological effectiveness. This standard was designed to simulate extensive leaching by natural precipitation. The first step was impregnation with demineralised water. The samples were stacked in a container, weighed down and vacuum impregnated (4 kPa) with demineralised water for 20 min and soaked for an additional 2 h. Samples were then immersed in water for 14 days and during this period water was replaced nine times. After the ageing process was completed, samples were dried in ambient conditions for 2 weeks prior to subsequent tests.

The decay test was performed according to the modified EN 113 [40] on treated and untreated beech and spruce samples. Disposable Petri dishes containing 20 mL of 4% potato dextrose agar (PDA, Difco, NJ, USA) were inoculated with 3 different fungi: one white rot fungi (*Trametes versicolor* (L.) Lloyd (ZIM L057)) and two brown rot fungi: *Gloeophyllum trabeum* (Pers.) Murrill (ZIM L018) and *Poria monticola* Murrill (ZIM L033). The fungal isolates originate from the fungal collection of the Biotechnical Faculty, University of Ljubljana, Slovenia. A plastic mesh was used to avoid direct contact between the samples and the medium. The assembled test dishes were then incubated at 25 °C and 80% RH for 12 weeks.

Samples of dimensions 8 mm × 25 mm × 25 mm were prepared and 5 replicates per fungus species were used for each type of treatment. The untreated beech and spruce wood samples served as reference wood species to assess the validity of the test. After incubation, the fungal mycelium was removed and the samples weighed for moisture content. After 24 h of drying at 103 °C, mass loss was determined gravimetrically.

2.5.2. Efficacy against Subterranean Termites

Subterranean termites belonging to the species *Reticulitermes grassei* Clément were captured in a pine forest in Sesimbra, Setúbal district of Portugal and were brought to the laboratory and kept in a conditioned room at 24 ± 1 °C and 8 ± 5% RH. Groups of 150 workers of termites were established in 200 mL glass jars with moistened sand (Fontainebleau sand and water; 4:1 v/v) as substrate. Three replicates (30 × 10 × 10 mm) per treatment were then placed in contact with the termites and the test run for four weeks at the described conditions. Maritime pine test specimens with the same dimensions were also included as internal virulence controls [42,43]. The initial moisture content of the blocks was measured in sets of three additional replicates per treatment and these values were used to determine the theoretical initial dry mass (IDM) of the exposed specimens (in all tests conducted). At the end of the trial, the final moisture content was recorded and the mass loss was obtained according to:

$$\text{Mass loss} = (\text{FDM} - \text{IDM})/\text{IDM} \quad (6)$$

where FDM is the oven-dry mass of the block at the end of the test. The survival of the termites was also recorded and all wood blocks were graded in terms of termite attack using the scale: 0 = no damage; 1 = attempted attack; 2 = slight damage; 3 = superficial and inner damage; 4 = heavy inner damage.

2.6. Statistical Analysis

Where applicable, a multivariate analysis of variance (ANOVA) was performed to determine any significant affect as a result of specific treatments and resulting tests. Significance established at the $p < 0.05$. Tukey's honest significance test was applied to find means that are significantly different from each other. Additionally, statistical analysis was undertaken using IBM SPSS Statistics V26.

3. Results and Discussion

3.1. Weight Uptakes

Figure 2 gives an overview of the weight changes as a result of the treatments undertaken. The results from experiments involving spruce produced weight gains from treatment with bicine or tricine, respectively. However, the combined chemical treatment of bicine with thermal treatment resulted in what appears to be an erroneous higher weight gain than that observed for treatment with bicine alone, though standard deviation could suggest similar results for S_Bi and S_Bi_HT, respectively. Since the thermal modification process resulted in a weight loss, it would be expected that similar thermal treatment following chemical treatment would result in some reduction in weight uptake unless some degree of interaction of bicine with components that would normally be lost during the thermal modification has occurred. The mass loss noted after thermal modification (1.30%) can be explained by the mild conditions applied during the experiments. This would seem to suggest that there is some retention of components because of the thermal process, which can react with the bicine. This contradicts the expected Maillard-type reaction, which is dependent on the presence of a reactive N-H group, usually in a primary amine, but suggested in work by Hauptmann et al., [3] as being possible in the secondary amine of tricine.

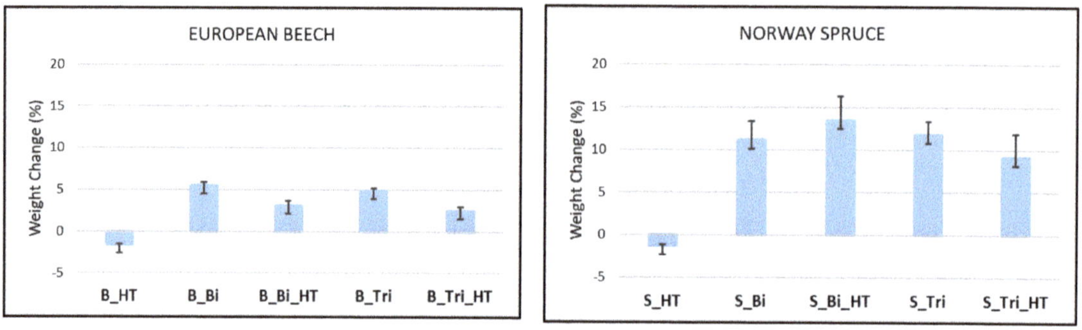

Figure 2. Weight changes as a result of treatments on beech (B) and spruce (S). Bi = bicine, Tri = tricine, HT = thermal modification.

3.2. Results from Physical Tests

3.2.1. Colour Changes

Evaluation of the colour changes as a result of the treatments can be seen in Tables 2 and 3 for European beech and Norway spruce, respectively.

The results for spruce indicated that treatment with bicine in conjunction with heat treatment had the greatest impact in terms of darkening (ΔL*) of samples, resulting in a value of −30.4. However, with beech, it was the corresponding treatment with tricine that produced the greatest darkening (ΔL* = −23.04), compared to the darkening by heat treatment alone, where ΔL* values of −12.4 and −11.1 were obtained for bicine and tricine, respectively. For spruce, ΔL* was lower for tricine with heat treatment (−15.7), but still significantly different from the heat treatment alone. Similar effects were noted for bicine when heat treated with beech. In general, a* and b* values remained relatively unchanged for all samples.

Table 2. Colour determinations of European beech (B) treated with bicine (Bi) and tricine (Tri), respectively. HT = thermal modification.

	B_C	B_HT	B_Bi	B_Bi_HT	B_Tri	B_Tri_HT
L* (s.d.)	76.00 (2.59)	64.94 (0.96)	64.86 (0.96)	63.14 (1.36)	64.72 (2.22)	52.96 (3.26)
a* (s.d.)	7.68 (0.18)	6.90 (0.27)	9.70 (0.27)	8.02 (0.32)	7.86 (0.62)	9.3 (0.60)
b* (s.d.)	8.90 (0.22)	7.68 (0.19)	10.20 (0.27)	8.20 (0.27)	7.94 (0.37)	8.40 (0.22)
ΔL*		−11.06	−11.14	−12.86	−11.28	−23.04
Δa*		−0.78	2.02	0.34	0.18	1.66
Δb*		−1.22	1.30	−0.70	−0.96	−0.50
ΔL* (%)		−14.55	−14.66	−16.92	−14.84	−30.32
Δa* (%)		−10.16	26.30	4.43	2.34	21.61
Δb* (%)		−13.71	14.61	−7.87	−10.79	−5.62
ΔE*		11.15	11.40	12.88	11.32	23.11

Table 3. Colour determinations of Norway spruce (S) treated with bicine (Bi) and tricine (Tri), respectively. HT = thermal modification.

	S_C	S_HT	S_Bi	S_Bi_HT	S_Tri	S_Tri_HT
L* (s.d.)	84.46 (1.08)	72.06 (1.33)	75.20 (1.96)	54.04 (1.61)	83.26 (0.22)	68.80 (1.61)
a* (s.d.)	6.90 (0.45)	8.20 (0.27)	8.40 (0.55)	9.52 (0.38)	7.00 (0.22)	8.20 (0.41)
b* (s.d.)	10.78 (0.38)	11.80 (0.35)	13.60 (0.42)	10.98 (0.18)	13.06 (0.40)	13.02 (0.40)
ΔL*		−12.40	−9.26	−30.42	−1.20	−15.66
Δa*		1.30	1.50	2.62	0.10	1.30
Δb*		1.02	2.82	0.20	2.28	2.24
ΔL* (%)		−14.7	−11.0	−36.0	−1.40	−18.5
Δa* (%)		18.8	21.7	38.0	1.40	19.1
Δb* (%)		9.50	26.20	1.90	21.20	6.90
ΔE*		12.51	9.80	30.53	2.58	15.73

As previously described [9], the change as a result of thermal modification may be explained by a darkening of lignin components alone, whereas results here indicate some additional processes, albeit seemingly differing between spruce and beech for bicine and tricine, respectively. This suggests that some form of additional chemical reaction assumed to be a Maillard process has occurred as a result of the combination of treatments, with a p-value for the interaction effect of heat treatment and choice of respective chemicals of 0.021, which was possibly as a result of the formation of polymerisation products such as melanoidins [24].

Based on the results in Tables 2 and 3, it would appear that the combined effect of chemical treatment and thermal modification had differing impacts on the beech and spruce, respectively. For softwoods, the treatment with bicine produced the most significant colour change on thermal modification ($p = 0.068$ for S_Bi, but $p < 0.001$ for S_Bi_HT), whilst tricine was most significant with thermally treated beech. The result for bicine and heat treatment for spruce would appear to confirm the retention of material suggested from weight uptakes in Figure 2, given that the retained materials show greater colouration, typical of a Maillard-type reaction occurring.

3.2.2. Contact Angle Measurements

Studies into how treated beech and spruce samples adsorb a droplet of water over a period of time are shown in Table 4. It can be seen that the thermal modification, albeit under mild conditions, had a significant effect on how quickly a droplet was adsorbed into the wood surface. Increasing the thermal treatment temperatures from those in this study (165 °C) to typical thermal modification conditions (above 190 °C) would further increase the hydrophobicity of the wood surface and help maintain the droplet stability. However, operating the thermal processing at these higher temperatures would result in thermal decomposition of the bicine and tricine.

Table 4. Overview of contact angle studies of a water droplet over time. B = beech, S = spruce, Bi = bicine, T = tricine, HT = thermal modification, C = control.

		Time (Seconds)					
		1 s	**10 s**	**20 s**	**30 s**	**45 s**	**60 s**
Beech	B_C	70.71	57.45	53.93	51.47	48.19	45.97
	B_HT	99.78	90.17	84.49	81.66	78.07	75.28
	B_Bi	87.04	70.98	65.27	61.53	57.80	55.80
	B_Bi_HT	101.83	86.24	79.78	75.70	71.73	68.95
	B_Tri	99.13	83.06	77.64	75.31	72.26	70.43
	B_Tri_HT	113.17	102.75	96.96	93.04	89.23	86.50
Spruce	S_C	106.05	99.87	97.99	96.31	94.74	93.86
	S_HT	112.79	110.24	109.55	109.08	108.33	107.69
	S_Bi	86.16	69.90	63.04	59.79	58.04	56.64
	S_Bi_HT	102.58	87.81	82.87	79.85	76.88	75.44
	S_Tri	102.53	86.50	81.27	77.93	74.42	72.39
	S_Tri_HT	108.10	101.79	99.16	97.88	96.47	95.16

It is interesting to note the effect of impregnating bicine and tricine into the wood specimens. For spruce, there is a significant reduction in immediate contact angle, with the droplet being more rapidly adsorbed into the wood surface over the 60-s duration. However, there is a slight increase in droplet stability when beech is treated with bicine ($p < 0.001$). When combined with thermal treatment, bicine treatments are equivalent to the thermal treatment alone, whilst tricine treatment followed by thermal modification resulted in a significant increase in droplet stability (and therefore hydrophobicity) for beech ($p < 0.05$). The combined salt and thermal treatments on spruce yielded results similar to those of the thermal treatment alone. The slightly better results for tricine may be partly described by its different solubility level to bicine (18 g per litre compared to 160 g per litre for bicine). Whilst this did not appear to cause any problems in preparation of the treatment sample, it may be possible that once deposited within the cell wall, the tricine resulted in a coating that slightly enhanced the hydrophobic nature of the wood surface. The potential of subsequent reaction with reduced sugars during the thermal modification process may further enhance this effect.

3.2.3. Dynamic Vapour Sorption (DVS)

The equilibrium moisture contents of samples (Table 5) mainly followed the patterns observed for low temperature thermal modification processes, with the slight decreases noted matching those previously reported under similar conditions [44].

Table 5. Average equilibrium moisture content (EMC) values of combined milled samples at 95% relative humidity (RH). B = beech, S = spruce, Bi = bicine, Tri = tricine, HT = thermal modification, C = control.

Beech	$EMC_{95\%\ RH}$ (%)	Spruce	$EMC_{95\%\ RH}$ (%)
B_C	23.40	S_C	23.79
B_HT	19.98	S_HT	19.62
B_Bi	26.95	S_Bi	22.96
B_Bi_HT	25.36	S_Bi_HT	20.92
B_Tri	24.54	S_Tri	20.41
B_Tri_HT	21.58	S_Tri_HT	20.86

Significant differences were noted in the EMC values for samples treated with bicine and tricine, respectively, for both beech and spruce. For spruce, all samples exhibited a lower EMC than the control sample (with the exception of the samples undergoing heat treatment alone), with these values further lowered following respective heat treatments.

The individual sorption curves show typical patterns for sorption and desorption, though there are slight differences in the hysteresis values of samples, representing the EMC difference at a particular RH. However, as pointed out by Frederiksson and Thybring [45], care must be taken in analysing sorption hysteresis, since the common method, as in this study, depends on the determination of an absorption isotherm followed by desorption. The desorption step, based on the definition by Frederiksson and Thybring [45], is a scanning isotherm, since the isotherm has not been obtained from either a dry or water-saturated state. The hysteresis graphs, shown in Figure 3 for spruce, would seem to suggest that the treatments with bicine and tricine alone resulted in similar results. However, when thermal modification was subsequently undertaken, the hysteresis curve for S_Bi_HT closer matched that of S_HT, whereas the effect of tricine seemed to have the greater effect on the hysteresis results. Similar experiments with beech resulted in almost identical hysteresis curves for all samples related to bicine treatment, whilst for tricine alone there was a similar reduced hysteresis, but when combined with thermal treatment resulted in a curve matching those of the control and thermal treatment, respectively.

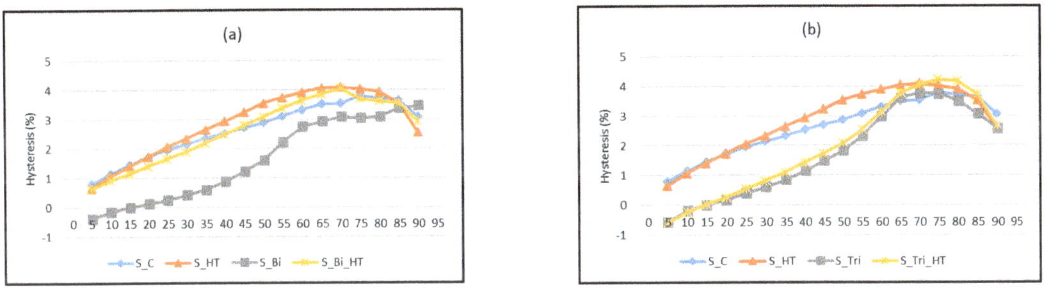

Figure 3. Hysteresis results for spruce, showing (**a**) effects of bicine treatment and (**b**) effects of tricine treatment, respectively.

From these results, it would appear there is a reduction in the hysteresis of spruce samples treated with bicine or tricine. This effect is then lost when applying a thermal treatment to the bicine specimens. However, to determine the actual effects, it would be necessary to undertake analysis from water-saturated specimens, as previously suggested [45], instead of running a standard adsorption–desorption analysis between 0 and 95% RHs.

3.3. Mechanical Testing

3.3.1. Mechanical Performance Testing

The results of the mechanical tests are shown below (Table 6), in terms of Modulus of Elasticity by three-point bending and compression testing, Modulus of Rupture (MOR) values and Compressive strength (Fm) for both beech and spruce.

For beech, the MOE from three-point bending tests suggested that very little variation was observed, with maximum reductions compared to control samples being around 7.5%. However, compressive MOE values revealed a more significant ($p < 0.001$) reduction, with up to 60% reduction when either bicine or tricine was added. In comparison, the compressive MOE for thermally modified beech only reduced by approximately 5%. MOR values were found to be fairly consistent for samples where thermal modification was not applied. Thus, control samples (B_C) had MOR value of 182.7 N/mm^2, whereas treatments with bicine and tricine led to values of 178.4 and 175.4 N/mm^2, respectively. Greater changes were noted when looking at thermally modified beech samples, whereby MOR values for samples treated with tricine and subsequently thermally modified (B_Tri_HT) were around 13% lower than for thermal modification alone (B_HT) and samples treated with bicine and thermally modified (B_Bi_HT). In all cases with beech, values of Fm were higher than those of the reference samples, suggesting that the use of bicine or tricine had a positive effect on Fm. Brinell hardness (HB) results failed to show any general trends.

Table 6. Overview of mechanical properties of bicine (Bi)- and tricine (Tri)-treated wood. B = beech, S = spruce, HT = thermal modification, C = control. Standard deviation in brackets ().

		Three-Point Bending Test		Compression Test		
	Group	MOE (N/mm^2)	MOR (N/mm^2)	MOE (N/mm^2)	Fm (N/mm^2)	HB (N/mm^2)
Beech	B_C	16,030 (1213)	182.70 (18.68)	20,100 (1120)	99.73 (5.40)	30.06 (3.45)
	B_HT	14,750 (1732)	162.60 (34.63)	19,100 (6140)	105.43 (7.77)	29.73 (4.83)
	B_Bi	15,800 (812)	178.44 (24.74)	7800 (920)	108.37 (4.68)	30.18 (4.66)
	B_Bi_HT	15,956 (1138)	159.78 (21.90)	7200 (920)	106.44 (6.50)	35.29 (5.52)
	B_Tri	16,120 (2420)	175.40 (40.32)	7300 (1060)	108.24 (5.01)	35.45 (4.69)
	B_Tri_HT	14,790 (1931)	140.21 (34.31)	8100 (740)	110.84 (9.41)	30.84 (4.83)
Spruce	S_C	14,410 (1799)	106.87 (9.72)	19,400 (970)	88.73 (11.61)	16.77 (1.92)
	S_HT	12,019 (4761)	93.66 (25.51)	21,000 (1250)	88.49 (12.38)	16.74 (2.37)
	S_Bi	12,040 (935)	97.72 (15.10)	18,600 (1710)	96.39 (10.35)	17.14 (2.97)
	S_Bi_HT	13,440 (1751)	93.95 (9.62)	19,000 (1490)	85.70 (10.10)	15.77 (2.30)
	S_Tri	12,823 (1860)	92.93 (32.09)	8000 (1250)	88.73 (7.31)	20.18 (2.39)
	S_Tri_HT	13,805 (1756)	87.39 (26.73)	8400 (520)	97.98 (7.77)	19.35 (2.18)

For spruce, MOE values from three-point bending tests showed that thermal modification (S_HT) realised a 14% reduction compared with control samples (S_C). However, direct comparison of the effects of bicine and tricine showed increases in MOE compared with the impregnated samples alone. MOR values generally showed the thermal modification process resulted in lower values to the non-thermally modified analogues. Tukey's HSD Test for compressive MOE values showed that, for spruce, only tricine had a significant effect on values measured ($p < 0.05$). Fm values showed no clear trends, though the results for tricine and thermal modified samples (S_Tri_HT) were considerably higher than the non-thermally modified samples. Similarly, the effects of tricine (both S_Tri and S_Tri_HT) realised higher HB values than any of the other treatments considered herein.

3.3.2. HEMI Tests

Earlier work [46,47] suggested that the main benefit of HEMI testing was for identifying the effects of treatments on the brittleness of materials. Other than ANOVA showing the obvious statistical significant difference between samples undergoing heat treatment and those that did not, results obtained for beech (Table 7) and spruce (Table 8) showed that there were no clear benefits of any specific treatment when considering the resistance to impact milling (RIM). However, there appeared to be a slight influence resulting from treating spruce with tricine (both as a stand-alone treatment and in conjunction with thermal treatment) and with treatment of beech with tricine and heat treatment (S_Bi was shown to be statistically different from ANOVA analysis when considering resultant F values, with $p < 0.001$). The effect of reduced temperature heat treatment showed little effect on the samples, as would be expected, since studies [46] showed that increasing the thermal modification temperature had no significant effect on the RIM values. These similarities in RIM values also suggested no brittleness as a result of treating with bicine or tricine, respectively.

Table 7. HEMI test results for beech (B). Bi = bicine, Tri = tricine, HT = thermal modification, C = control.

	B_C	B_HT	B_Bi	B_Bi_HT	B_Tri	B_Tri_HT
F (s.d.)	0.89 (0.23)	1.55 (0.18)	0.82 (0.22)	1.82 (0.69)	1.41 (0.64)	1.57 (0.62)
I (s.d.)	55.63 (2.39)	50.44 (1.80)	57.24 (1.93)	53.87 (4.10)	55.37 (3.85)	55.22 (2.64)
RIM (s.d.)	88.24 (0.72)	86.45 (0.43)	88.70 (0.41)	87.10 (1.41)	87.79 (1.35)	87.63 (0.90)

Table 8. HEMI test results for spruce (S). Bi = bicine, Tri = tricine, HT = thermal modification, C = control.

	S_C	S_HT	S_Bi	S_Bi_HT	S_Tri	S_Tri_HT
F (s.d.)	2.49 (0.66)	3.65 (0.60)	3.08 (0.68)	3.79 (0.98)	4.04 (1.13)	5.63 (1.26)
I (s.d.)	28.65 (4.55)	27.86 (1.73)	29.15 (4.68)	24.32 (2.61)	29.64 (4.64)	25.86 (3.94)
RIM (s.d.)	80.30 (1.33)	79.22 (0.66)	79.98 (1.48)	78.24 (1.17)	79.38 (1.86)	77.24 (1.74)

In all the treatments, it can be seen that the percentage of fine material present after the HEMI test was higher following the application of heat treatment ($p < 0.05$), any minor change in I values could result in greater fragmentation to smaller pieces.

3.4. Chemical Tests

3.4.1. VOC Analysis

Qualitative analysis of samples of collected extractives by gas chromatography–mass spectrometry (GC-MS) showed, as expected, the untreated controls exhibited typical VOCs: 3-carene, α-pinene, β-pinene, α-thujene, β-terpinene and formic acid in both beech and spruce, with an additional peak for acetaldehyde present in spruce. Table 9 gives an overview of the major VOCs detected from samples. Results from experiments combining bicine and heat treatment showed a large number of minor peaks, which could not be assigned probable structures.

Table 9. Principal compounds detected by GC-MS of volatile organics from beech (B) and spruce (S) reactions involving bicine (Bi) and tricine (Tri). HT = thermal modification.

VOC Detected	Formula	Mol. Wt.	B_Bi	B_Bi_HT	B_Tri	B_Tri_HT	S_Bi	S_Bi_HT	S_Tri	S_Tri_HT
Formic Acid	CH_2O_2	46.03					√			
Acetaldehyde	C_2H_4O	44.05	√	√	√	√	√	√	√	√
Tetraacetyl-d-xylonic nitrile	$C_{14}H_{17}NO_9$	343.29	√	√			√	√		
3N-(7-acetamido-[1,2,4]triazolo [4,3-b][1,2,4]triazol-3-yl)acetamide	$C_7H_9N_7O_2$		√				√	√		
Deoxyspergualin	$C_{17}H_{37}N_7O_3$	387.50	√				√	√		
Tetrahydro-4h-pyran-4-ol	$C_5H_{10}O_2$	102.13	√				√	√		
Oxiranemethanol	$C_3H_6O_2$	74.08	√	√	√	√	√	√	√	√
O-methylisourea	$C_2H_6N_2O$	74.08	√	√	√	√	√	√	√	√
Propane	C_3H_8	44.10	√	√	√	√	√	√	√	√

From results shown in Table 9, there appears to be more indication of reactions using bicine, though it is uncertain how identified compounds such as tetraacetyl-d-xylonic nitrile, 3N-(7-acetamido-[1,2,4]triazolo[4,3-b][1,2,4]triazol-3-yl)acetamide and deoxyspergualin can be derived, particularly from impregnation with both spruce and beech, where a maximum temperature of 60 °C should have limited any fragmentation of the reagent, whilst being below anticipated temperatures for reactions to occur. Despite this, it is probable that that tetraacetyl-d-xylonic nitrile is a result of an interaction between the chemical reagents and a hemicellulose fragment, whilst the presence of tetrahydro-4h-pyran-4-ol could also result from the breakdown of the hemicellulosic components in wood, though it is interesting that this only occurred at low temperatures in bicine-treated beech, and for the bicine treatment and bicine-high temperature treatment of spruce.

The presence of all the components identified to date did not provide any additional insight into possible reaction mechanisms at this time, other than the likelihood of probable incorporation of nitrogenous moieties from the bicine or tricine into hemicellulosic fragments.

3.4.2. Infrared Spectrometry

A detailed investigation into the FT-IR spectrometry of samples has previously been reported [48]. This showed that both the infrared spectra and their derivatives for the control (S_C and B_C) and the thermal treated (S_HT and B_HT) samples did not indicate any significant differences occurring as a result of the thermal process. However, the spectra for treatments involving bicine or tricine showed the presence of the chemicals, though the peaks were found to overlap with those from the wood at higher wavenumbers (3700 to 2500 cm^{-1}).

Analysis at lower wavenumbers, along with the respective second derivative spectra, offered more insight into possible interactions. These results allowed the observation of the peaks with bicine treatments described in Table 10, based on spectra shown in Figure 4. Comparative peaks from tricine experiments were less pronounced, probably due to the lower weight uptake of samples.

The use of chemometric methods such as principal component analysis and hierarchal cluster analysis [48] suggested the occurrence of reactions between the functional groups from bicine and tricine and those present in wood structure. In the case of bicine, this could suggest that a chain displacement from bicine to create the Amadori-type products and activation of the N via a quaternary ammonium intermediate is typical of a Maillard reaction. However, assessment of the spectra alongside VOC analysis did not give conclusive evidence of such mechanisms.

Table 10. Selection of peaks showing significant differences in 2nd derivate FTIR spectra for reactions involving bicine.

Wavenumber (cm^{-1})	Peak Identification
1740	Carbonyl/carboxyl stretching vibration
1647	Carbonyl/carboxyl stretching vibration
1493	Methyl or methylene deformation vibration
1465	Methyl or methylene deformation vibration
1422	Methyl or methylene deformation vibration
1402	OH stretching
1317	OH deformation
1267	C-N stretching vibration in amine groups
1208	C-N stretching vibration in amine groups
1166	C-O groups stretching vibration
1118	C-O groups stretching vibration
1074	C-O groups stretching vibration
1045	C-O groups stretching vibration
1024	C-O groups stretching vibration

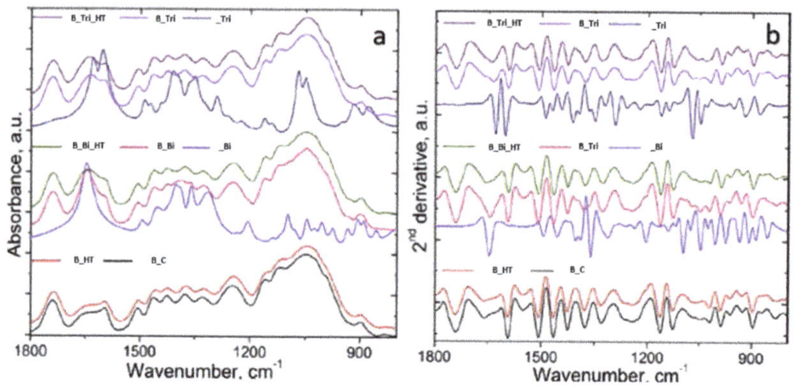

Figure 4. *Cont.*

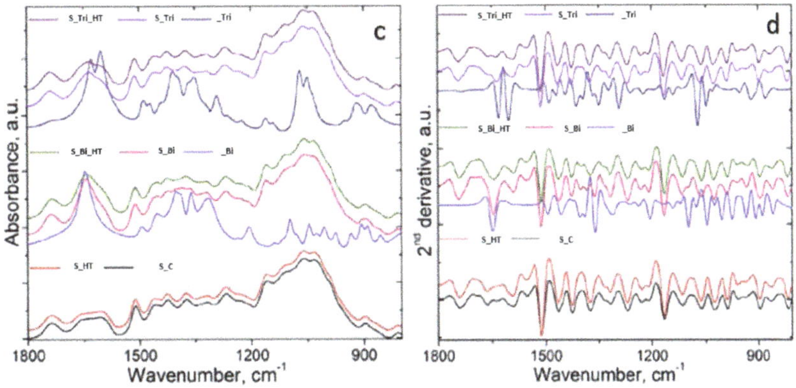

Figure 4. FTIR and 2nd derivative spectra of beech and spruce samples. (**a**) FT-IR spectra beech samples; (**b**) 2nd derivative spectra for beech samples; (**c**) FT-IR spectra spruce samples; (**d**) 2nd derivative spectra for spruce samples.

3.5. Biological Efficacy

3.5.1. Effects against Fungal Decay

Results from experiments with bicine and tricine, both individually and in combination with a reduced thermal modification, did not provide any discernible variation in effects against the three fungal species studied in these trials. However, there were significant variations in degradation studies depending on undertaking the prerequisite EN84 accelerated weathering prior to decay [41]. This is clear from Figure 5 for beech and Figure 6 for spruce, respectively, outlining analysis of decay with *Gloeophyllum trabeum*, *Poria monticola* and *Trametes versicolor*, respectively.

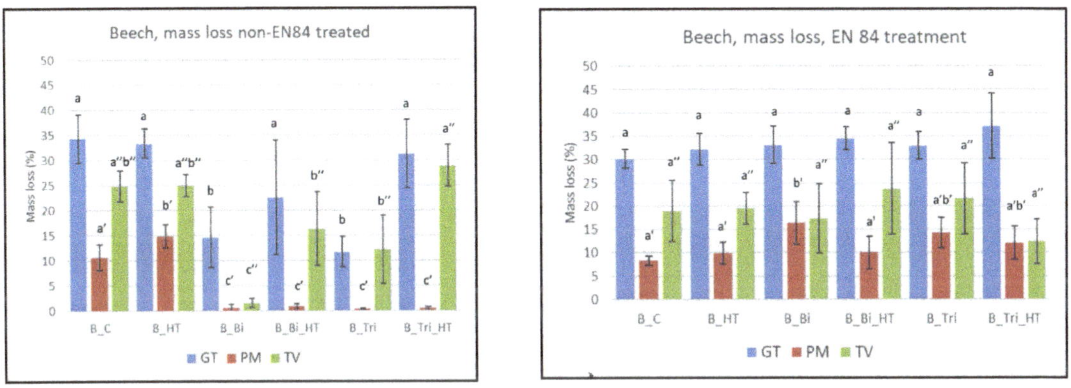

(**a**) (**b**)

Figure 5. Mass loss of beech as a result of fungal decay, (**a**) without EN84 weathering; (**b**) with EN84 weathering. GT—*G. trabeum*, PM—*P. monticola*, TV—*T. versicolor*. Tukey analysis groups shown above each bar graph.

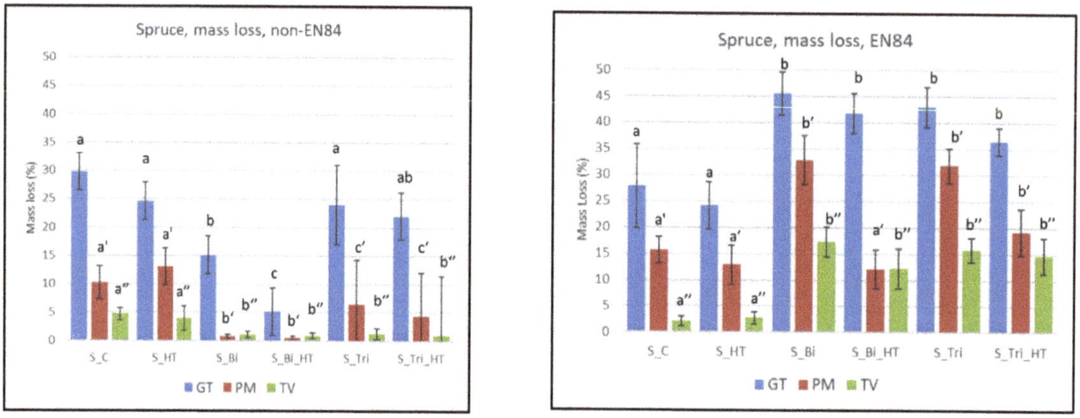

(a) (b)

Figure 6. Mass loss of spruce as a result of fungal decay, (**a**) without EN84 weathering; (**b**) with EN84 weathering. GT—*G. trabeum*, PM—*P. monticola*, TV—*T. versicolor*. Tukey analysis groups shown above each bar graph.

ANOVA analysis of results of non-EN84 accelerated ageing of both beech and spruce showed there were significant differences for all the fungi ($p < 0.05$). From Figures 5 and 6, it can be seen that the EN84 accelerated ageing of samples had a significant effect on the effectiveness of treatments considered in this study. Whereas the spruce control (S_C) and beech controls resulted in similar levels of decay irrespective of EN84, values for other treatments varied considerably, with the greatest effects noted for treatments of both spruce and beech involving bicine or tricine in non-EN84 tests, particularly when assessing with *P. monticola*. However, decay rates were similar or worse where EN84 ageing was carried out prior to decay studies. Similarly, when spruce underwent decay with *G. trabeum* with bicine (S_Bi and S_Bi_HT) without EN84, significant decreases in decay were noted, though these were found to result in greater decay rates when EN84 was carried out. A similar effect was noted with beech (B_Bi).

In an attempt to ascertain why the non-EN84 treated samples experienced considerable reduction in decay levels, moisture content assessments were also carried out on these samples. Results (Figures 7 and 8) suggest that the presence of bicine or tricine as part of the treatment regime significantly affected the moisture content levels noted for the various experimental regimes. Such increases were, however, not noted when looking at equilibrium moisture content measurements (Table 5). The exceptionally high moisture contents noted for some of the non-EN84 treated samples during fungal decay tests may result in water logging of samples, which in turn diminishes the fungal activity during the trial. However, these high moisture contents appear to be within the functional limits for the fungal species used in these trials. This suggests that the presence of the non-leached treatments may have a direct effect on the degree of biological attack by selected fungi.

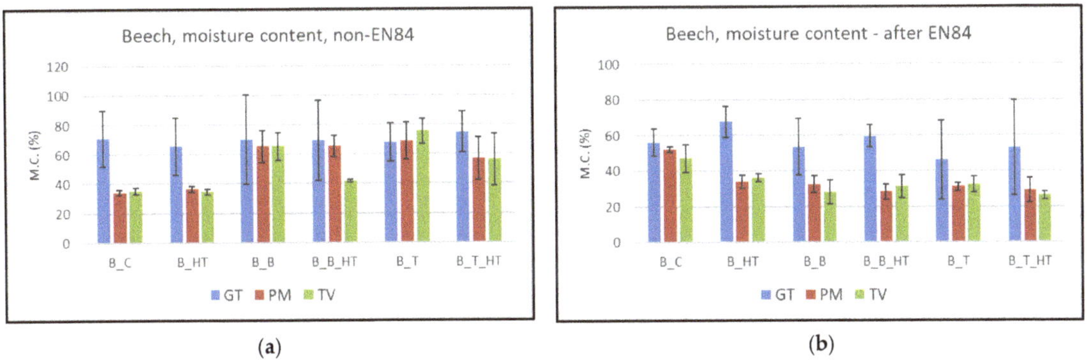

Figure 7. Moisture content of beech measured during fungal decay, (**a**) without EN84 weathering; (**b**) with EN84 weathering. GT—*G. trabeum*, PM—*P. monticola*, TV—*T. versicolor*.

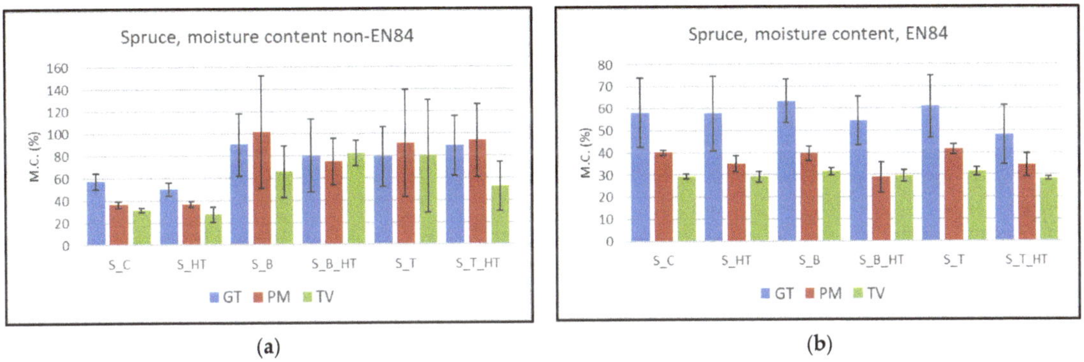

Figure 8. Moisture content of spruce measured during fungal decay, (**a**) without EN84 weathering; (**b**) with EN84 weathering. GT—*G. trabeum*, PM—*P. monticola*, TV—*T. versicolor*.

3.5.2. Termite Testing

As shown in Figure 9, the survival of the termites and the mass loss showed some differences between the treatments, although a high variability of the results was observed for some treatments. The durability of thermal modified wood is recognised as low and the results of the present work are in accordance with that perception. This low durability fits in well with previously reported data [49,50].

When considering the degree of termite attack of treated samples, it was found that S_B_HT had the lowest attack grade (2.7), followed by S_B, S_Tri and B_Bi, each of which had a grade of 3.0. The treatment B_Tri was found to have a very slight improvement (3.7) over control samples, which, along with the remaining treatments, had a grade of 4.0. In terms of termite survival rate, termites exposed to beech (untreated control and treated wood), though with high variability of results, did not show significant differences among them ($F = 1.83$, $p = 0.181$). For spruce wood (untreated control and treated wood), significant differences ($F = 6.28$, $p = 0.004$) were observed. Termites exposed to spruce treated with Bi_HT and Tri scored significantly lower survival rates than termites that fed on untreated controls (both pine and spruce) and heat-treated spruce (S_HT), with specific p values shown in Table 11.

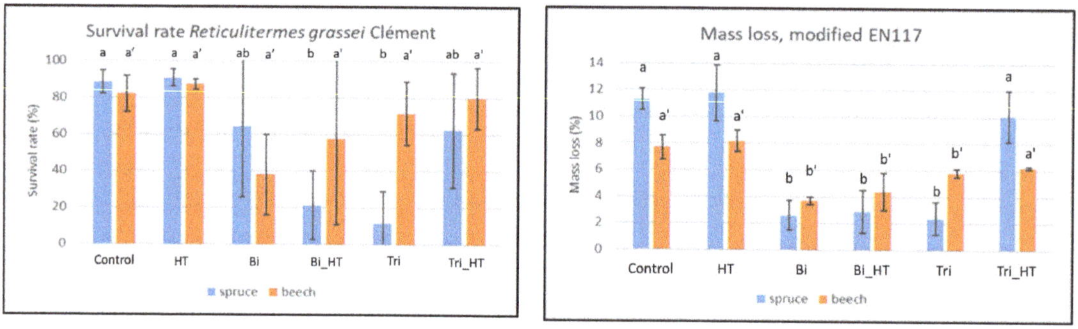

(**a**) (**b**)

Figure 9. Overview of termite tests with various treatments. (**a**) Survival rates of *Reticulitermes grassei* Clément and (**b**) mass losses of samples following a modified EN117 experiment [42]. Tukey analysis groups shown above each bar graph.

Table 11. Significant results ($p < 0.05$) of the Tukey's honestly significant difference test (HSD) for termite survival in untreated and treated spruce (non-significant p values are not shown in this table).

		_HT	_Bi	_Bi_HT	_Tri	_Tri_HT	S_C
Heat treatment	_HT	x					
Bicine and oven dry	_Bi		x				
Bicine, oven dry and heat treatment	_Bi_HT	0.028		x			
Tricine and oven dry	_Tri	0.012			x		
Tricine, oven dry and heat treatment	_Tri_HT					x	
Spruce control	S_C			0.035	0.015		x

For mass loss, termite consumption was significantly different among beech (untreated control and treated wood) (F = 14.40 $p < 0.001$); termites consumed less B_Bi and B_Bi_HT when compared with control and B_HT (all $p < 0.001$). The same significant differences were only found for tricine-treated beech without HT ($p = 0.031$). For spruce, significant differences among untreated and treated spruce were also observed (F = 26.96, $p < 0.001$); wood treated with S_Bi and S_Tri showed significantly lower mass losses relatively to the group of untreated spruce controls (all $p < 0.001$). S_Bi_HT-treated wood showed significantly lower mass loss than untreated control of spruce, heat treated spruce and THT ($p < 0.001$).

In addition to these mass loss data, a more detailed investigation into the effects of bicine and tricine as a result of the treatments detailed herein on the digestive protists within termite guts has been published [51]. Whilst both bicine and tricine are recognised as Good's buffers and have similar properties (e.g., pH levels), it is interesting to compare similar treatments between the two wood species. When bicine is introduced without subsequent thermal treatment, termite survival is considerably lower for beech than spruce. However, when subsequent thermal treatment was applied, termite survival rates were lower in spruce. This could suggest some mode of action between the bicine and pine as a result of the thermal treatment and not just the effect of the chemical alone.

Given the fact that EN84 treatment was not undertaken prior to exposure of these tests, there may be an increased impact of the unleached bicine or tricine within the wood samples having a direct effect on termite feeding habits and potential toxic effects due to reductions in specific protists within the termite guts [51].

4. Conclusions

The use of bicine and tricine as part of an enhanced thermal modification process has been considered. The results herein were part of establishing the feasibility of such combined chemical/thermal modification processes for treating wood. However, the thermal stability of the selected compounds resulted in the need for a reduced thermal modification temperature, which would have expected impacts on the effectiveness of the thermal modification on its own.

Studies using colour, FT-IR spectroscopy and VOC analysis seem to suggest there is some level of chemical interaction between the treatments and the wood resulting from the trials undertaken, particularly through the presence of compounds such as tetraacetyl-d-xylonic nitrile, though no clear evidence of exactly what the mechanism is has been determined to date. The use of principal component analysis with FT-IR studies confirmed specific peaks were a direct result of chemical reactions.

The hypothesis of combining chemical and thermal reactions of wood resulting in equivalent or better mechanical properties were not borne out in these studies, with three-point bending and compression tests being reduced. There were slight improvements in some Brinell hardness data, though this may be a direct result of the increased density resulting from the uptake of bicine or tricine. The lack of improvements may also be a direct result of the thermal instability of the bicine and tricine.

The results from fungal and termite attack of samples suggested that significant improvements were possible for experiments where the EN84 weathering had not been undertaken. Whilst moisture content levels for fungal decay tests on non-weathered samples suggested the possibility of water logging occurring, corresponding samples tested under standard hysteresis conditions did not show such significant uptakes of moisture. When samples underwent EN84 weathering prior to testing, the beneficial effects were not noted. This indicated that any benefits were lost when the bicine or tricine was leached from the samples. Thus, it would seem logical that further means of entrapment within the wood cell wall structure needs to be considered in future work.

Despite the inconclusive results from this study, there are sufficient indications to suggest some degree of reaction is occurring. Part of the issue with the process undertaken in this study was the use of an open system, whereby any potential intermediate moieties resulting from Maillard-type reactions were volatilised before subsequent reactions could occur. This could potentially be overcome through the use of a closed system reactor, thereby allowing these intermediate groups to undergo further reactions. In addition, the risk of thermal degradation of the key reagents in this study (bicine and tricine) at temperatures typically employed for more effective thermal modification (between 180 and 210 °C) may limit the degree of reactivity encountered in this study, even though reduced heat treatment temperatures (160 °C) were employed. Further consideration into the use of reagents that are more stable at elevated temperatures may enable the full effect of Maillard-type reactions to be explored in more detail.

Author Contributions: In this manuscript, the activities of the following persons are hereby summarized: conceptualization, D.J., M.H. (Miha Humar) and D.K.; methodology, D.K., M.H. (Miha Hočevar), C.-M.P., C.B. and L.N.; validation, M.H. (Miha Humar), A.Z., M.-C.P. and G.O.; formal analysis, D.J., D.K. and M.H. (Miha Humar), S.F.C.; investigation, D.J., M.H. (Miha Humar), L.N. and C.-M.P.; writing—original draft preparation, D.J.; writing—all authors; visualization, D.J. and M.H. (Miha Humar); statistical analysis, L.N.; supervision, D.J. and D.S.; project administration, D.J. and D.S.; funding acquisition, D.J. All authors have read and agreed to the published version of the manuscript.

Funding: This research was funded by COST Action FP1407 "Understanding wood modification through an integrated scientific and environmental impact approach (ModWoodLife)", which provided a Short-Term Scientific Mission (STSM) grant to D.J. to initiate this work.

Institutional Review Board Statement: Not applicable.

Informed Consent Statement: Not applicable.

Data Availability Statement: Not applicable.

Acknowledgments: The involvement of several co-authors as a result of discussions and interaction at both COST FP1407 "Understanding wood modification through an integrated scientific and environmental impact approach (ModWoodLife)" and COST FP1303 "Performance of bio-based building materials" is hereby acknowledged. Additional support to D.J. and D.S. through the project "Advanced research supporting the forestry and wood-processing sector's adaptation to global change and the 4th industrial revolution", OP RDE (Grant No. CZ.02.1.01/0.0/0.0/16_019/0000803) and CT WOOD—a centre of excellence at Luleå University of Technology supported by the Swedish wood industry—is also gratefully acknowledged (D.J. and D.S.). The termite work (L.N.) was conducted under LNEC P2I project ConstBio. Support of the Slovenian Research Agency (D.K., M.H. (Miha Hočevar), M.H. (Miha Humar) and A.Z.) in the frame of the programme P4-0015 is acknowledged as well.

Conflicts of Interest: There are no conflict of interest.

References

1. Ramage, M.H.; Burridge, H.; Busse-Wicher, M.; Fereday, M.G.; Reynolds, T.; Shah, D.U.; Wu, G.; Yu, L.; Fleming, P.; Densley-Tingleye, D.; et al. The wood from the trees: The use of timber in construction. *Renew. Sustain. Energy Rev.* **2017**, *68*, 333–359. [CrossRef]
2. Popescu, C.-M.; Hill, C.; Curling, S.; Ormondroyd, G.; Xie, Y. The water vapour sorption behaviour of acetylated birch wood: How acetylation affects the sorption isotherm and accessible hydroxyl content. *J. Mater. Sci.* **2014**, *49*, 2362–2371. [CrossRef]
3. Hauptmann, M.; Gindl-Altmutter, W.; Hansmann, C.; Bacher, M.; Rosenau, T.; Liebner, F.; D'Amico, S.; Schwanninger, M. Wood modification with tricine. *Holzforschung* **2015**, *69*, 985–991. [CrossRef]
4. Jones, D.; Sandberg, D.; Goli, G.; Todaro, L. *Wood Modification in Europe: A State-of-the-Art about Processes, Products and Applications*; Firenze University Press: Florence, Italy, 2019; ISBN 978-88-6453-970-6.
5. Sandberg, D.; Kutnar, A.; Karlsson, O.; Jones, D. *Wood Modification Technologies. Principles, Sustainability, and the Need for Innovation*; CRC Press: Cambridge, UK, 2021; ISBN 9781138491779.
6. Herrera-Díaz, R.; Sepúlveda-Villarroel, V.; Pérez-Peña, N.; Salvo-Sepúlveda, L.; Salinas-Lira, C.; Llano-Ponte, R.; Ananías, R.A. Effect of wood drying and heat modification on some physical and mechanical properties of radiata pine. *Dry. Technol.* **2017**, *36*, 537–544. [CrossRef]
7. Korkut, S.; Korkut, D.S.; Kocaefe, D.; Elustondo, D.; Bajraktari, A.; akıcıer, N. Effect of thermal modification on the properties of narrow-leaved ash and chestnut. *Ind. Crops Prod.* **2012**, *35*, 287–294. [CrossRef]
8. González-Penã, M.M.; Curling, S.F.; Hale, M.D.C. On the effect of heat on the chemical composition and dimensions of thermally-modified wood. *Polym. Degrad. Stab.* **2009**, *94*, 2184–2193. [CrossRef]
9. González-Peña, M.M.; Hale, M.D.C. Colour in thermally modified wood of beech, Norway spruce and Scots pine. Part 1: Colour evolution and colour changes. *Holzforschung* **2009**, *63*, 385–393. [CrossRef]
10. Hill, C.A.S.; Jones, D. The dimensional stabilisation of Corsican pine sapwood by reaction with carboxylic acid anhydrides. *Holzforschung* **1996**, *50*, 457–462. [CrossRef]
11. Jebrane, M.; Harper, D.; Labbe, N.; Sebe, G. Comparative determination of the grafting distribution and viscoelastic properties of wood blocks acetylated by vinyl acetate or acetic anhydride. *Carbohydr. Polym.* **2011**, *84*, 1314–1320. [CrossRef]
12. Marfo, E.D.; Wereko, E.Y.; Larbi, K.O. Chemical modification of the tropical hardwood species, *Celtis mildbraedii* (ESA FUFUO), to improve its durability. *J. Wood Chem. Technol.* **2017**, *38*, 51–56. [CrossRef]
13. Kumar, A.; Richter, J.; Tywoniak, J.; Hajek, P.; Adamopoulos, S.; Šegedin, U.; Petrič, M. Surface modification of Norway spruce wood by octadecyltrichlorosilane (OTS) nanosol by dipping and water vapour diffusion properties of the OTS-modified wood. *Holzforschung* **2018**, *72*, 45–56. [CrossRef]
14. Gérardin, P. New alternatives for wood preservation based on thermal and chemical modification of wood—A review. *Ann. For. Sci.* **2016**, *73*, 559–570. [CrossRef]
15. Rowell, R.M. Acetylation of Wood—A Review. *Int. J. Lignocellul. Prod.* **2014**, *1*, 1–27.
16. Sandberg, D.; Kutnar, A.; Mantanis, G. Wood modification technologies—A review. *iForest* **2017**, *10*, 895–908. [CrossRef]
17. Popescu, M.-C.; Froidevaux, J.; Navi, P.; Popescu, C.-M. Structural modifications of *Tilia cordata* wood during heat treatment investigated by FT-IR and 2D IR correlation spectroscopy. *J. Mol. Struct.* **2013**, *1033*, 176–186. [CrossRef]
18. Hill, C.A.S.; Papadopoulos, A.N.; Payne, D. Chemical modification employed as a means of probing the cell-wall micropore of pine sapwood. *Wood Sci. Technol.* **2004**, *37*, 475–488. [CrossRef]
19. Özmen, N.; Çetin, N.S.; Tingaut, P.; Sèbe, G. A new route for the functionalisation of wood through transesterification reactions. *Eur. Polym. J.* **2006**, *42*, 1617–1624. [CrossRef]
20. Hill, C.A.S. *Wood Modification—Chemical, Thermal and Other Processes*; John Wiley & Sons: Chichester, UK, 2006.
21. Mai, C.; Militz, H. Modification of wood with silicon compounds. Inorganic silicon compounds and sol-gel systems: A review. *Wood Sci. Technol.* **2004**, *37*, 339–348. [CrossRef]
22. Ames, J.M. Applications of the Maillard reaction in the food industry. *Food Chem.* **1998**, *62*, 431–439. [CrossRef]

23. Manzocco, L.; Calligaris, S.; Mastrocola, D.; Nicoli, M.C.; Lerici, C.R. Review of non-enzymatic browning and antioxidant capacity in processed foods. *Trends Food Sci. Technol.* **2001**, *11*, 340–346. [CrossRef]
24. Echavarría, A.P.; Pagán, J.; Ibarz, A. Melanoidins formed by Maillard reaction in food and their biological activity. *Food Eng. Rev.* **2012**, *4*, 203–223. [CrossRef]
25. Kim, J.S.; Lee, Y. Activity of melanoidins from different sugar/amino acid model systems: Influence of the enantiomer type. *Amino Acids* **1999**, *36*, 465–474. [CrossRef] [PubMed]
26. Peeters, K.; Larnøy, E.; Kutnar, A.; Hill, C.A.S. An examination of the potential for the use of the Maillard reaction to modify wood. *Int. Wood Prod. J.* **2018**, *9*, 108–114. [CrossRef]
27. Hauser, C.; Müller, U.; Sauer, T.; Augner, K.; Pischetsrieder, M. Maillard reaction products as antimicrobial components for packaging films. *Food Chem.* **2014**, *145*, 608–613. [CrossRef]
28. Mohebby, B. Application of ATR Infrared Spectroscopy in Wood Acetylation. *J. Agric. Sci. Technol.* **2008**, *10*, 253–259.
29. González-Penã, M.M.; Hale, M.D.C. Rapid assessment of physical properties and chemical composition of thermally modified wood by mid-infrared spectroscopy. *Wood Sci. Technol.* **2011**, *45*, 83–102. [CrossRef]
30. Ferraz, A.U.; Baeza, J.; Rodrigues, J.; Freer, J. Estimating chemical composition of biodegraded pine and eucalyptus by DRIFT spectroscopy and multivariate analysis. *Bioresour. Technol.* **2000**, *74*, 201–212. [CrossRef]
31. Popescu, C.-M.; Popescu, M.-C.; Vasile, C. Characterization of fungal degraded lime wood by FT-IR and 2D IR correlation spectroscopy. *Microchem. J.* **2010**, *95*, 377–387. [CrossRef]
32. Popescu, C.-M.; Popescu, M.-C.; Vasile, C. Structural analysis of photodegraded lime wood by means of FT-IR and 2D IR correlation spectroscopy. *Int. J. Biol. Macromol.* **2011**, *48*, 667–675. [CrossRef]
33. Tolvaj, L.; Popescu, C.-M.; Molnar, Z.; Preklet, E. Dependence of the air relative humidity and temperature on the photodegradation processes of beech and spruce wood species. *BioResourses* **2016**, *11*, 296–305.
34. Rep, G.; Pohleven, F.; Košmerl, S. Development of industrial kiln for thermal wood modification by a procedure with an initial vacuum and commercialisation of modified Silvapro wood. In Proceedings of the Sixth European Conference on Wood Modification, Ljubljana, Slovenia, 17–18 September 2012; Jones, D., Militz, H., Petrič, M., Pohleven, F., Humar, H., Pavlič, M., Eds.; University of Ljubljana, Biotechnical Faculty, Department of Wood Science and Technology: Ljubljana, Slovenia; pp. 11–17.
35. *CEN. EN 310:1993*; Wood-Based Panels. Determination of Modulus of Elasticity in Bending and of Bending Strength. European Committee for Standardisation: Brussels, Belgium, 1993.
36. *ASTM D1037-99*; Standard Test Methods for Evaluating Properties of Wood-Base Fiber and Particle Panel Materials. ASTM: West Conshohocken, PA, USA, 1999.
37. *CEN. EN 1534:2020*; Wood Flooring and Parquet. Determination of Resistance to Indentation. Test Method. European Committee for Standardisation: Brussels, Belgium, 2020.
38. Rapp, A.O.; Brischke, C.; Welzbacher, C.R. Interrelationship between the severity of heat treatments and sieve fractions after impact ball milling: A mechanical test for quality control of thermally modified wood. *Holzforschung* **2006**, *60*, 64–70. [CrossRef]
39. *CEN. EN ISO 5223:2015*; Test Sieves for Cereals. European Committee for Standardisation: Brussels, Belgium, 2015.
40. *CEN. EN 113e*; Wood Preservatives—Test Method for Determining the Protective Effectiveness against Wood-Destroying Basidiomycetes. Determination of Toxic Values, Volume 28. European Committee for Standardisation: Brussels, Belgium, 2006.
41. *CEN. EN 84*; Wood Preservatives—Accelerated Ageing of Treated Wood Prior to Biological Testing—Leaching Procedure. European Committee for Standardisation: Brussels, Belgium, 2002.
42. *CEN. EN 117:2012*; Wood Preservatives. Determination of Toxic Values against Reticulitermes Species (European termites) (Laboratory Method). European Committee for Standardisation: Brussels, Belgium, 2012.
43. Esteves, B.; Ribeiro, F.; Cruz-Lopes, L.; Ferreira, J.; Domingos, I.; Duarte, M.; Duarte, S.; Nunes, L. Combined treatment by densification and heat treatment of maritime pine wood. *Wood Res.* **2017**, *62*, 373–388.
44. Humar, M.; Repič, R.; Kržišnik, D.; Lesar, B.; Cerc Korošec, R.; Brischke, C.; Emmerich, L.; Rep, G. Quality Control of Thermally Modified Timber Using Dynamic Vapor Sorption (DVS) Analysis. *Forests* **2020**, *11*, 666. [CrossRef]
45. Frederiksson, M.; Thybring, E.E. Scanning or desorption isotherms? Characterising sorption hysteresis of wood. *Cellulose* **2018**, *25*, 4477–4485. [CrossRef]
46. Brischke, C.; Rapp, A.O.; Welzbacher, C.R. High-energy multiple impact (HEMI)–test Part 1: A new tool for quality control of thermally modified timber. In Proceedings of the 37th Annual Conference, International Research Group on Wood Protection, Tromsø, Norway, 18–22 June 2006.; IRG/WP 06-20346.
47. Welzbacher, C.R.; Brischke, C.; Rapp, A.O. High-energy multiple impact (HEMI)–test—Part2: A mechanical test for the detection of fungal decay. In Proceedings of the 37th Annual Conference, International Research Group on Wood Protection, Tromsø, Norway, 18–22 June 2006; IRG/WP 06-20339.
48. Popescu, C.-M.; Jones, D.; Kržišnik, D.; Humar, M. Determination of the effectiveness of a combined thermal/chemical wood modification by the use of FT–IR spectroscopy and chemometric methods. *J. Mol. Struct.* **2020**, *1200*, 127133. [CrossRef]
49. Mayes, D.; Oksanen, O. *ThermoWood Handbook*; Finnish ThermoWood Association: Helsinki, Finland, 2003.
50. Salman, S.; Thévenon, M.F.; Pétrissans, A.; Dumarçay, S.; Candelier, K.; Gérardin, P. Improvement of the durability of heat-treated wood against termites. *Maderas. Cienc. Tecnol.* **2017**, *19*, 317–328. [CrossRef]
51. Duarte, S.; Nunes, L.; Kržišnik, D.; Humar, M.; Jones, D. Influence of Zwitterionic Buffer Effects with Thermal Modification Treatments of Wood on Symbiotic Protists in *Reticulitermes grassei* Clément. *Insects* **2021**, *12*, 139. [CrossRef]

forests

Article

Effect of Thermo-Hydro-Mechanical Treatment on Mechanical Properties of Wood Cellulose: A Molecular Dynamics Simulation

Feiyu Ouyang and Wei Wang *

College of Engineering and Technology, Northeast Forestry University, Harbin 150040, China; fayeoy@nefu.edu.cn
* Correspondence: vickywong@nefu.edu.cn; Tel.: +86-133-1361-3588

Abstract: Based on molecular dynamics, a water and cellulose model was constructed to provide more theoretical support for the behavior characteristics of cellulose properties in thermo-hydro-mechanical treatment. In this paper, dynamic simulations were carried out under the NPT ensemble at 4, 5.5, 8, and 12 MPa, respectively. Moreover, we analyze the effects on the mechanical properties of wood cellulose in terms of the hydrogen bond numbers, small molecule diffusion coefficients, end-to-end distances, and mechanical parameters of the water–cellulose model. The results indicate that the densification of the water–cellulose model gradually increases with increasing pressure. The effect of pressures on mechanical properties is mainly due to the formation of massive hydrogen bonds within the cellulose chain and between water and cellulose. This is reflected in the fact that water molecules are more difficult to diffuse in the cellulose, which therefore weakens the negative effect of large amounts of water on the cellulose. The increase in end-to-end distance represents the stiffness of the cellulose chains being strengthened. The mechanical parameters indicate an increase in wood stiffness to resist deformation better, while reducing tensile properties at the same time. The dynamic simulation results in this paper can well correspond to macroscopic experiments.

Keywords: thermo-hydro-mechanical treatment; molecular dynamics; wood cellulose; mechanical properties

Citation: Ouyang, F.; Wang, W. Effect of Thermo-Hydro-Mechanical Treatment on Mechanical Properties of Wood Cellulose: A Molecular Dynamics Simulation. *Forests* **2022**, *13*, 903. https://doi.org/10.3390/f13060903

Academic Editors: Tomasz Krystofiak and Pavlo Bekhta

Received: 19 May 2022
Accepted: 6 June 2022
Published: 9 June 2022

1. Introduction

As the only renewable green material, wood is used extensively because of its wide coverage of raw materials and strong maneuverability. The hygroscopic expansion of wood materials can form bonding interfaces, but at the same time, they are also vulnerable to corrosion cracking and dimensional deformation, which reduces the utilization rate of wood. Therefore, natural wood needs to be modified [1]. Heat treatment is the most common method used in wood modification [2]. Since the concept of heat treatment was first proposed by Stamm [3] in 1946, this method has been widely studied and continuously modified to better improve wood utilization, such as dimensional stability, moisture absorption, durability, etc. [4].

Hydrothermal treatment uses water vapor or liquid water as the heat transfer medium and modifies wood in the range of 160–240 °C [5]. In general, as the temperature of hydrothermal treatment increases, the dimensional stability and durability of wood increases significantly, but it will cause the loss of mechanical properties of wood [6]. This was confirmed by the study of Camiyani Black Pine, where the compression strength of Black Pine was reduced by 27% after 10 h treatment at 180 °C [7], while in the M. Gaff et al. study, the highest reduction in the impact bending strength of oak and spruce reached 32.2% and 39.8%, respectively [8].

To reduce the influence of hydrothermal treatment on the mechanical properties of wood, thermo-hydro-mechanical (THM) treatment can be used to modify wood [9]. Thermo-hydro-mechanical wood treatment is based on a combination of high temperatures, moisture, and mechanical loading, which leads to modified wood [10]. It has been demonstrated that THM treatment can better improve the mechanical strength of wood compared to dry-heating conditions [11]. Bao et al. applied 4–12 MPa of pressure while THM densified hybrid poplar wood at 160 °C [12]. The modulus of elasticity increased significantly compared to the wood without high pressure. Moreover, this study showed a positive correlation with the mechanical properties of the wood as the densification degree of wood enhanced.

By calculating the dynamic evolution of molecules, molecular dynamics can help us understand their motion. Cellulose is the primary component of wood. The study of cellulose movements by molecular dynamics assists us in both verifying the changes of wood in macroscopic experiments and predicting the impact of changes in conditions on the properties of wood. Wood cellulose belongs to higher plant cellulose, whose crystalline form is monoclinic crystal [13]. Cellulose is composed of crystalline and amorphous regions. The cellulose chains are disordered in the amorphous region. It is obviously observed that there are many pore structures in the system of cellulose [14], which causes a high frequency of interaction between water and cellulose in the amorphous region. Du et al. investigated the mechanical properties of amorphous cellulose by constructing an amorphous region from a cellulose chain with a degree of polymerization of 20 [15]. The mechanical properties obtained were similar to the experimental values. Therefore, this study focuses on the amorphous region of cellulose.

The accuracy of molecular dynamics simulation depends on the selection of force fields. Zhang et al. used the GLYCAM force field to study the effect of temperature on the thermal response of crystalline cellulose [16]. Chen et al. investigated the structural stability and thermal motion of natural cellulose in different potential hydrogen bonding patterns using the Gromos56Acarbo force field [17]. Wang et al. used the PCFF force field to simulate the cellulose at high temperatures, which investigated the effects of different temperatures on the mechanical properties of cellulose [18]. Meanwhile, it is also proved that the PCFF force field is suitable for the amorphous region of cellulose.

Available studies on molecular dynamics simulation mainly focused on the effects of temperature and transfer medium on wood heat treatment. The research on the effect of THM treatment on the properties of wood is limited to macroscopic properties such as wood size, hardness, and strength. However, the mechanism between heat and pressure and solid wood with high moisture has not been well explained. Exploring the effects of the combined mechanisms of pressure, temperature, and moisture on wood is crucial to improving wood utilization.

In this paper, a cellulose model was conducted under certain humidity in the controlled environment of THM treatment, which is used to analyze the micro-mechanism associated with the effect of THM treatment on the mechanical properties of the water–cellulose model that was studied. Moreover, we analyzed this model in terms of its hydrogen bond numbers, small molecule diffusion coefficients, and end-to-end distances. The main purpose of this paper is to study the changes in the macroscopic properties of wood in the THM treatment while providing stronger theoretical support for wood thermo-hydro-mechanical modification.

2. Materials and Methods

2.1. Model Establishment

Material Studios software (version 2020, Accelrys, San Diego, CA, USA) was used to simulate a cellulose chain with a degree of polymerization of 20. In order to investigate the effect of humidity on cellulose, a composite model with water contents of 25% was constructed [12]. The model was constructed using the method proposed by Theodorou et al. for the construction of amorphous polymers [19]. The Ewald method was adopted to charge addition. The Van der Waals addition method used the atom-based method. The water–cellulose composite model was established with a target density of 1.5 g/cm^3 [20], which consists of a cellulose chain with 125 water molecules, as shown in Figure 1. The water molecules are displayed in ball-and-stick mode and the cellulose molecule is displayed in stick mode. The hydrogen and oxygen atoms are represented with the white and yellow spheres, respectively.

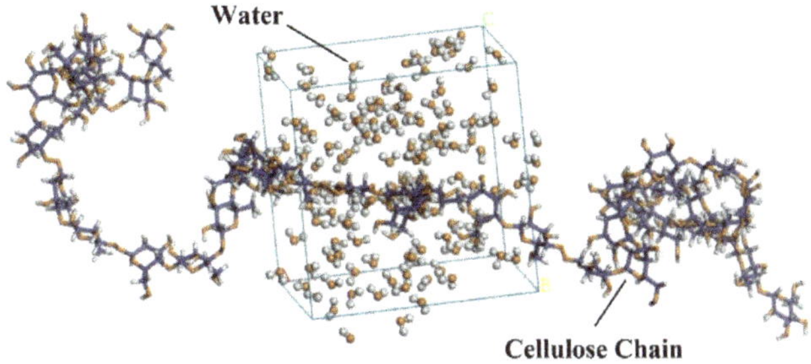

Figure 1. Water–cellulose composite model.

2.2. Dynamic Simulation

All simulations in this paper were based on the PCFF force field [18]. To minimize the energy of the system before the molecular dynamics simulation, the model must first completely relax and minimize its energy. First, the Smart algorithm was used to perform 5000 steps of operations, after which the model was geometrically optimized. Then we performed a dynamic relaxation at 433.15 K under the NVT ensemble for 1000 ps. One frame was output every 5000 steps. Huang et al. has proved that these parameters are reliable [21]. After all these steps, the system energy tended to be stable.

Based on the previous complete relaxation, we performed a 1000 ps under the NVT ensemble with a temperature of 433.15 K as pre-heating. After that, a 1000 ps dynamic simulation was performed under the NPT ensemble. The temperature was set to 433.15 K and the pressure was set to 4, 5.5, 8, and 12 MPa for simulation, respectively. The model with dynamic simulations at atmospheric pressure of 0.1 MPa as the untreated control group was prepared as well. Bao et al. tested the selection of temperature and pressure, demonstrating that the effect of the pressurized hydrothermal treatment on cellulose could be well explored [12]. The electronic effect was controlled by the Ewald method [22], the van der Waal force was calculated by the atom-based method [23], the temperature was controlled by the Anderson method [24], and the pressure was controlled by the Berendsen method [25].

3. Results and Discussion

3.1. Model Characterization

3.1.1. System Energy

In molecular dynamics, temperature and energy will change with time, which can be used as criteria for determining whether a system is in equilibrium. The variation of the water–cellulose system's energy with time is shown in Figure 2a. The system equilibrium can be expressed by the energy convergence parameter δE [20], which is defined in Equation (1):

$$\delta E = \frac{1}{N}\sum_{i=1}^{N}\left|\frac{E_i - E_0}{E_0}\right| \quad (1)$$

where N is the total number of steps of the simulation, E_0 represents the system's initial energy, and E_i represents the energy after the simulation i steps.

When $\delta E \leq 0.001 - 0.003$, the system tends to be in equilibrium, while the simulation results are reliable. After analysis, the δE for the water–cellulose model was 0.0015 at the last 200 ps.

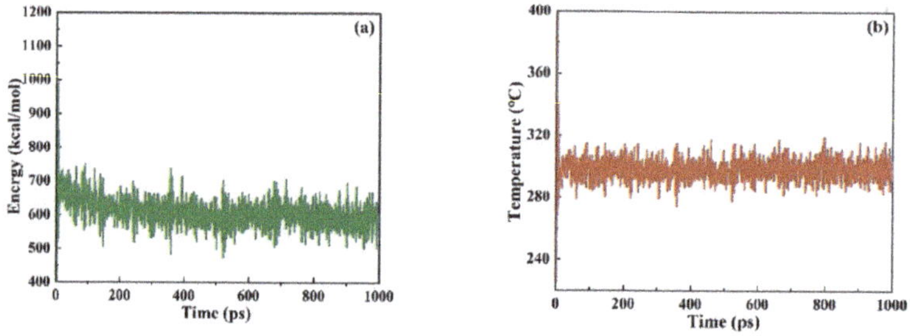

Figure 2. Energy and temperature variation of the model: (**a**) Energy–Time; (**b**) Temperature–Time.

The temperature–time variation of the water–cellulose composite model is shown in Figure 2b. The temperature, as shown in the figure, varied slightly over time while remaining within ±25 K, which indicated that the system was in equilibrium after energy relaxation.

As a result of geometric optimization and energy relaxation, the energy of the water–cellulose composite model is stable, and the subsequent dynamic simulation results are trustworthy.

3.1.2. Lattice Parameters and Density

The lattice cell of the water–cellulose model is a cube. The cell parameters show the length of the cell, which can be used to characterize the cell size. The variation in pressure affects the cell size as well as density, and the variation of the cell parameters under different pressures is shown in Table 1.

Table 1. Cell parameters of water–cellulose model.

Pressure (MPa)	Cell Parameters (Å)		
	The Length	The Width	The Height
Untreated	24.86	24.86	24.86
4	22.83	22.83	22.83
5.5	22.79	22.79	22.79
8	22.71	22.71	22.71
12	22.69	22.69	22.69

The density of the model reflects the densification degree of cellulose chains under different pressures, which can be used to analyze the mechanical properties of wood cellulose [26]. The density of the water–cellulose composite model after hydrothermal treatment under high pressure is reported in Table 2.

Table 2. Density of water–cellulose model.

Pressure (MPa)	Density (g/cm^3)		
	Final	Average	Std. Dev.
Untreated	0.979	0.978	0.020
4	1.242	1.264	0.014
5.5	1.264	1.270	0.015
8	1.268	1.280	0.014
12	1.285	1.288	0.013

Where Final represents the density of the model at the end of the dynamic simulation and Average is the average density after dynamic simulation. Std. Dev. is the standard deviation, which indicates the dispersion degree of the density at different moments. The lower the standard deviation, the more reliable the density value.

Analyzing the data in Table 1, the cell parameters decreased from 24.86 to 22.69. The cell parameters decrease as the pressure increases, which indicates that the cell is compressed under high pressure. From Table 2, it can be inferred that the average density of the water–cellulose model increases with increasing pressure. The cellulose volume decreases and the average density increases significantly under high pressure, which demonstrates that the densification degree of wood can be greatly enhanced.

3.2. Hydrogen Bond Analysis

Glucose is the minimum component unit of cellulose, which contains multiple hydroxyl groups, hence the strong hydrogen bond in cellulose [27]. The hydrogen bond network is directly related to the mechanical properties of materials. Hydrogen bonds can be formed within the cellulose chain, between cellulose chains, as well as between cellulose and water molecules. The number of various hydrogen bonds under different pressures can be clearly seen in Table 3.

Table 3. Number of hydrogen bonds in the water–cellulose model.

Pressure (MPa)	Number of Hydrogen Bonds			
	Between Cellulose Chains	Between Water Molecular	Between Water–Cellulose	Total
Untreated	39	56	20	115
4	59	134	150	343
5.5	62	130	160	352
8	64	128	167	359
12	70	127	173	370

Water and cellulose form more hydrogen bonds than cellulose chains, which may contribute to moisture diffusion into the cellulose system to form hydrogen bonds. Significant quantities of moisture may weaken the original hydrogen bonding network, making the wood susceptible to fracture.

Table 3 demonstrates that the THM treatment significantly increased the number of hydrogen bonds of the THM cellulose in comparison with those of the untreated model. Table 3 also determines that hydrogen bonding between cellulose chains is higher at high pressure, which is probably owing to the increased cellulose densification caused by high pressure. The volume of cellulose decreases and intermolecular contacts boost, making it easier to form hydrogen bonds [28]. The formation of intermolecular hydrogen bonds not only enhances the structural stability, but also mitigates the negative effect of water intrusion on the mechanical properties of cellulose.

3.3. The Molecular Diffusion Coefficient

The diffusion of small molecules in polymers can be characterized by the diffusion coefficient. The size of the diffusion coefficient can be expressed by the motion range of small molecules in the system. The greater the diffusion coefficient, the more likely it is that water molecules will react with cellulose chains. The investigation of the diffusion coefficient of water molecules contributes to the movement of cellulose chains under high pressure [29]. The mean square displacement (MSD) is one of the important parameters to describe the motion of molecular chains, which is usually used to describe the centroid displacement of molecular chains within a section of time, and can be obtained from Equation (2):

$$\mathrm{MSD} = \sum_{i=1}^{n}\left\langle\left|\vec{r_i}(t) - \vec{r_i}(0)\right|^2\right\rangle \tag{2}$$

where $\vec{r_i}(t)$ represents the displacement of the molecule i at time t; $\vec{r_i}(0)$ represents the initial displacement. The diffusion coefficient D of water molecules can be calculated through the Einstein equation [30], which is related to MSD, represented by Equation (3):

$$D = \frac{1}{6n}\lim_{t\to\infty}\frac{d}{dt}\sum_{i=1}^{n}\left\langle\left|\vec{r_i}(t) - \vec{r_i}(0)\right|^2\right\rangle \tag{3}$$

when t is sufficiently large, the diffusion coefficient can be solved as Equation (4),

$$D = \frac{k}{6} \tag{4}$$

where k represents the slope obtained by simulating the MSD curve with the least square method. Analyzing the water–cellulose model produced the MSD curves of water molecules under various pressures.

It can be seen from Figure 3 that the MSD curve of water molecules linearly correlated with time. The MSD curves were simulated to obtain the slope k and the correlation coefficient r^2, as well as the diffusion coefficients D at various pressures, which are shown in Table 4.

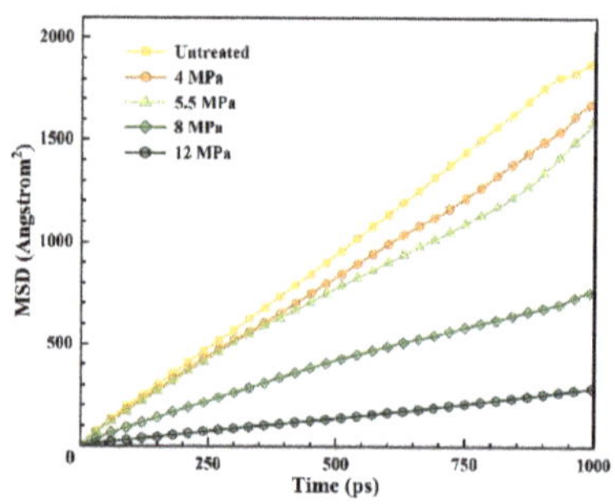

Figure 3. MSD of water–cellulose model.

Table 4. Diffusion coefficients of water–cellulose model under different pressures.

Pressure (MPa)	k	D	R-Square
Untreated	1.9112	0.3185	0.9993
4	1.6162	0.2694	0.9989
5.5	1.4387	0.2398	0.9952
8	0.7320	0.1220	0.9967
12	0.2752	0.0459	0.9983

From the analysis of Table 4, the diffusion coefficient of the pressurized model was smaller than those of the untreated model. Furthermore, the diffusion coefficient of water molecules decreases as the pressure increases, implying that water molecules move less flexibly in cellulose, as summarized in Table 4. As a strong polar molecule, water easily forms hydrogen bonds with the cellulose chain, thus inhibiting water movement in cellulose [31], as clearly reflected by the increased number of hydrogen bonds formed between water and cellulose.

The interaction between water and cellulose is increased as a result of the massive formation of intermolecular hydrogen bonds [32], while the diffusion coefficient of water molecules decreases as the adsorption capacity of cellulose to water increases. The rapid diffusion of water in cellulose accelerates the thermal movement of molecules, which is the primary reason for wood fracture. THM treatment reduces the diffusion of water molecules, which improves the stability of molecular chain movement and the deformation resistance of wood.

3.4. End-to-End Distance

End-to-end distance is the distance between the ends of a polymer chain, describing the degree of molecular curl. The curling of the cellulose polymer chain is owing to the non-bonding force in the molecular chain, such as hydrogen bond and van der Waals force, so that the energy of the whole system reaches the lowest value to ensure the stability of the system, which conforms to the entropy increase principle [33]. The end-to-end distance of the water–cellulose model is shown in Figure 4, and the cellulose presents a crimp state.

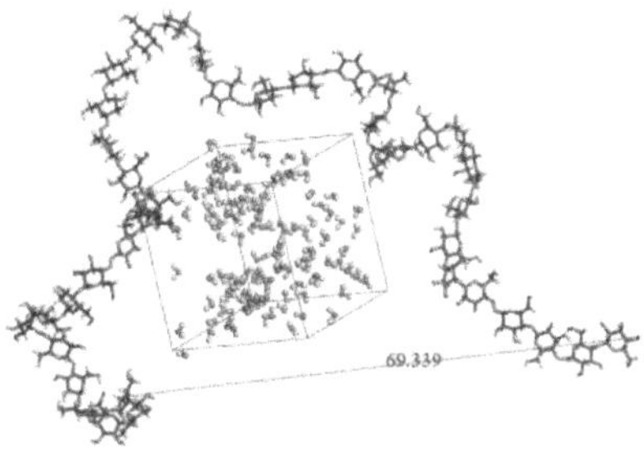

Figure 4. End-to-end distance of water–cellulose model.

In this paper, end-to-end distance was used to characterize the stiffness of the cellulose chain. The higher stiffness of the cellulose chain at the microscopic level strengthens the deformation resistance of the wood, which means it is harder to deform due to external forces [34]. The Flory's characteristic ratio [35] is defined as follows:

$$C = \frac{\langle {h_0}^2 \rangle}{Nl^2} \tag{5}$$

where h_0 is end-to-end distance, N denotes the number of cellulose chain molecules, and l denotes the length of a single bond. The larger value of C indicates the better stiffness of the cellulose chain. Given that N and l remain unchanged in this paper, the change in C is only related to h_0.

The end-to-end distance data at each pressure were extracted to obtain the graph shown in Figure 5.

Comparing the untreated model with the THM groups, end-to-end distance is demonstrated in Figure 5. With the enhancement of pressure, the amplitude of end-to-end distance keeps increasing. Hence, higher pressure lowered the curl degree of the cellulose. Large amounts of hydrogen bonds are formed, which makes a great contribution to the increasing end-to-end distance and increases the difficulty of rotation and displacement of cellulose chains. Above all, the equilibrium conformation of molecular chains depends more on the interaction energy rather than the conformation entropy. Non-bonding interaction of hydrogen bonds between water and cellulose is also the reason why cellulose chains are stretched [36]. As a result, the increase in end-to-end distance means that cellulose chain stiffness is strengthened, and its deformation resistance will also be enhanced.

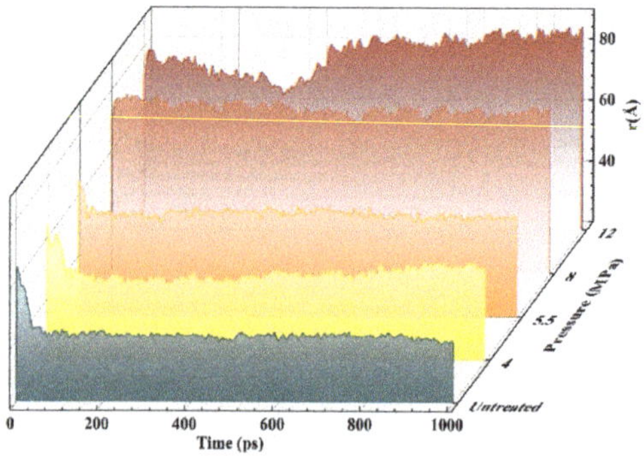

Figure 5. End distance distribution curve of cellulose chain.

3.5. Mechanical Properties

The mechanical parameters were obtained from the dynamic simulation results at different pressures, which were able to characterize the mechanical properties of the wood. Hooke's law governs the stress–strain behavior of solid linear–elastic materials, so that the mechanical properties of cellulose can be calculated by Equation (6):

$$[C_{ij}] = \begin{bmatrix} \lambda+2\mu & \lambda & \lambda & 0 & 0 & 0 \\ \lambda & \lambda+2\mu & \lambda & 0 & 0 & 0 \\ \lambda & \lambda & \lambda+2\mu & 0 & 0 & 0 \\ 0 & 0 & 0 & \mu & 0 & 0 \\ 0 & 0 & 0 & 0 & \mu & 0 \\ 0 & 0 & 0 & 0 & 0 & \mu \end{bmatrix} \tag{6}$$

where λ and μ are known as Lamé constants and are often applied to compute mechanical parameters, such as Young's modulus (E), shear modulus (G), Poisson's ratio (γ), and so on. The formulas are as follows:

$$E = \frac{\mu(3\lambda+2\mu)}{\lambda+\mu} \tag{7}$$

$$G = \mu \tag{8}$$

$$K = \lambda + \frac{2}{3\mu} \tag{9}$$

$$\gamma = \frac{\lambda}{2(\lambda+\mu)} \tag{10}$$

As shown in Figure 6, each mechanical parameter of the water–cellulose model under different pressures to that of the untreated model was calculated to quantify the influence of the THM process.

Young's modulus (E) is the ratio of stress to the corresponding strain of the material within the elastic limit, which is commonly applied to indicate whether the material is prone to deformation. A large value indicates that wood is more resistant to deformation. The shear modulus (G), as the ratio of shear stress to strain, also represents the stiffness of the material.

It can be seen from Figure 6a that the Young's modulus and shear modulus increase with increasing pressure, which indicates that wood stiffness increases. The results showed that the E and G value of the THM model with 12 MPa increased by 12.0381 GPa and 4.4378 GPa compared with those of the untreated model, respectively. Both Young's modulus and shear modulus increases that are caused by the enhancing densification degree of wood have been investigated [37]. The increase in hydrogen bonds between cellulose chains also increases the van der Waals force of the inter-chain contact [31]. The enhancement of the intermolecular force makes the entire cellulose system form a denser structure, resulting in the decrease in Young's modulus and shear modulus. In consequence, the wood has a stronger stiffness to resist deformation. It is also consistent with the conclusion that pressurized hydrothermal treatment can improve the Young's modulus and shear modulus of wood, which Bao et al. [12] studied.

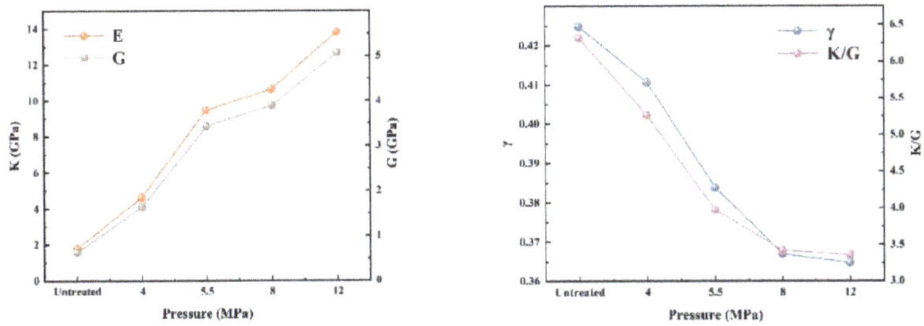

(a) (b)

Figure 6. Mechanical parameter of water–cellulose model (**a**) Young's modulus and shear modulus; (**b**) Poisson's ratio and the ratio of volume variable to shear variable.

Poisson's ratio (γ) and the ratio of volume variable to shear variable (*K/G*) are used as the elastic constants to describe the transverse deformation of the material. The greater the value, the better the flexibility of the material. The results shown in Figure 6b excluded the increasing pressure between γ and *K/G* values, showing that the flexibility of wood is weakened. An updated review by Yu et al. confirms the reduction in wood toughness attributed to the degradation of cellulose at high temperatures [38].

Different application scenarios of wood have different requirements for its performance. Compared with other industries, wood has higher requirements of stiffness when used as a building material. Pressurized hydrothermal treatment can significantly improve the stiffness and deformation resistance of wood, allowing it to be used in applications that require high wood stiffness.

4. Conclusions

This paper studies the effect of pressurized hydrothermal treatment on the mechanical properties of wood cellulose. Model parameters, hydrogen bond numbers, molecule diffusion coefficients, end-to-end distances, and mechanical parameters of the water–cellulose model were analyzed. The conclusions obtained are as follows:

1. High pressure improves the densification degree of wood. The total number of hydrogen bonds in the system, the number of hydrogen bonds between cellulose chains, and the combined hydrogen bonds of water and cellulose are on the rise, which improves the structural stability of cellulose. Reduced diffusion coefficient of water molecules in the cellulose indicates that high pressure weakens the cracking of the wood caused by the bulk diffusion of moisture.
2. The stiffness of the polymer chain can reflect the macroscopic mechanical properties of the material from the microscopic level. The analysis found that the increase in pressure caused the continuous increase in the end-to-end distance of the cellulose chain and the decrease in the degree of curling, which was related to the formation of many hydrogen bonds between the cellulose chains. The increase in the end-to-end distance also means that the stiffness of the cellulose chain increases, indicating that the wood can have better deformation resistance at the macro level.
3. The mechanical properties of the water–cellulose model were studied, and the Young's modulus, shear modulus, Poisson's ratio, and *K/G* value were calculated and analyzed.

With the increase in the pressure, the Young's modulus and the shear modulus were in an upward trend, and the increase was very large, which means that the deformation resistance and rigidity of the wood have been greatly improved. Moreover, Poisson's ratio and *K/G* value decreased with the increase in pressure, and the toughness of the wood decreased. The pressurized hydrothermal treatment of wood can significantly improve its rigidity and deformation resistance, and what is more, its utilization rate in applications requiring high rigidity.

Author Contributions: Conceptualization, F.O.; methodology, F.O.; software, F.O.; validation, F.O. and W.W.; formal analysis, F.O.; investigation, F.O.; resources, F.O.; data curation, F.O.; writing—original draft preparation, F.O.; writing—review and editing, F.O.; visualization, F.O.; supervision, F.O.; project administration, F.O.; funding acquisition, W.W. All authors have read and agreed to the published version of the manuscript.

Funding: This research was funded by the Fundamental Research Funds for the Central Universities, grant number 2572019BL04, and the Scientific Research Foundation for the Returned Overseas Chinese Scholars of Heilongjiang Province, grant number LC201407.

Institutional Review Board Statement: Not applicable.

Informed Consent Statement: Not applicable.

Data Availability Statement: Data available upon request from the corresponding author.

Conflicts of Interest: The authors declare no conflict of interest.

References

1. Esteves, B.; Pereira, H. Wood modification by heat treatment: A review. *BioResources* **2009**, *4*, 370–404. [CrossRef]
2. Pelaez-Samaniego, M.R.; Yadama, V.; Lowell, E.; Espinoza-Herrera, R. A review of wood thermal pretreatments to improve wood composite properties. *Wood Sci. Technol.* **2013**, *47*, 1285–1319. [CrossRef]
3. Stamm, A.J.; Burr, H.K.; Kline, A.A. Staybwood—Heat-stabilized wood. *Ind. Eng. Chem.* **1946**, *38*, 630–634. [CrossRef]
4. Mohebby, B.; Ilbeighi, F.; Kazemi-Najafi, S. Influence of hydrothermal modification of fibers on some physical and mechanical properties of medium density fiberboard (*MDF*). *Holz Als Roh-Und Werkst.* **2008**, *66*, 213–218. [CrossRef]
5. Lee, S.H.; Ashaari, Z.; Lum, W.C.; Halip, J.A.; Ang, A.F.; Tan, L.P.; Chin, K.L.; Tahir, P.M. Thermal treatment of wood using vegetable oils: A review. *Constr. Build. Mater.* **2018**, *181*, 408–419. [CrossRef]
6. Xu, J.; Zhang, Y.; Shen, Y.; Li, C.; Wang, Y.; Ma, Z.; Sun, W. New perspective on wood thermal modification: Relevance between the evolution of chemical structure and physical-mechanical properties, and online analysis of release of VOCs. *Polymers* **2019**, *11*, 1145. [CrossRef]
7. Gündüz, G.; Korkut, S.; Korkut, D.S. The effects of heat treatment on physical and technological properties and surface roughness of Camiyanı Black Pine (*Pinus nigra Arn. subsp. pallasiana var. pallasiana*) wood. *Bioresour. Technol.* **2008**, *99*, 2275–2280. [CrossRef]
8. Gaff, M.; Kačík, F.; Gašparík, M. Impact of thermal modification on the chemical changes and impact bending strength of European oak and Norway spruce wood. *Compos. Struct.* **2019**, *216*, 80–88. [CrossRef]
9. Fang, C.-H.; Mariotti, N.; Cloutier, A.; Koubaa, A.; Blanchet, P. Densification of wood veneers by compression combined with heat and steam. *Eur. J. Wood Wood Prod.* **2012**, *70*, 155–163. [CrossRef]
10. Zhan, T.; Lu, J.; Jiang, J.; Peng, H.; Li, A.; Chang, J. Viscoelastic Properties of the Chinese Fir (*Cunninghamia lanceolata*) during Moisture Sorption Processes Determined by Harmonic Tests. *Materials* **2016**, *9*, 1020. [CrossRef]
11. Willems, W.; Altgen, M. Hygrothermolytic wood modification. process description and treatment level characterisation. *Wood Mater. Sci. Eng.* **2019**, *15*, 213–222. [CrossRef]
12. Bao, M.; Huang, X.; Jiang, M.; Yu, W.; Yu, Y. Effect of thermo-hydro-mechanical densification on microstructure and properties of poplar wood (*Populus tomentosa*). *J. Wood Sci.* **2017**, *63*, 591–605. [CrossRef]
13. Atalla, R.H.; Vanderhart, D.L. Native cellulose: A composite of two distinct crystalline forms. *Science* **1984**, *223*, 283–285. [CrossRef] [PubMed]
14. Meier, R.; Maple, J.; Hwang, M.-J.; Hagler, A. Molecular modeling urea-and melamine-formaldehyde resins. 1. A force field for urea and melamine. *J. Phys. Chem.* **1995**, *99*, 5445–5456. [CrossRef]
15. Du, D.; Tang, C.; Yang, L.; Guo, L.; Qiu, Q. Molecular dynamics simulation on the distribution and diffusion of different sulfides in oil-paper insulation systems. *J. Mol. Liq.* **2020**, *314*, 113678. [CrossRef]
16. Zhang, Q.; Bulone, V.; Ågren, H.; Tu, Y. A molecular dynamics study of the thermal response of crystalline cellulose Iβ. *Cellulose* **2011**, *18*, 207–221. [CrossRef]
17. Chen, P.; Nishiyama, Y.; Putaux, J.-L.; Mazeau, K. Diversity of potential hydrogen bonds in cellulose I revealed by molecular dynamics simulation. *Cellulose* **2014**, *21*, 897–908. [CrossRef]

18. Wang, W.; Ma, W.; Wu, M.; Sun, L. Effect of Water Molecules at Different Temperatures on Properties of Cellulose Based on Molecular Dynamics Simulation. *BioResources* **2022**, *17*, 269–280. [CrossRef]
19. Theodorou, D.N.; Suter, U.W. Detailed molecular structure of a vinyl polymer glass. *Macromolecules* **1985**, *18*, 1467–1478. [CrossRef]
20. Allen, M.P.; Tildesley, D.J. *J Computer Simulation of Liquids*; Clarendon Press: Oxford, UK, 1987.
21. Huang, S.; Sun, Y.; Xu, Y.; Meng, L. Molecular dynamics simulation of the effect of ammonia molecules on the performance of cellulose Iβ. *Acta Polym. Sin.* **2014**, *14*, 188–193.
22. Ewald, P. Evaluation of optical and electrostatic lattice potentials. *Ann. Phys.* **1921**, *64*, 253–287. [CrossRef]
23. Andersen, H.C. Molecular dynamics simulations at constant pressure and/or temperature. *J. Chem. Phys.* **1980**, *72*, 2384–2393. [CrossRef]
24. Andrea, T.A.; Swope, W.C.; Andersen, H.C. The role of long ranged forces in determining the structure and properties of liquid water. *J. Chem. Phys.* **1983**, *79*, 4576–4584. [CrossRef]
25. Berendsen, H.J.; Postma, J.v.; Van Gunsteren, W.F.; DiNola, A.; Haak, J.R. Molecular dynamics with coupling to an external bath. *J. Chem. Phys.* **1984**, *81*, 3684–3690. [CrossRef]
26. Wang, X.; Tu, D.; Chen, C.; Zhou, Q.; Huang, H.; Zheng, Z.; Zhu, Z. A thermal modification technique combining bulk densification and heat treatment for poplar wood with low moisture content. *Constr. Build. Mater.* **2021**, *291*, 123395. [CrossRef]
27. Nishiyama, Y.; Langan, P.; Chanzy, H. Crystal structure and hydrogen-bonding system in cellulose Iβ from synchrotron X-ray and neutron fiber diffraction. *J. Am. Chem. Soc.* **2002**, *124*, 9074–9082. [CrossRef]
28. Taghiyari, H.R.; Bayani, S.; Militz, H.; Papadopoulos, A.N. Heat treatment of pine wood: Possible effect of impregnation with silver nanosuspension. *Forests* **2020**, *11*, 466. [CrossRef]
29. David, G.; Orszag, S.A. *Numerical Analysis of Spectral Methods: Theory and Applications*; Society for Industrial and Applied Mathematics: Philadelphia, PA, USA, 1977.
30. Einstein, A. Zur elektrodynamik bewegter körper. *Ann. Phys.* **1905**, *4*, 891–921. [CrossRef]
31. Du, D.; Tang, C.; Zhang, J.; Hu, D. Effects of hydrogen sulfide on the mechanical and thermal properties of cellulose insulation paper: A molecular dynamics simulation. *Mater. Chem. Phys.* **2020**, *240*, 122153. [CrossRef]
32. Zhang, N.; Shen, Z.; Chen, C.; He, G.; Hao, C. Effect of hydrogen bonding on self-diffusion in methanol/water liquid mixtures: A molecular dynamics simulation study. *J. Mol. Liq.* **2015**, *203*, 90–97. [CrossRef]
33. Liao, R.-J.; Xiang, B.; Yang, L.-J.; Tang, C.; Sun, H.-G. Study on the thermal aging characteristics and bond breaking process of oil-paper insulation in power transformer. In Proceedings of the Conference Record of the 2008 IEEE International Symposium on Electrical Insulation, Vancouver, BC, Canada, 9–12 June 2008; pp. 291–296.
34. Zhang, B.; Zhu, S.; Guo, T. *Modern Polymer Science*; Beijing Chemical Industry Press: Beijing, China, 2006.
35. Flory, P.J.; Volkenstein, M. *Statistical Mechanics of Chain Molecules*; Wiley: Hoboken, NJ, USA, 1969.
36. Ren, Z.; Guo, R.; Zhou, X.; Bi, H.; Jia, X.; Xu, M.; Wang, J.; Cai, L.; Huang, Z. Effect of amorphous cellulose on the deformation behavior of cellulose composites: Molecular dynamics simulation. *RSC Adv.* **2021**, *11*, 19967–19977. [CrossRef] [PubMed]
37. Sözbir, G.D.; Bektas, I.; Ak, A.K. Influence of combined heat treatment and densification on mechanical properties of poplar wood. *Maderas. Cienc. Tecnol.* **2019**, *21*, 481–492. [CrossRef]
38. Yu, Y.; Zhang, F.; Zhu, S.; Li, H. Effects of high-pressure treatment on poplar wood: Density profile, mechanical properties, strength potential index, and microstructure. *BioResources* **2017**, *12*, 6283–6297. [CrossRef]

forests

MDPI

Article

Physical and Mechanical Properties Performance between Untreated and Treated with CCA Treatment at Different Age Groups of Fast-Growing Acacia Hybrid of Sarawak

Gaddafi Ismaili [1,*], Ellyne Enduat [1], Nur Syahina Yahya [1], Fanthy Moola Malek [1], Noor Azland Jaimudin [1], Khairul Khuzaimah Abdul Rahim [2], Mohd Effendi Wasli [3], Meekiong Kalu [3], Iskanda Openg [4], Ahmad Nurfaidhi Rizalman [5], Jack Liam [6] and Biha Razali [6]

1 Department of Civil Engineering, Faculty of Engineering, Universiti Malaysia Sarawak, Kota Samarahan 94300, Sarawak, Malaysia
2 Sarawak Forestry Corporation, Kuching 93250, Sarawak, Malaysia
3 Department of Plant Resource Science and Management, Faculty Resource Science & Technology, Universiti Malaysia Sarawak, Kota Samarahan 94300, Sarawak, Malaysia
4 Faculty of Civil Engineering, Universiti Teknologi MARA, Kota Samarahan 94300, Sarawak, Malaysia
5 Department of Civil Engineering, Faculty of Engineering, Universiti Malaysia Sabah, Kota Kinabalu 88400, Sabah, Malaysia
6 Sarawak Forest Department, East Wing Bangunan Baitul Makmur II, Medan Raya, Petra Jaya, Kuching 93050, Sarawak, Malaysia
* Correspondence: igaddafi@unimas.my

Abstract: An effort was carried out to fully utilise fast-growing Acacia hybrid usage in the timber engineering field; however, the research data are still lacking. This paper aims to evaluate the physical and mechanical properties performance between untreated (control) and treated with 10% copper chrome arsenic of Acacia hybrid collected from Daikin Plantation Sdn. Bhd. Bintulu, Sarawak at air-dry condition at different age groups using the small clear method. Mechanical properties test refers to shear parallel to grain (tangential and radial directions), cleavage (tangential and radial directions), and tension parallel to grain test. Meanwhile, the physical properties test refers to moisture content (MC) and density test. After treatment, mechanical properties increase with an average of 13.67%; meanwhile, moisture content decreased with an average of 0.58% or 0.09% MC, and density slightly increased with an average of 0.44% or 0.002 g/cm^3. Results indicate that 10-year-old Acacia hybrid exhibits the highest strength values in shear parallel to the grain, tension parallel to the grain, and cleavage, followed by 13-year-old and 7-year-old. Treated samples in the tangential direction performed better with consistent mean results than that of the untreated samples, while radial direction gave a high average strength increment when treated.

Keywords: Acacia hybrid; fast-growing; copper chrome arsenic; physical and mechanical properties

Citation: Ismaili, G.; Enduat, E.; Yahya, N.S.; Malek, F.M.; Jaimudin, N.A.; Abdul Rahim, K.K.; Wasli, M.E.; Kalu, M.; Openg, I.; Rizalman, A.N.; et al. Physical and Mechanical Properties Performance between Untreated and Treated with CCA Treatment at Different Age Groups of Fast-Growing Acacia Hybrid of Sarawak. *Forests* **2022**, *13*, 1969. https://doi.org/10.3390/f13121969

Academic Editors: Tomasz Krystofiak and Pavlo Bekhta

Received: 15 October 2022
Accepted: 16 November 2022
Published: 22 November 2022

1. Introduction

Timber was used throughout the history of mankind and provided humans with a broad range of building products and construction materials [1]. It is the most sustainable construction material, as it is renewable and absorbs carbon dioxide as it grows [1]. Malaysia is one of the leading producers of the world's good quality timbers, which are very highly demanded all over the globe [2]. However, over the year, Malaysia was unable to accommodate the huge demand for timber, especially primary hardwood timber, due to a shortage of timber resources. In conjunction with that, Malaysia introduced *Acacia mangium* for forest plantation species due to its fast-growing rate. Sarawak's effort, by planting the fast-growing species tree, began in the 1980s, and includes plantations of *Acacia mangium* with the largest forest plantation area in the country. However, the effort plantations grew and turned out to be prone to several diseases [3].

In a serious effort to meet the current and future raw material demand, the Acacia hybrid was introduced in Sarawak. Acacia hybrid is a cross-breed of *Acacia mangium* × *Acacia auriculiformis* and tends to possess many desirable features, such as less tapering, straight bole, fast growth, and heart rot resistance [4]. It was first reported in Sabah in the late 1970s. The wood properties of the Acacia hybrid are similar to those of *A. mangium* [5]. Its morphological traits, such as flower colour, pod aspect, leaf shape and size, bark aspect, and wood density, are generally an even mixture between its pure parent species [6]. However, it also differs from its parents in several ways. The tree has many small and light branches that can be easily pruned, and the main stem is not as straight as that of *A. mangium*, but it is much straighter than the main stem of *A. auriculiformis* [5]. It is reported that the Acacia hybrid shows more excellent resistance to diseases, a higher growth rate and better adaptation to different soil types than the parental species [7]. In Malaysia, heart rot disease was frequently observed in *A. mangium*, but it was never reported in Acacia hybrid [8]. Acacia hybrid also showed the ability to improve soil properties both physically and chemically [9]. The research on improved species with fast-growing time, durable and high engineering properties, viz., mechanical and physical properties, were carried out to perceive the suitability and potential of the species. Research conducted by Choong et al. [10] found that the improvement of *Acacia mangium* species to the Acacia hybrid improved the wood quality. The Acacia hybrid is more durable and less susceptible to heart rot disease [11]. The density of the Acacia hybrid was reported as slightly higher; the shape of the log is almost completely round and not susceptible to termite attack [4]. Study in determining the engineering properties of Sarawak species, especially fast-growing timber, to determine its strength grouping still at the beginning stage. There are more species to be explored and timber engineering research approaches to be done before the timber engineering properties of the species can be introduced in engineering design or other related fields for its various utilisation and application. Thus this has led to lacking knowledge in timber properties and design among architects and engineers [2]. Little is known about the mechanical and physical properties of the Acacia hybrid, especially in the timber engineering field, and the utilisation of this species is only limited to furniture, pulp, and paper industries. It is an alternative to evaluate the mechanical properties of Acacia hybrid and introduce this species as a building material in the timber engineering field. As mentioned by Treza et al. [12], fast-growing wood is an alternative solution for replacing the function of broadleaf plants as a material for floor, furniture, interior elements, and as structural components. Additionally, this research information will be used to utilise timber species more efficiently and effectively. Wood's strength properties are classified according to its moisture content, density, durability, and grain direction [13]. Timber's strength properties are essential to determining timber applications in real life, whether suitable for furniture, outdoor, or structural. The stronger the strength of the timber, the heavier the loading it will be able to withstand.

This study was conducted to evaluate and compare the behaviours of Acacia hybrid's engineering properties between untreated and treated samples under 10% of the copper chrome arsenic (CCA) treatment at different age groups using a small clear method. Copper chrome arsenic (CCA) treatment was used in this study since this treatment is widely used in the timber industry in the country [14]. Due to the environmental impact of timber engineering applications, understanding the treated timber for durability is very important. To achieve the aim of this study, several objectives should be considered. Firstly, evaluate the behaviours of physical and mechanical strength performance between untreated and treated Acacia hybrid in different age groups at air-dry conditions. Secondly, compare the mechanical properties of Acacia hybrid in different age groups and grain directions. Third, identify the age group with the best performance in mechanical and physical properties treated or untreated.

2. Materials and Methods

The sample for the experiment used was Acacia hybrid species, and it was taken from Daikin Plantation Sdn. Bhd. located at Bintulu, Sarawak (Figure 1a). There are three age groups of Acacia hybrid collected, which are the age groups of 7-year-old, 10-year-old, and 13-year-old. The selected trees must be free from any decay, termites, and defects. However, there is an exception for the knots on the trees. The trees were cut approximately 30 cm above the ground, with the diameter at breast height over the bark of trees approximately 20 cm at the height of 1.3 m from the root collar.

(a) (b) (c)

Figure 1. Timber collection process. (**a**) The plot number were marked on Acacia hybrid logs at Daikin Plantation. (**b**) The logs were sawn and trimmed into plank size. (**c**) The timber was stacked according to age groups in airdry room.

All the collected logs were transferred to Samling Plywood Bintulu Sdn. Bhd. for the sawing process. The processed samples were in plank size and were transferred to Sarawak Forestry Corporation, Kuching (Figure 1b). All of the processed green condition samples were appropriately stacked and allowed for the natural air-drying process placed in a shed air-drying room at AFSID Sarawak Forestry Corporation. The samples were stacked according to their age group. The natural air-drying process was chosen for this study to minimise the strength of the wood failures, such as various splits and cracks [15], as shown in Figure 1c. This air-drying process for this species took about one year to reach an approximate moisture content of 19%. The air-drying process depends on the type of species. Ismaili [16] conducted a study on *Acacia mangium* where the air-drying process took more than nine months to reach 19% moisture content. The plank's moisture content (MC) under natural air-drying needs to be checked using electric moisture metres before cutting it into test samples. The moisture content (MC) must be less than 19% for the test at air-dry condition.

Two types of samples were prepared for each age group shear parallel to grain test, tension parallel to grain test, and cleavage test. These samples were prepared in small clear sizes under untreated and treated samples at air-dry condition. It was understood that timber is a heterogeneous or nonhomogeneous material; thus, the most suitable sample to be tested was suggested using a small clear sample, which is defect-free [17–20]. The samples are prepared in accordance with Tan et al. [21], which is adopted from BS 373:1957 [22]. The samples for shear (Figure 2a), cleavage (Figure 3a), and tensile (Figure 4a) are cut from a stick with a cross-sectional dimension of 30 mm × 30 mm. The sticks are yielded from the flitches by reaping the 2.5 m of log sample using a band saw. Due to its hygroscopic behaviour [18,23,24], to prevent the dry condition of timber, the flitches are left in the air-conditioned room and air-dried to ensure its moisture content is less than 19%. Samples will be prepared in solid untreated and solid treated, with different age groups (Figure 5). A total of 3600 samples for untreated and treated samples were prepared for these three tests. The mechanical properties test was designated with 1200 samples, with 40 samples for each test and 2400 samples for the physical properties test, with 200 samples for each test.

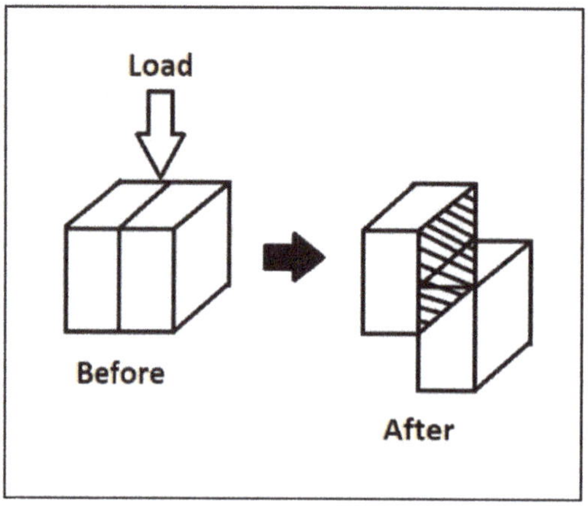

(a)

(b)

Figure 2. Shear parallel to grain test. (**a**) Shear parallel to grain test sample. (**b**) Measuring of the shear parallel to grain of an Acacia hybrid sample with universal testing machine Instron 5569, Norwood, MA, USA.

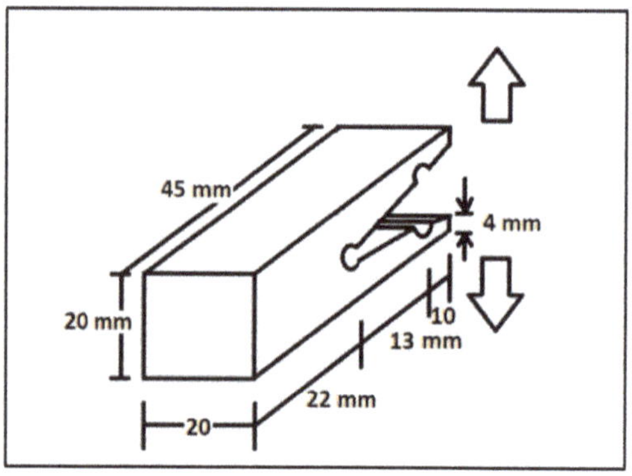

(a)

(b)

Figure 3. Cleavage test. (**a**) Cleavage test sample. (**b**) Measuring of the cleavage of an Acacia hybrid sample with universal testing machine Instron 5569, Norwood, MA, USA.

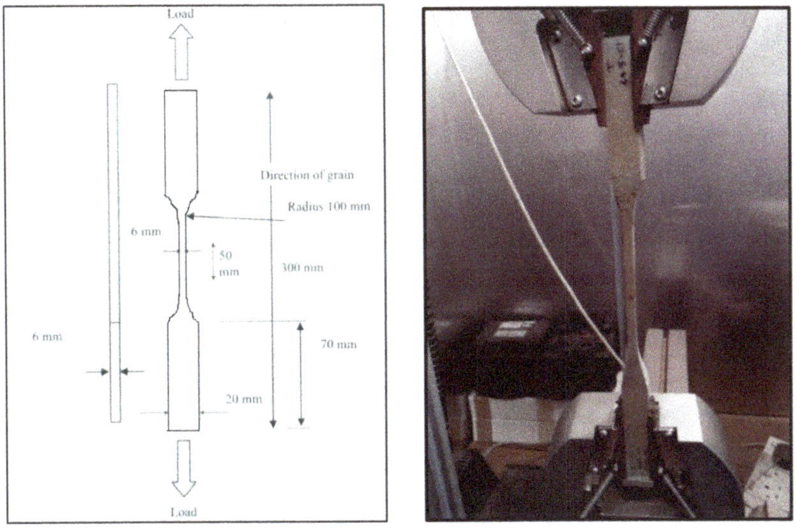

(a) (b)

Figure 4. (**a**) Tension parallel to grain test sample. (**b**) Measuring of the tension parallel to grain of an Acacia hybrid sample with universal testing machine Instron 5569, Norwood, MA, USA.

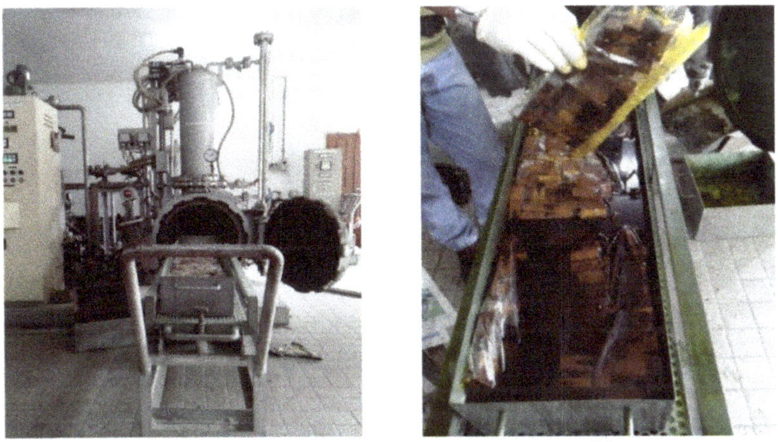

(a) (b)

Figure 5. Timber treatment process. (**a**) Samples are placed in a treatment cylinder. (**b**) Samples were treated by using copper chrome arsenic (CCA).

2.1. Moisture Content Determination

The moisture content is according to the BS 373: 1957 [22] and Tan et al. [21]. Samples were dried in an oven with a temperature of 103 ± 3 °C for 24 h, or until a constant weight was obtained. In order to achieve accurate results, the sample should be kept in a desiccator before reweighing. The samples were weighed again in order to obtain the oven-dry weight with MC ≤ 19% by using an electronic balance to obtain the initial weight of Acacia hybrid samples. Samples of 1200 test pieces of size 20 × 20 mm were used. The moisture content was calculated using the formula given:

$$MC = (\text{Initial weight} - \text{Oven-dried weight (g)})/(\text{Oven-dried weight (g)}) \times 100\% \quad (1)$$

2.2. Density Determination

According to BS 373: 1957 [22], a density determination test was conducted using the moisture content determination test samples. The prepared samples' dimensions are measured using vernier calipers to obtain the green volume of the Acacia hybrid. A total of 1200 samples were dried in an oven with a temperature of 103 ± 3 °C for 24 h or until a constant weight was obtained. The samples were weighed to obtain oven-dry weight with MC ≤ 19%. The basic density was calculated using the formula given:

$$\text{Density} = (\text{Oven-dried weight (g)})/(\text{Green volume of sample (cm}^3\text{)}) \quad (2)$$

2.3. Treatment Process

The treatment process of the sample using copper chrome arsenic (CCA) is according to Tan et al. [21]. A total of 1800 samples of Acacia hybrid at three different age groups were treated using copper chrome arsenic (CCA) treatment, as shown in Figure 5a,b. Test samples underwent a full-cell process using 10% of the copper chrome arsenic (CCA) solution to achieve maximum absorption into the wood. The process was started by performing the initial vacuum, where −85 kPa was subjected to the samples for 1 h. The preservative pumped into the treatment cylinder with the vacuum was maintained at −85 kPa. Then, 200 psi (14 bars) of hydraulic pressure was subjected to samples for 2 h. The copper chrome arsenic (CCA) solution was drained out from the treatment cylinder after the pressure period was completed. The final vacuum was subjected to samples at −85 kPa for 30 min to remove excess copper chrome arsenic (CCA) solution in the samples.

2.4. Mechanical Testing

All testing was conducted based on BS 373: 1957 [22] and Tan et al. [21]. This study has three types of mechanical strength tests: shear parallel to grain, tension parallel to grain, and cleavage test. The tests were performed using the Instron 5569 universal testing machine to determine the samples' strength value. The shear parallel to grain test was carried out by a constant loading speed at 0.6 mm per minute on the samples. This test aims to measure the maximum dividing load by the cross-section area when the load is applied parallel to grain direction of timber. The samples are prepared with dimensions 20 mm × 20 mm × 20 mm. The shear failures are shown along a plane parallel to the tangential and radial directions. Shear stress at maximum load in megapascal (MPa) is calculated by using the formula as follows:

$$\text{Shear stress at maximum load} = (\text{Force (N)})/(\text{Area (mm}^2\text{)}). \quad (3)$$

The cleavage test aims to measure the resistance of timber to splitting based on the failure caused by the maximum load on the test samples. The samples are prepared with a dimension of 20 mm × 20 mm × 45 mm. The load is applied to the cleavage sample at a constant crosshead speed of 2.5 mm per minute to give a failure along the tangential and radial surface. The test layout is shown in Figure 3a,b. The maximum splitting load in N/mm is calculated by using the formula given:

$$\text{Load per mm of width to resist splitting} = (\text{Max. load (N)})/(\text{breath (mm)}). \quad (4)$$

The tension parallel to grain test is conducted to determine the maximum loadable to cater by sample before tensile failure occurs. This test will give the maximum load and maximum tensile stress exerted by the samples. In correlation with that, the samples are subjected to load at constant head speed until the samples break. Both ends of the test samples are held by toothed and self-aligning grips, as shown in Figure 4a,b. The samples are a break in 1.5 to 2 min from the start of loading. The failure of tension parallel to grain happened at the minimum cross-section of the test sample. Tensile stress at maximum load in MPa is calculated by using the formula given:

$$\text{Tensile stress at maximum load} = ((\text{Force (N)})/((\text{Area (mm}^2)). \quad (5)$$

3. Results

From this study, Tables 1 and 2 show the mean value with a 95% coefficient interval of physical and mechanical properties taken during the experiment accordingly to the condition of the samples that are untreated and treated. The results of the Acacia hybrid at an untreated condition show that the 7-year-old age group sample recorded a higher mean value of 15.56% MC with 95% coefficient interval value of 0.04, followed by the 13-year-old sample that recorded 15.52% MC with 95% coefficient interval value of 0.04, and the 10-year-old sample that recorded 15.44% MC with 95% coefficient interval value of 0.04. The results of the Acacia hybrid under the treated condition show a similar pattern, where the 7-year-old age group sample recorded a higher mean value of 15.48% MC with 95% coefficient interval value of 0.04, followed by the 13-year-old sample that recorded 15.45% MC with a 95% coefficient interval value of 0.03, and the 10-year-old sample that recorded 15.32% MC with 95% coefficient interval value of 0.04. After the samples were treated with 10% copper chrome arsenic, the moisture content value for the Acacia hybrid at air-dry condition from each testing sample at different age groups showed a decrement with an average of 0.58% or 0.09% MC. It was recorded that moisture content in the age group 10-year-old sample decreased more moisture content with 0.76% or 0.12% MC, followed by the 7-year-old sample with 0.52% or 0.08% MC, and the 13-year-old sample with 0.45% or 0.07% MC, which can be clearly observed in Figure 6. The decreasing moisture content from untreated to treated samples was observed in this study. The study by Ferreira et al. [25] shows that CCA treatment increased the moisture content from an untreated to treated sample of plywood panel from *Pinus taeda* wood. However, previous work by Epmeier et al. [26] and Bruno et al. [27] has similar results to this study, where pine species were treated by furfurylation treatment, showing that the moisture content decreased when treated.

Table 1. Summary of the mean mechanical properties of Acacia hybrid.

Condition		Unit	Controlled			Treated		
Age Groups			7	10	13	7	10	13
Mechanical properties	Shear parallel to grain (tangential)	Nos	40	40	40	40	40	40
		MPa	17.06	18.18	17.61	17.52	18.59	17.67
		SD	1.55	1.85	2.62	1.85	1.86	2.17
		CV %	9.1	10.18	14.91	10.55	10	12.3
		SE	0.25	0.29	0.42	0.29	0.29	0.34
		95% CI	0.50	0.59	0.84	0.59	0.59	0.70
	Shear parallel to grain (radial)	Nos	40	40	40	40	40	40
		MPa	16	17.46	16.62	16.78	17.92	16.88
		SD	2.06	2.09	2.64	1.74	1.92	2.2
		CV %	12.86	11.95	15.88	10.38	10.71	13.06
		SE	0.33	0.33	0.42	0.28	0.30	0.35
		95% CI	0.66	0.67	0.84	0.56	0.61	0.71
	Cleavage (tangential)	Nos	40	40	40	40	40	40
		N/mm	15.82	17.66	16.19	16.69	18.12	16.86
		SD	2.68	2.52	1.98	2.27	3.16	2.1
		CV %	16.94	14.29	12.21	13.62	17.43	12.43
		SE	0.42	0.40	0.31	0.36	0.50	0.33
		95% CI	0.85	0.80	0.63	0.72	1.00	0.66

Table 1. *Cont.*

Condition Age Groups		Unit	Controlled 7	Controlled 10	Controlled 13	Treated 7	Treated 10	Treated 13
Mechanical properties	Cleavage (Radial)	Nos	40	40	40	40	40	40
		N/mm	13.67	15	14.19	14.61	16.41	15.41
		SD	2.57	2.09	2.16	2.75	2.75	1.68
		CV %	18.78	13.9	15.18	18.79	16.77	10.91
		SE	0.41	0.33	0.34	0.43	0.44	0.27
		95% CI	0.81	0.66	0.68	0.87	0.87	0.53
	Tension parallel to grain	Nos	40	40	40	40	40	40
		MPa	141.8	148.99	144.19	158.4	176.44	160.26
		SD	24.26	22.33	17.49	17.56	27.08	11.94
		CV %	17.11	14.99	12.13	11.08	15.35	7.45
		SE	3.84	3.53	2.77	2.78	4.28	1.89
		95% CI	7.67	7.06	5.53	5.55	8.56	3.78

Table 2. Summary of the mean moisture content and density of Acacia hybrid.

Condition Age Groups		Unit	Controlled 7	Controlled 10	Controlled 13	Treated 7	Treated 10	Treated 13
Physical properties	Moisture content	Nos	200	200	200	200	200	200
		%	15.56	15.44	15.52	15.48	15.32	15.45
		SD	0.31	0.27	0.28	0.31	0.29	0.20
		CV %	2.01	1.74	1.80	1.99	1.87	1.32
		SE	0.02	0.02	0.02	0.02	0.02	0.01
		95% CI	0.04	0.04	0.04	0.04	0.04	0.03
	Density	Nos	200	200	200	200	200	200
		g/cm^3	0.503	0.479	0.507	0.503	0.482	0.511
		SD	0.09	0.09	0.08	0.08	0.08	0.08
		CV %	17.03	18.83	16.67	15.91	17.37	16.42
		SE	0.01	0.01	0.01	0.01	0.01	0.01
		95% CI	0.01	0.01	0.01	0.01	0.01	0.01

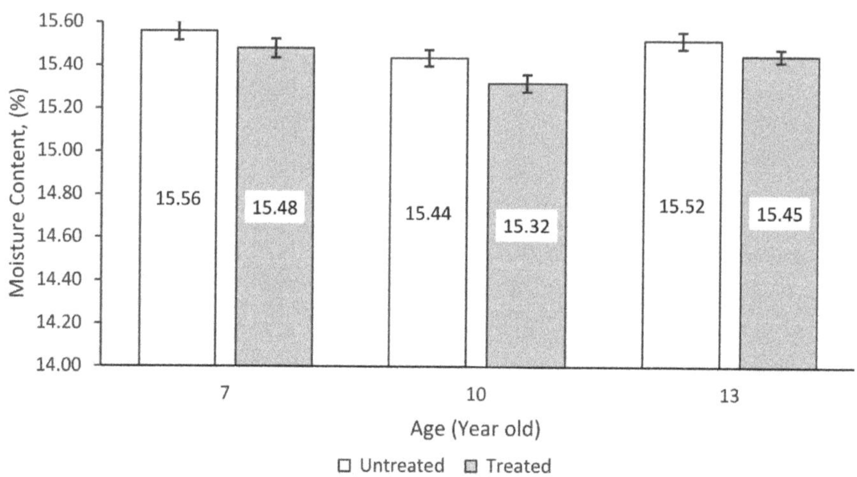

Figure 6. Moisture content mean values of Acacia hybrid sample compared within the age groups, grain direction, and treatment condition at air-dry condition.

For density, untreated Acacia hybrid shows that the 13-year-old sample recorded higher mean value 0.507 g/cm^2 with a 95% coefficient interval value of 0.01, followed by the 7-year-old sample, which recorded 0.503 g/cm^2 with 95% coefficient interval value of 0.01, and the 10-year-old sample, which recorded 0.479 g/cm^2 with 95% coefficient interval value of 0.01. The mean results of the Acacia hybrid under treated condition show a similar pattern as of untreated Acacia hybrid, with the 13-year-old age group sample recording a higher mean value of 0.511 g/cm^2 with a 95% coefficient interval value of 0.01, followed by the 7-year-old age group sample, which recorded 0.503 g/cm^2 with a 95% coefficient interval value of 0.01, and the 10-year-old age group sample recorded a lower mean value of 0.482 g/cm^2 with s 95% coefficient interval value of 0.01. As a result of the decrement in moisture content value, the density of samples increased after the sample was treated. The density result observed for each testing sample in different age groups is in the range of 0.479 g/cm^3 to 0.507 g/cm^3 for untreated and 0.482 g/cm^3 to 0.511 g/cm^3 for untreated samples. It was recorded that 13-year-old density increased more by 0.78% or 0.004 $g/cm^{3,}$ followed by the 10-year-old increase by 0.62% or 0.003 g/cm^3, and the 7-year-old age group sample remained unchanged, which can be clearly observed in Figure 7. From these results, we can observe that the reduction percentage in treated samples' moisture content led to an increase in density. According to Ismaili [16], the moisture content is reported to greatly influence the mechanical strength of the timber. The more the timber dried, the greater the timber's strength. This shows that changing moisture content directly influences the basic density, and this is supported by the study conducted by Alik and Naohiro [28]. The increment of strength is due to the reduction in moisture content in timber because of the shortening and, consequently, strengthening of hydrogen bonds linking together the microfibrils in the timber [17,29]. Therefore, high basic density has low moisture content and vice versa. Although the timber density is relatively reflected in the strength of the timber, it should not be the definitive measurement of its strength [19].

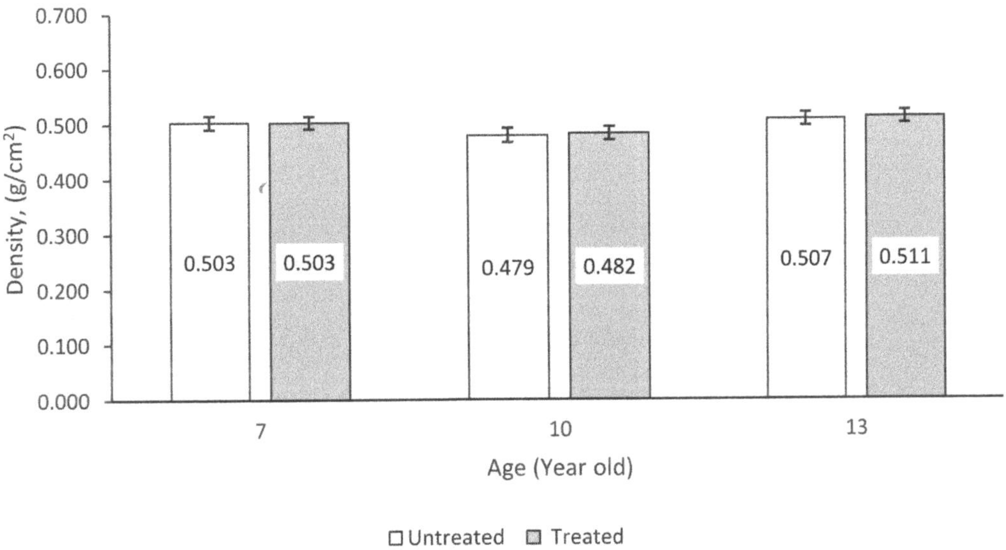

Figure 7. Density mean values of the Acacia hybrid sample compared within the age groups, grain direction, and treatment condition at air-dry condition.

3.1. Shear Parallel to Grain

As shown in Table 1, the mean value was observed with a 95% coefficient interval. For a shear parallel to grain tangential direction, the 10-year-old group exhibited the highest shear strength for both untreated and treated samples. For the untreated 10-year-old sample, the mean value was recorded as 18.18 MPa with a 95% coefficient interval value of 0.5 and the shear strength increased by 2.26% or 0.41 MPa to 18.59 MPa with a 95% coefficient interval value of 0.59. This was followed by the untreated 13-year-old and 7-year-old samples. The untreated 13-year-old sample was recorded with a mean value of 17.61 MPa with a 95% coefficient interval value of 0.84 and slightly increased by 0.34% or 0.06 MPa to achieve 17.67 MPa with a 95% coefficient interval value of 0.7 when treated. The 7-year-old untreated sample recorded a mean value of 17.06 MPa with a 95% coefficient interval value of 0.5 and increased by 2.7% or 0.46 MPa higher than the 10-year-old to achieve 17.52 MPa with a 95% coefficient interval value of 0.59 when treated.

The results in shear parallel to grain radial direction show a similar pattern, where the untreated 10-year-old sample recorded a higher mean value of 17.46 MPa with a 95% coefficient interval value of 0.67. The shear strength increased by 2.63% or 0.46 MPa to achieve 17.92 MPa with a 95% coefficient interval value of 0.61 when being treated. This was followed by an untreated 13-year-old sample with a mean value of 16.62 MPa with a 95% coefficient interval value of 0.84, and shear strength increased by 1.56% or 0.26MPa to achieve 16.88 MPa when treated. Meanwhile, the untreated 7-year-old recorded 16 MPa with a 95% coefficient interval value of 0.66, and a huge increment was spotted with 4.88% or 0.78 MPa to achieve 16.78 MPa with a 95% coefficient interval value of 0.56 when the sample was treated. From this result, it can be concluded that the CCA treatment improves the shear strength parallel to grain for both grain directions when samples were treated for all age groups, which can be observed clearly in Figure 8. The previous study by Andy [30] also shows a similar trend, where southern pine shear strength increased when treated with CCA. The outcome from this study also shows a similar strength pattern for each age group to the study conducted by Alik et al. [31], where the highest shear strength value is the 10-year-old group, followed by the 13-year-old and 7-year-old group.

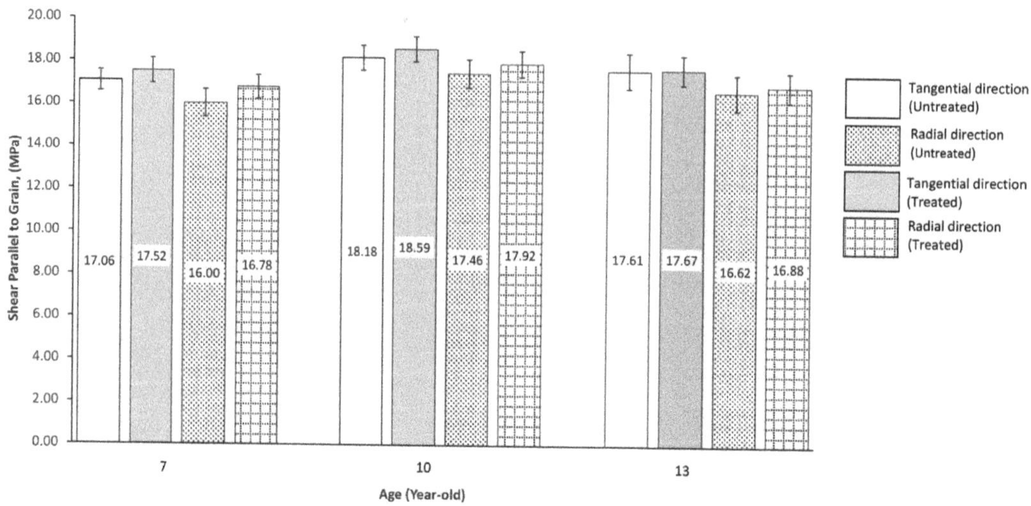

Figure 8. Shear parallel to grain mean values of Acacia hybrid sample comparing within the age groups, grain direction and treatment condition at air-dry condition.

From observation, the mean values obtained from this study show that shear parallel to grain tangential direction gives higher shear strength than the radial direction, and this finding was supported by Ismaili et al. [3] and Gaddafi [18] in their research. Compared to the percentage difference of strength increment, shear parallel to grain at tangential direction increased by 1.76% or 0.31 MPa after treatment, which is slightly higher than at radial direction with an increment of 1.43% or 0.19 MPa. The result also shows an improvement in shear strength in both directions, especially in the radial direction. This means the treatment process improved the shear parallel to the grain of the Acacia hybrid species, especially in the radial direction. As Ismaili et al. [3] mentioned in their research, this could be due to differences in the proportion of major wood constituents, such as cellulose, hemicelluloses, and lignin present in the woods or differences in extractive contents. Riyanto and Gupta [32] also explained that, although ring orientation was strong, inconsistent factors affected shear strength parallel to the grain. Moreover, other factors affect shear strength, such as moisture content. According to Table 2, the 10-year-old Acacia hybrid sample exhibited the lowest moisture content, where moisture content is reported to significantly influence timber's mechanical strength [28,31,33,34]. Madsen [35], in his study, found that moisture content significantly affects shear strength. According to Jamil [36], density is greatly influenced by the amount of moisture content in the timber. The presence of moisture in the wood not only increases the mass, but also the volume of timber. From this study, the 7- and 13-year-old age group sample has higher density compared to the 10-year-old age group sample. Similarly, the percentage of moisture content for both age groups was also higher compared to 10 years old. However, compared to mechanical or strength properties, the 10-year-old Acacia hybrid sample recorded higher values for all tests. Thus, it is agreed that the density of wood should not be the definite measurement of its strength [37]. McKenzie [33] reported that heartwood tissue's presence provides mechanical rigidity and strength to timber. From this study, the age group of 10 years old possessed the highest strength properties value as compared to other age groups. Therefore, from this study, 10-year-old timber has more heartwood tissue presence than 13-year-old and 7-year-old timber. From this study, the highest mean value of shear parallel to grain is from the 10-year-old age group. Further analysis was conducted using statistically significant difference ($p < 0.5$) to observe the error bars between the 10-year-old age group and other age groups, grain directions, and treatment conditions of Acacia hybrid samples, as shown in Figure 8. The observation will later be computed with the number of frequencies in percentage having the condition statistically significant difference ($p < 0.5$), and the results will be tabulated in Table 3. For a shear parallel to grain at tangential direction, the untreated 10-year-old sample has a statistically significant difference ($p < 0.5$) with the 7-year-old ($p = 0.0044$). A similar pattern was observed for the 10-year-old Acacia hybrid untreated sample at a radial direction has a statistically significant difference ($p < 0.5$) with the 7-year-old ($p = 0.0023$) sample. The treated 10-year-old Acacia hybrid sample in the tangential direction also has a statistically significant difference ($p < 0.5$) when compared with the 7-year-old ($p = 0.0112$) and 13-year-old ($p = 0.0432$) treated samples at the tangential direction. Meanwhile, shear parallel to grain in the radial direction, indicated that the 10-year-old Acacia hybrid treated sample has a significant difference with the 7-year-old ($p = 0.0067$) and 13-year-old ($p = 0.0275$) samples. Statistically significant difference analysis observed no significant difference ($p = 0.3157$) between the treated 10-year-old sample compared with the untreated 10-year-old Acacia hybrid of shear parallel to grain in the tangential direction sample. Similarly, the treated 10-year-old sample compared with the untreated 10-year-old sample has no significant difference for shear parallel to grain in radial direction. The analysis shows that a 10-year-old Acacia hybrid sample at shear parallel to grain yield a 60% statistically significant difference ($p < 0.05$) when compared with other age groups, grain directions, and treatment conditions, which can be clearly observed in Table 3. Therefore, concrete evidence from this analysis shows that the 10-year-old Acacia hybrid of shear parallel to grain sample has a statistically significant difference ($p < 0.05$), especially in the tangential direction at the treated condition.

Table 3. ANOVA on the effect of age groups, treatment conditions, and grain directions with 10-year-old untreated and treated for shear parallel to grain of Acacia hybrid samples at air-dry condition.

Age Group	Treatment Condition	Shear Parallel to Grain Direction	Age Groups	Treatment Condition (UT, TD)	Significant Difference p-Values
10	UT	T	7	UT	0.0044 *
			10	TD	0.3157 ns
			13	UT	0.2687 ns
		R	7	UT	0.0023 *
			10	TD	0.3040 ns
			13	UT	0.1191 ns
	TD	T	7	TD	0.0112 *
			13	TD	0.0432 *
		R	7	TD	0.0067 *
			13	TD	0.0275 *
Number of frequencies having condition statistically significant difference ($p < 0.05$), %					60.0

T—Tangential direction; R—Radial direction; UT—Untreated (Control); TD—Treated; ns—not significant; * Statistically significant difference at 95% ($p < 0.05$).

3.2. Cleavage

Figure 9 was observed with a 95% coefficient interval for cleavage tangential direction. It was reported that the untreated 10-year-old sample exhibited the highest cleavage strength of 17.66 N/mm, with a 95% coefficient interval value of 7.06, followed by the 13-year-old and 7-year-old samples, with a mean value of 16.19 N/mm with a 95% coefficient interval value of 0.563 and 15.82 N/mm, with a 95% coefficient interval value of 0.85, respectively. The cleavage strength increased when samples were treated. This can be observed in the 10-year-old age group sample, which achieved a higher mean value of 18.12 N/mm with a 95% coefficient interval value of 1 with 2.60% or 0.46 N/mm increment after the samples were treated. A similar pattern was observed in the treated 13-year-old sample with an increment of 4.14% or 0.67 N/mm to achieve 16.86 N/mm with a 95% coefficient interval value of 0.66. This is followed by a treated 7-year-old sample with an increment of 5.50% or 0.87 N/mm to achieve 16.69 N/mm with a 95% coefficient interval value of 0.72. These results are supported by the findings from Alik et al. [31], which also stated that the 10-year-old age group sample exhibited the highest cleavage strength at air-dry condition, followed by the 13-year-old and 7-year-old age group samples. This gave an average increment after the sample was treated with 4.08% or 0.66 N/mm.

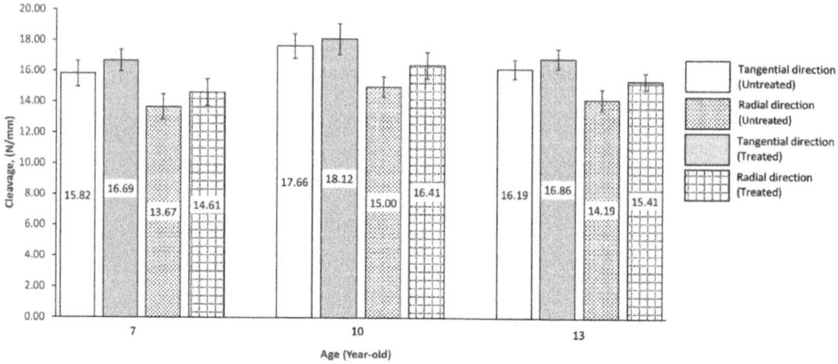

Figure 9. Cleavage strength values of Acacia hybrid sample compared within the age groups, grain direction, and treatment condition at the air-dry condition.

For cleavage in the radial direction, the same trend was observed correspondingly with the results in the tangential direction, where the untreated 10-year-old sample recorded a higher mean value of 15 N/mm with a 95% coefficient interval value of 0.66 followed by the 13-year-old and 7-year-old sample, each recorded with 14.19 N/mm with a 95% coefficient interval value of 0.68 and 13.67 N/mm with a 95% coefficient interval value of 0.81, respectively. For the treated sample, the 10-year-old was also recorded with a higher mean value with an increment of 9.40% or 1.41 N/mm to achieve 16.41 N/mm with a 95% coefficient interval value of 0.87, followed by the 13-year-old and 7-year-old age groups, each recorded with an increment of 8.6% or 1.22 N/mm to achieve 15.41 N/mm with a 95% coefficient interval value of 0.53 and an increment of 6.88% or 0.94 N/mm to achieve 14.61 N/mm with a 95% coefficient interval value of 0.87, respectively. It can be observed that the cleavage results show a high percentage of increment in radial direction after the sample was treated with 8.33% or 1.19 N/mm compared with the tangential direction, which is 4.08% or 0.67 N/mm. However, comparing the results obtained between the mean value of cleavage in both radial and tangential for untreated and treated samples, the mean value obtained in the tangential direction for all age groups was higher than in the radial direction. From the results, it can be concluded that the 10-year-old age group sample had the highest mean value for both untreated and treated at cleavage in the tangential direction. This was confirmed by Moya and Muñoz [38] in their research, where cleavage in the tangential direction gives a higher strength mean value compared with the radial direction. According to Wallis [34], the lowest value cleavage in the radial direction compared with the tangential direction was due to the relationship between air-dry density and the cleavage strength of timber along the fibres.

The study indicated that the highest mean value of cleavage is from the 10-year-old age group. When the results were analysed using statistically significant difference ($p < 0.05$), a similar pattern as in the shear parallel to grain results was observed in Table 4. The statistically significant difference ($p < 0.5$) analysis will be carried out with the highest mean value obtained by the 10-year-old sample compared with other age groups for both grain directions, untreated, and treated samples. For the untreated Acacia hybrid of cleavage in the tangential direction sample, the 10-year-old sample has a significant difference ($p < 0.5$) when compared with the 7-year-old ($p = 0.0023$) and 13-year-old ($p = 0.0046$). Meanwhile, for the untreated Acacia hybrid of cleavage in the radial direction sample, the 10-year-old sample has a significant difference ($p < 0.5$) when compared with untreated 7-year-old ($p = 0.0128$) and treated 10-year-old ($p = 0.0117$) Acacia hybrid sample. Furthermore, for the treated Acacia hybrid of cleavage in the tangential direction sample, the 10-year-old sample has a statistically significant difference ($p < 0.5$) when compared with the 7-year-old ($p = 0.0227$) and 13-year-old ($p = 0.0383$). Meanwhile, for the treated Acacia hybrid of cleavage in the radial direction sample, the 10-year-old sample significantly differs only from the 7-year-old ($p = 0.0045$) sample. Statistically significant difference analysis observed no significant difference ($p = 0.4788$) between the treated 10-year-old sample compared with the untreated 10-year-old Acacia hybrid of cleavage in the tangential direction sample. However, there is a significant difference ($p = 0.0117$) between treated 10-year-old samples compared with the untreated 10-year-old Acacia hybrid of cleavage in the radial direction sample. The analysis shows that the 10-year-old Acacia hybrid for the cleavage sample yields a 70% statistically significant difference ($p < 0.05$) when compared with other age groups, grain directions, and treatment conditions, which can be clearly seen in Table 4. Therefore, it can be concluded that the 10-year-old Acacia hybrid of the cleavage sample has a statistically significant difference ($p < 0.05$), especially in the tangential direction at the treated condition.

Table 4. ANOVA on the effect of age groups, treatment conditions, and grain directions with 10-year-old untreated and treated cleavage of Acacia hybrid samples at air-dry condition.

Age Group	Treatment Condition	Cleavage to Grain Direction	Age Groups	Treatment Condition (UT, TD)	Significant Difference p-Values
10	UT	T	7	UT	0.0023 *
			10	TD	0.4788 ns
			13	UT	0.0046 *
		R	7	UT	0.0128 *
			10	TD	0.0117 *
			13	UT	0.0921 ns
	TD	T	7	TD	0.0227 *
			13	TD	0.0383 *
		R	7	TD	0.0045 *
			13	TD	0.0527 ns
Number of frequencies having condition statistically significant difference ($p < 0.05$), %					70.0

T—Tangential direction; R—Radial direction; UT—Untreated (Control); TD—Treated; ns—not significant; * Statistically significant difference at 95% ($p < 0.05$).

3.3. Tension Parallel to Grain

As shown in Figure 10, the mean results observed with a 95% coefficient interval reported that the 10-year-old has the highest strength mean value for the untreated sample with 148.99 MPa with a 95% coefficient interval value of 0.8, followed by the 13-year-old and 7-year-old samples with 144.19 MPa with a 95% coefficient interval value of 5.53 and 141.80 MPa with a 95% coefficient interval value of 7.67, respectively. When the sample was treated, the treated 10-year-old sample also recorded with the highest mean value increased by 18.42% or 27.45 MPa to achieve 176.44 MPa with a 95% coefficient interval value of 8.56, followed by 13-year-old and 7-year-old age group samples, which each recorded increments of 11.15% or 16.07 MPa to achieve 160.26 MPa with a 95% coefficient interval value of 3.78 and increment of 11.71% or 16.6 MPa to achieve 158.40 MPa with a 95% coefficient interval value of 5.55, respectively. The treated 10-year-old sample shows excellent increment value with an average mean difference of 10.7% or 17.11 MPa more compared to both the 13-year-old and 7-year-old samples. As expected, untreated samples in all age groups increased their strength value when treated. It can be observed that the average increment in strength value from the untreated sample to the treated sample for all age groups increased with the average increment in strength value of 13.8% or 20.04 MPa, whereby the 10-year-old age group sample is more prominent when being treated. These results, supported by findings from Alik et al. [31], stated that the 10-year-old age group sample exhibited the highest strength at the air-dry condition, followed by the 13-year-old and 7-year-old age group.

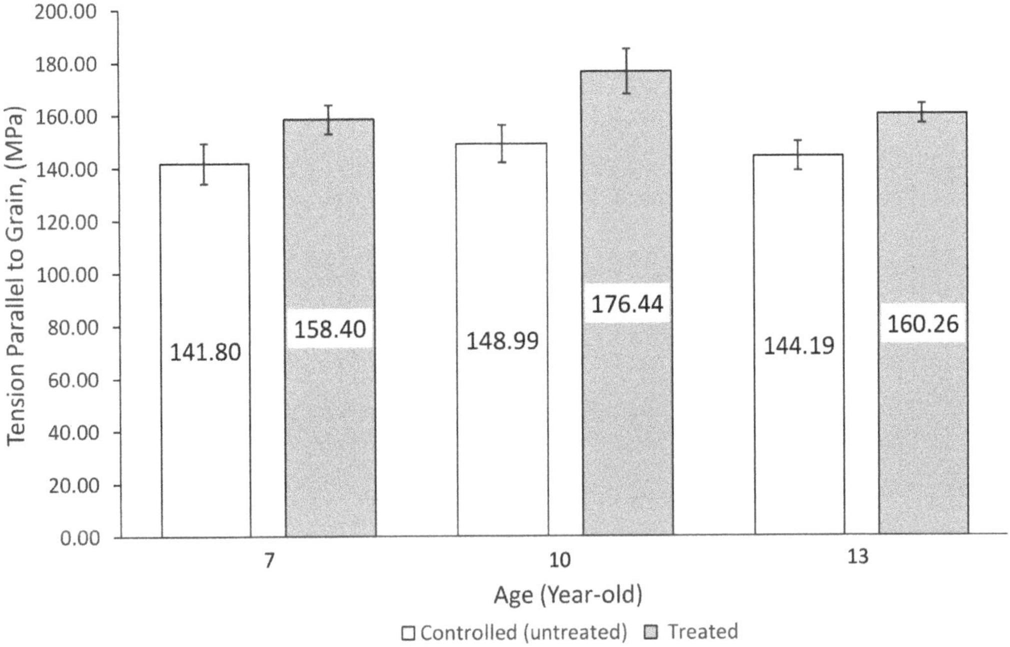

Figure 10. Tension strength values of Acacia hybrid sample compared within the age groups, grain direction, and treatment condition at the air-dry condition.

Similarly, this study's 10-year-old age group sample also recorded the highest tensile strength value. Thus, the statistically significant difference ($p < 0.5$) analysis will be carried out with the highest mean value obtained by the 10-year-old age group sample compared with other age groups, untreated, and treated samples. For the untreated Acacia hybrid sample, the 10-year-old age group sample has no statistically significant difference ($p < 0.5$) when compared with 7-year-old and 13-year-old age group samples. However, for untreated 10-year-old Acacia hybrid tensile sample has a statistically significant difference ($p < 0.5$) when compared with the treated 10-year-old ($p = 0.0000$) sample. Meanwhile, for the treated Acacia hybrid sample, the 10-year-old age group sample has a statistically significant difference ($p < 0.5$) when compared with 7-year-old ($p = 0.0007$) and 13-year-old ($p = 0.0009$) samples. Therefore, from this analysis, the 10-year-old age group Acacia hybrid of the tensile sample yields a 60 % statistically significant difference ($p < 0.05$) when compared with other age groups and treatment conditions, which can be observed in Table 5. From the study conducted, we can observe that the 10-year-old age group sample constantly recorded with higher strength value as compared with the 7-year-old and 13-year-old age group samples, although the density of the 10-year-old age group sample was slightly lower with an average of 5.03%. However, the result contradicts the study conducted by Alik and Kuroda [28], where basic density is a strong indicator correlated to mechanical strength properties. Therefore, this revealed that timber density should not be the definitive measurement of its strength [19].

Table 5. ANOVA on the effect of age groups and treatment conditions with 10-year-old Acacia hybrid untreated and treated for tensile samples at air-dry condition.

Age Group	Treatment Condition	Age Groups	Treatment Condition (UT, TD)	Significant Difference p-Values
10	UT	7	UT	0.1719 ns
		10	TD	0.0000 *
		13	UT	0.2879 ns
	TD	7	TD	0.0007 *
		13	TD	0.0009 *
Number of frequencies having condition statistically significant difference ($p < 0.05$), %				60.0

T—Tangential direction; R—Radial direction; UT—Untreated (Control); TD—Treated; ns—not significant; * Statistically significant difference at 95% ($p < 0.05$).

4. Conclusions

Based on the mean results obtained in this study for the moisture content, density, shear parallel to grain, cleavage, and tensile properties of Acacia hybrid at different age groups under 10% of the copper chrome arsenic (CCA) treatment at air-dry condition, the following conclusions are made.

1. The moisture content for the Acacia hybrid decreases when the sample is treated with an average 0.58% or 0.09% MC, whereas the treated 10-year-old sample recorded a higher mean value due to a decrement of 0.76% or 0.118% MC. The density of Acacia hybrid shows that, as a result of the sample being treated, the mean density increased with an average of 0.44% or 0.002 g/cm^3, whereas the treated 13-year-old sample recorded a higher increment value of 0.74% or 0.004 g/cm^3. Similarly, mechanical properties also increased with an average 6.18% increment of strength value when treated. Tension value recorded a huge average increment of 13.67% when treated.
2. The tangential direction gives a high mean value in both untreated and treated samples for shear parallel to grain and cleavage. However, in terms of the percentage difference of strength increment from untreated to treated samples, radial direction samples show significant improvement where it recorded a higher percentage increment with an average of 3.02% or 0.5 MPa and 8.29% or 1.19 MPa for shear parallel to grain and cleavage, respectively. The 10-year-old age group sample showed a significant increment of strength in both the radial and tangential direction when treated, followed by 13-year-old and 7-year-old age group.
3. Findings revealed that the mechanical and physical properties of Acacia hybrid obviously perform better in the 10-year-old age group, followed by the 13-year-old and 7-year-old age groups. It is demonstrated that the mechanical properties of Acacia hybrid serve better after being treated using CCA treatment.
4. It is recommended that the air-dried 10-year-old Acacia hybrid is the most suitable with consistent physical property values and mechanical strength values in the shear parallel to grain at the tangential direction, cleavage in the tangential direction, and tension when treated with copper chrome arsenic (CCA) treatment. However, the radial direction selection can also be considered due to its greater percentage difference in strength increment when being treated.

To predict the treated sample's physical and mechanical properties under 10% chrome arsenic (CCA) treatment at air-dry condition, it can be calculated using the mean value of the untreated sample using the equation in Table 6 below.

Table 6. Calculation of mean value of engineering properties of treated Acacia hybrid sample under 10% chrome arsenic (CCA) treatment at air-dry condition using the mean value of the untreated sample.

Engineering Properties	Mean Value for Untreated Sample		Unit	Age of Sample Factor, Y_i			Mean Value for Treated Sample with 10% CCA
				Y_7	Y_{10}	Y_{13}	
Physical properties (Air-dry)	MC_C	Moisture content	%	0.995	0.992	0.996	$MC_C \times Y_i$
	D_C	Density	g/cm^3	0.999	1.006	1.007	$D_C \times Yi$
Strength Properties (Air-dry)	$S//T_C$	Shear Parallel to Grain (Tangential)	MPa	1.027	1.023	1.003	$S//T_C \times Yi$
	$S//R_C$	Shear Parallel to Grain (Radial)	MPa	1.049	1.026	1.016	$S//R_C \times Yi$
	T_C	Tension	MPa	1.117	1.184	1.111	$T_C \times Yi$
	CLT_C	Cleavage (Tangential)	N/mm	1.055	1.026	1.041	$CLT_C \times Yi$
	CLR_C	Cleavage (Radial)	N/mm	1.069	1.094	1.086	$CLR_C \times Yi$

Author Contributions: Conceptualisation, G.I.; Data curation, G.I. and E.E.; investigation, G.I. and E.E.; methodology, G.I. and N.S.Y.; resources, N.A.J. and K.K.A.R.; validation, G.I. and I.O.; visualisation, F.M.M. and A.N.R.; writing—original draft, G.I. and E.E.; writing—review and editing, M.E.W. and M.K.; project administration, J.L. and B.R. All authors have read and agreed to the published version of the manuscript.

Funding: This research was funded by the Ministry of High Education of Malaysia (MOHE), Fundamental Research Grant Scheme (FRGS) (FRGS/1/2020/WAB03/UNIMAS/02/1).

Institutional Review Board Statement: Not applicable.

Informed Consent Statement: Not applicable.

Data Availability Statement: The data presented in this study are available on request from the corresponding author. Due to privacy concerns, the data is not publicly available.

Acknowledgments: The authors would like to gratefully acknowledge everyone involved in this project, Sarawak Forestry Corporation, Sarawak Forest Department, Samling Plywood Bintulu Sdn. Bhd., Daikin Plantation Sdn. Bhd. and Universiti Malaysia Sarawak. Thank you for the guidance and support.

Conflicts of Interest: The authors declare no conflict of interest.

References

1. Davies, H.E. *Timber: Structure, Properties. Conversion and Use*, 7th ed.; Food Product Press: New York, NY, USA, 1996.
2. Jumaat, M.Z.; Abdul Rahim, A.H.; Othman, J.; Razali, F.M. Timber engineering research and education in Malaysia. In Proceedings of the 9th World Conference on Timber Engineering 2006, WCTE, Portland, ON, USA, 6–10 August 2006.
3. Ismaili, G.; Abu Bakar, B.H.; Duju, A.; Abdul Rahim, K.K.; Openg, I. Domination of Grain Bearing on The Strength Properties of Engkabang Jantong as Fast Growing Timber in Sarawak. *Iran. J. Energy Environ.* **2013**, *4*, 311–315. [CrossRef]
4. Jusoh, I.; Zaharin, F.A.; Adam, N.S. Wood quality of Acacia hybrid and second-generation *Acacia Mangium*. *BioResources* **2014**, *9*, 150–160. [CrossRef]
5. Kijkar, S.; Somyos, K. *Part II, Species Description*; Association of South-East Asian Nations (ASEAN) Forest Tree Seed Centre: Chiang Mai, Thailand, 2003.
6. Sein, C.C.; Mitlöhner, R. *Acacia Mangium Wild: Ecology and Silviculture in Vietnam*; Center for International Forestry Research: Bogor Barat, Indonesia, 2011.
7. Pinso, C.; Nasi, R. The Potential Use of Acacia Mangium x Acacia Auriculiformis Hybrid in Sabah. In *Breeding Technologies for Tropical Acacias*; Carron, L.T., Aken, K.M., Eds.; No. 37; Australian Centre for International Agricultural Research; Goanna Print Pty Ltd.: Canberra, Australia, 1992; pp. 17–21.
8. Barry, K. *Heartrots in Plantation Hardwoods in Indonesia and Australia*; ACIAR Technical Reports No. 51e; Sun Photoset Pty Ltd.: Brisbane, Australia, 2002.
9. Ng, C.H.; Lee, S.L.; Ng, K.K.S.; Muhammad, N.; Ratnam, W. Mating system and seed variation of Acacia hybrid (*A. mangium* x *A. auriculiformis*). *J. Genet.* **2009**, *88*, 25–31. [CrossRef] [PubMed]
10. Choong, C.Y.; Yong, S.Y.C.; Cheong, P.L.; Pang, S.L.; Harikrishna, J.A.; Mat Isa, M.N.; Wickneswari, R. Analyses of ESTs Generated from Inner Bark Tissue of Acacia Hybrid. In Proceedings of the Acacia Research in Malaysia, Seminar on Current Updates on Acacia Genomics and Breeding 2005, Putrajaya, MA, USA, 12 July 2005.
11. Sukganah, A.; Wickneswari, R.; Choong, C.Y. Full length cDNA for Cinnamoyl-CoA Reductase (*CCR*) and Caffeic Acid O-Methyltransferase (*COMT*) in *Acacia mangium* Willd. X Acacia auriculiformis Cunn. Ex Benth. Hybrid. In Proceedings of the

Acacia Research in Malaysia, Seminar on Current Updates on Acacia Genomics and Breeding 2005, Putrajaya, MA, USA, 12 July 2005.
12. Treza, C.J.; Hiroatsu, F.; Didit, N. The influence of High-Temperature and -Pressure Treatment on Physical Properties of *Albizia falcataria* Board. *Forest* **2022**, *13*, 239.
13. Ismaili, G.; Roslani, R.; Abdul Rahim, K.K.; Duju, A.; Ismaili, Z.; Openg, I.; Ilias, M.N. Behaviour of Strength Properties of Fast-Growing Endospermum diadenum Species at Different Distance from the Vortex. *Mater. Sci.* **2017**, *23*, 254–259. [CrossRef]
14. Ling, W.C.; Wong, A.H.H. *Tropical In-Ground Durability of Structural Sarawak Hardwoods Impregnated to High Retention with CCA-salts, CCA-oxide and FCAP after 20 Years Exposure*; The International Research Group On Wood Protection, Wood Protecting Chemicals: Bangalore, India, 2005.
15. Forest Products Laboratory. *Air Drying of Lumber*; General Technical Report—117; USDA Forest Service: Madison, WI, USA, 1999.
16. Ismaili, G. Engineering Properties of Fast-Growing Indigenous Timber in Sarawak Compare to *Acacia mangium*: Aras. *Int. J. Sci. Mod. Eng.* **2015**, *3*, 8–12.
17. Ismaili, G.; Abu Bakar, B.H.; Abdul Rahim, K.K. Basic and grade stress for some timber in Sarawak. *J. Civ. Eng. Sci. Technol.* **2011**, *2*, 35–38. [CrossRef]
18. Gaddafi, I. Engineering Properties of Selected Sarawak Fast-Growing Indigenous Timber Species. Ph.D. Thesis, Universiti Sains Malaysia, Pulau Pinang, Malaysia, 2012.
19. Ismaili, G.; Abdul Rahim, K.K.; Duju, A.; Openg, I.; Ismaili, Z. Strength Classification of Aras as Fast-Growing Indigenous Species Timber in Sarawak. *Appl. Mech. Mater.* **2014**, *695*, 617–621. [CrossRef]
20. Pearson, R.G.; Kloot, N.H.; Boyd, J.D. *Timber Engineering Design Handbook*; Jacaranda Press Pty Ltd.: Melbourne, Australia, 1962.
21. Tan, Y.E.; Lim, N.P.T.; Gan, K.S.; Wong, T.C.; Lim, S.C.; Thilagawathy, M. *Testing Methods for Plantation Grown Tropical Timbers. ITTO Project on Improving Utilisation and Value Adding of Plantation Timbers from Sustainable Sources in Malaysia*; FRIM: Selangor Malaysia, 2010.
22. *BS 373:1957*; Methods of Testing Small Clear Specimens of Timber. British Standard Institution: London, UK, 1957.
23. Winandy, J.E. *Wood Properties*; Arntzen, C.J., Ed.; Encyclopaedia of Agricultural Science; Academic Press: Orlando, FL, USA, 1994.
24. Nur Emilia Azira, K.; Gaddafi, I. Strength Performance of Hevea Brasiliensis at Structural Size. *Int. Adv. Res. J. Sci. Eng. Technol.* **2015**, *2*, 144–147.
25. Ferreira, B.S.; Silva, J.V.F.; Campos, C.I. Static bending strength of heat treated and chromated copper arsenate-treated plywood. *BioResources* **2017**, *12*, 6276–6282. [CrossRef]
26. Epmeier, H.; Johansson, M.; Kliger, R.; Westin, M. Bending creep performance of modified timber. *Holz Roh-Werkst* **2007**, *65*, 343–351. [CrossRef]
27. Bruno, E.; Lina, N.; Helena, P. Properties of furfurylated wood (*Pinus pinaster*). *Eur. J. Wood Prod.* **2011**, *69*, 521–525.
28. Alik, D.; Naohiro, K. The relationship between basic density and mechanical strength properties of some Sarawak timber. In Proceedings of the TRTTC/STA Forest Product Seminar '96, TRTTC/STA, Kuching, Sarawak, 12–13 March 1996; pp. 117–126.
29. Desch, H.E. *Timber Its Structure Properties and Utilisation*, 6th ed.; Timber Press: Forest Glove, ON, USA, 1981.
30. Andy, W.C.L. Effect of CCA-Treating and Air-drying on the Properties of Southern Pine Lumber and Plywood. *Wood Fiber Sci.* **1985**, *17*, 209–213.
31. Alik, D.; Lai, J.K.; Yang, M.C.; John, S. Properties and utilisation of three age groups of *Acacia mangium* planted in Sarawak 2015. In Proceedings of the International Science and Nature Congress 2015 (ISNaC), Kuala Lumpur, Malaysia, 21–23 September 2015.
32. Riyanto, D.S.; Gupta, R. Effect of Ring Angle on Shear Strength Parallel to Grain of Wood. *J. Forest Product* **1995**, *46*, 87–92.
33. McKenzie, W.M.C. *Design of Structural Timber*; McGraw-Hill: London, UK, 2000; pp. 2–3.
34. Wallis, N.K. *Australian Timber Handbook*; Halstead Press: Sydney, Australia, 1970.
35. Madsen, B.; Janzen, W.; Zwaagstra, T. *Moisture Effects in Lumber*; Structural Research Series Report No. 27; Department of Structural Engineering, U.B.C: Vancouver, BC, Canada, 1980.
36. Mohd Jamil, A.W. Strength Assessment of Structural Size Malaysia Timbers. MSc, University of Malaya, Kuala Lumpur, Malaysia. 2013. Available online: http://studentsrepo.um.edu.my/8642/7/jamil.pdf (accessed on 7 October 2022).
37. Ping, X. The Mechanical Properties and Stability of Radiata Pine Structural Timber. Ph.D. Thesis, University of Canterbury, Christchurch, New Zealand, 2000.
38. Moya, R.; Muñoz, F. Physical and Mechanical Properties of Eight Fast-Growing Plantation Species in Costa Rica. *Trop. Forest Sci.* **2010**, *22*, 317–318.

forests

MDPI

Article

Multi-Objective Optimization and Analysis of Mechanical Properties of Coir Fiber from Coconut Forest Waste

Shaofeng Ru, Can Zhao and Songmei Yang *

Mechanical and Electrical Engineering College, Hainan University, Haikou 570228, China
* Correspondence: yangsongmei@hainanu.edu.cn

Abstract: As a type of natural fiber with excellent elongation, coir fiber has been applied in a wide range of fields. To ensure superb performance, coir fiber is usually treated with alkali before being applied. Previous studies paid little attention to the multiple alkali treatment of coir fiber; however, this study focuses on its influence on the mechanical properties of coir fiber and conducts multi-objective optimization and analysis of the tensile strength, elastic modulus and elongation of coir fiber. Our objective is the comprehensive enhancement of the mechanical properties of coir fiber. In this study, the experimental design is based on the Box-Behnken design method, and three treatment parameters were selected for the study, namely NaOH concentration, treatment time and treatment temperature. Analysis of variance (ANOVA) was adopted to analyze the experimental data, and response surface methodology (RSM) was used to investigate how the treatment factors interact with each other and affect the responses values. To improve the tensile strength, elastic modulus and elongation of coir fiber simultaneously, the experimental parameters were optimized. The results showed that the optimal values of NaOH concentration, treatment time and treatment temperature were 4.12%, 15.08 h and 34.21 °C, respectively. Under these conditions, the tensile strength of coir fiber was 97.14 MPa, the elastic modulus was 2.98 GPa and the elongation was 29.35%, which were 38.28%, 39.91% and 25.59% higher than that of untreated coir fiber, respectively. Furthermore, scanning electron microscopy (SEM), thermogravimetric analysis (TGA-DTG), Fourier-transform infrared spectroscopy (FTIR) and X-ray diffraction (XRD) were used to analyze the changes in surface, weight loss, composition and crystallinity of coir fiber treated with alkali under optimum conditions compared with untreated coir fiber to obtain a deeper insight into the influential mechanisms of alkali treatment.

Keywords: coir fiber; alkali treatment; parameter optimization; mechanical properties

Citation: Ru, S.; Zhao, C.; Yang, S. Multi-Objective Optimization and Analysis of Mechanical Properties of Coir Fiber from Coconut Forest Waste. *Forests* **2022**, *13*, 2033. https://doi.org/10.3390/f13122033

Academic Editors: Tomasz Krystofiak and Pavlo Bekhta

Received: 30 October 2022
Accepted: 29 November 2022
Published: 30 November 2022

1. Introduction

As people's awareness of environmental protection increases, synthetic fibers with a negative impact on the ecological environment are phased out. In development models prioritizing a green environment, people are looking for alternatives to synthetic fibers and are looking to natural fibers as options [1]. Natural fibers are attracting attention because of their environmental friendliness and renewable nature, as well as their low cost and good mechanical properties, accounting for their wide application [2]. Earlier studies on natural fibers were mostly in the textile field [1,3–5]; recently, there have been more studies on natural fiber composites [6,7]. Saikrishnan et al. studied the mechanical properties, including tensile strength and flexural strength, of ramie/kenaf fiber composites [8]. Sumesh et al. studied the friction and wear properties of sisal/pineapple fiber composites [9]. Natural fiber composites are advocated in the automotive and construction industries because of their environmental friendliness and superb mechanical properties. It is expected that the demand for natural fibers will increase in the future, involving more fields [10,11].

There are many natural fibers that are frequently used at present, such as the flax fiber, banana fiber and coir fiber, which is a kind of fiber obtained from waste coconut shells.

About 650,000 tons of waste coconut shells are produced globally every year. Extracting coir fiber for applications also makes use of waste coconut shells. Coir fiber is mainly composed of cellulose, hemicellulose, lignin and pectin, among other substances [12]. Among them, cellulose and lignin account for a large proportion. Cellulose is the most crucial component of coir fiber and positively affects its elongation, while excessively high content of lignin will damage the toughness of the fiber. In addition, cellulose is hydrophilic, and its hydroxyl groups facilitate the bonding of fibers with the matrix when composites are produced. Pectin and other impurities attached to the fiber surface will hinder the bonding of the fiber with the matrix, making it easy to fall off. Therefore, the purpose of coir fiber treatment is to remove pectin and other impurities and to partially remove lignin, hemicellulose and other non-cellulose substances to improve the properties of the fiber and the interfacial bonding of the fiber and the matrix [13,14]. Conventional methods to treat natural fibers include alkali treatment, radiation treatment, acid anhydride treatment, coupling agent treatment, and others. Many natural fiber treatments have been studied by researchers. Hestiawan et al. compared the effects of alkali treatment, silane treatment and alkali-silane treatment on the tensile properties and interfacial shear strength of fan palm fibers. The results showed that the tensile strength of alkali–treated and alkali-silane–treated fibers increased compared with untreated fibers, while that of silane–treated fibers decreased in comparison. In addition, the interfacial shear strength of the fibers treated in the above three ways rose compared with untreated ones [15]. Loong et al. treated flax fiber with acetic anhydride to investigate the effect on the properties of flax fiber epoxy composites. They found significant differences between the tensile strength and modulus of the composites and the untreated samples and reported that treatment of flax fiber can enhance interfacial adhesion of the flax fiber composites [16]. Using sodium chlorite solution, Ma et al. removed lignin from corn stalk fibers to enhance the friction wear properties and interfacial bonding of resin-based friction materials [17]. Alkali treatment is the most frequently used treatment for natural fibers, with the advantages of low cost and easy operation. In addition, alkali treatment can effectively improve the surface roughness of natural fibers, allowing closer combination with the matrix materials. This is an important reason for its selection. There are different requirements for conditions to treat different fibers, such as alkali concentration, treatment time and treatment temperature. Different treatment conditions will exert different effects on the same fiber. Valášek et al. explored the changes in the mechanical properties and surface microstructure of coir and abaca fibers after immersion in NaOH solution for different periods of time [18]. Frącz et al. used 2%, 5% and 10% NaOH solutions for surface modification of flax and hemp fibers, respectively, exploring their effects on the properties of the composites. It was shown that most of the properties of the hemp fiber composites were improved as concentration of the alkali solution increased. On the other hand, treatment of the flax fiber composites with a higher concentration of NaOH solution deteriorated most of their properties [19]. As mentioned earlier, alkali treatment conditions are closely related to the properties of the treated fiber, and unreasonable treatment conditions will degrade their properties. The purpose of fiber treatment is to ensure the optimal performance of the fiber, which requires accurately selecting factors influencing the alkali treatment.

Coir fiber possesses good mechanical properties, making it an object of great interest to researchers [20]. Although the strength of coir fiber is not dominant compared with other hemp fibers, it has the highest elongation compared with any known natural fiber [21,22]. As elongation is an advantageous property of coir fiber, the study of elongation is essential. Previous studies on the alkali treatment of natural fibers showed that the required alkali treatment conditions for optimum tensile strength, optimum elastic modulus, and optimum elongation are not identical [18]. Previous research on alkali treatment of natural fibers failed to focus on this aspect, and the mechanical properties of fibers obtained by the alkali treatment of fibers under non-optimized treatment conditions may not be improved in a holistic way. If a certain mechanical property of the fiber is not significantly improved after alkali treatment, there will be shortcomings in this aspect of the fiber. As a result, it is of great importance to improve the comprehensive mechanical properties of the fiber. The present study differs from other studies on coir fiber alkali treatment in that the tensile strength, elastic modulus and elongation are optimized and analyzed to obtain the treatment conditions under which these properties are optimally improved simultaneously. The tensile strength, elastic modulus and elongation of coir fiber can be improved more effectively by alkali treatment under the optimized NaOH concentration, treatment time and treatment temperature, so that the mechanical properties of coir fiber can be comprehensively improved in comparison with those before the treatment. The optimization of alkali treatment conditions is necessary. The experiment involves a multi-objective simultaneous optimization technology whose design is based on Box-Behnken. The data are derived from the tensile strength, elastic modulus and elongation of coir fibers treated under different NaOH concentrations, times and temperatures. Design-Expert software 11 was used to analyze and obtain the regression equation and obtain the optimization results. Finally, the properties of coir fiber under the optimal conditions were compared with untreated samples. The mechanism of its influence on coir fiber was analyzed with a scanning electron microscopy, thermogravimetric analysis, Fourier-transform infrared spectroscopy and X-ray diffraction.

2. Materials and Methods

2.1. Materials

The waste coconut shells used in the experiments were collected from mature coconuts grown in Hainan Province, China. The coir fiber was extracted from the waste in the coconut forest. Figure 1 shows the coconut forest, waste coconut shell and coir fiber. The chemical used to treat the coir fibers was sodium hydroxide (NaOH) solution, which was purchased from Phygene Biotechnology, Fuzhou.

Figure 1. Coconut forest, waste coconut shell and coir fibers.

2.2. Preparation of Coir Fibers

The coir fibers for experimental use were selected by hand to ensure that they were similar in shape and diameter. Prior to alkali treatment, coir fibers were cleaned with distilled water to remove impurities attached to the surface; they were then were cut into 50 mm after drying. The coir fibers were divided into 17 groups, each undergoing the treatment illustrated in Figure 2 under different treatment conditions. The coir fibers were immersed in a beaker containing the specified concentration of NaOH solution, and the beaker was placed in a multi-purpose incubator at a required temperature and for a required period for alkali treatment of the fibers. After the treatment, the fibers were immersed in distilled water for 2 h, followed by repeated washing with distilled water until all NaOH was removed from their surface. Finally, the coir fibers were dried at 60 °C for 3 h and stored in sealed bags for subsequent analyses.

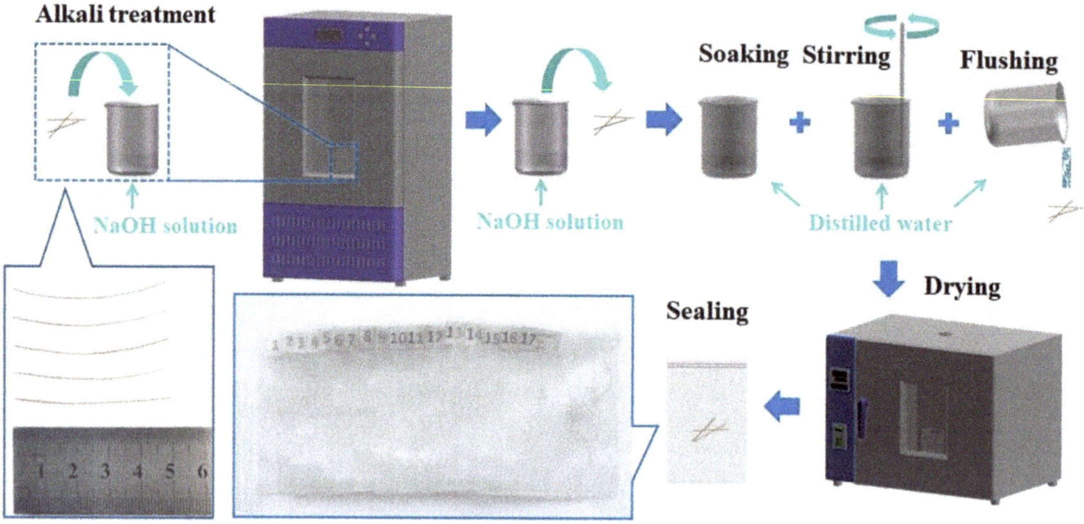

Figure 2. Preparation process of coir fiber samples.

2.3. Experimental Design

Response surface methodology (RSM) is a mathematical and statistical technique used to analyze experimental data for response optimization. RSM is one of the most frequently used optimization tools in experimental design. According to the impact of treatment conditions on alkali treatment, NaOH concentration, treatment time and treatment temperature were selected as variables in the study [11,14,18], and the experimental scheme was selected using a three-factor Box-Behnken design using three levels. The factors and levels are shown in Table 1. The diameter and post-break length of each fiber were measured with the stereomicroscope to obtain accurate experimental data, such as tensile strength and elongation. Each fiber was measured once in both mutually perpendicular directions to take the average value as the diameter. Then, Design-Expert software 11 was employed to analyze the experimental data and optimize the alkali treatment conditions, including NaOH concentration, time and temperature. This method can analyze the influential trend of alkali treatment on the tensile strength, elastic modulus and elongation of coir fiber through the response surface to obtain the best conditions for alkali treatment in combination with regression equations.

Table 1. Factors and levels of experiments.

Levels	Factors		
	NaOH Concentration (%)	Time (h)	Temperature (°C)
	A	B	C
−1	2	2	20
0	6	12	40
1	10	22	60

2.4. Testing of Coir Fiber

2.4.1. Tensile Test

Tensile tests were conducted on coir fiber using an Electronic Universal Testing Machine (3343, INSTRON, Boston, MA, USA) with a crosshead speed of 3 mm/min to compare and analyze the changes in mechanical properties of coir fibers in 17 groups treated under different conditions. The mechanical properties tested in this experiment included tensile strength, elastic modulus and elongation. Specifically, the coir fiber sample needed to be fixed on the upper and lower collets of the instrument through a special fixture. The fiber sample was kept at the centerline of the collet and clamped to avoid fiber slip. The test was repeated five times for each group of fiber samples to obtain the average value as the test result.

2.4.2. Scanning Electron Microscopy

The morphologies of the untreated and treated coir fibers were investigated with a scanning electron microscopy (SEM) (Verios G4 UC, Thermoscientific, Waltham, MA, USA). To make it conductive, the sample of coir fiber was sprayed with gold to make its surface a thin gold palladium layer. Surface and cross-sections of the fibers were observed and analyzed at a voltage of 5 kV to understand microscopic changes.

2.4.3. Thermogravimetric Analysis

The weight changes of untreated and treated coir fiber specimens were studied with the thermogravimetric analyzer (Q600, TA, New Castle, DE, USA). The coir fibers were cut into 1 mm segments before the test, and their specimens were put into the crucible. The test studied the changes in weight loss and derivative weight of coir fibers in testing temperature ranging from 30 °C to 500 °C. The whole test was conducted in nitrogen, and the heating rate was maintained at 10 °C/min.

2.4.4. Fourier-Transform Infrared Spectroscopy

The untreated and treated samples of coir fiber were studied and analyzed with the Fourier-Transform Infrared Spectrometer (FTIR) (T27, Bruker, Billerica, Germany). An appropriate amount of coir fiber powder was mixed and ground with potassium bromide (KBr). The test samples were prepared by the tablet pressing method. The data were recorded in the wave numbers ranging from 400 to 4000 cm^{-1} at a resolution of 4 cm^{-1}. The infrared spectra of each sample could be obtained after scanning.

2.4.5. X-ray Diffraction

The crystallinity index of untreated and treated samples of coir fiber was measured by an X-ray Diffractometer (XRD) (Smart Lab, Rigaku, Tokyo, Japan). Before the determination, the coir fiber was broken with a pulverizer to prepare powder samples, all of which were scanned in the range 2θ of 5°–60° at the scanning speed of 5°/min. The relative crystallinity index of each fiber sample was calculated according to the obtained spectral data and the Segal empirical method [23].

$$CrI = \frac{I_{002} - I_{am}}{I_{002}} \times 100\% \quad (1)$$

where CrI is crystallinity index, I_{002} is the maximum intensity of 002 lattice diffraction plane at a 2θ close to 22° and I_{am} is the intensity diffraction of amorphous materials at a 2θ close to 18°.

3. Results and Discussion

3.1. Tensile Test Results and Analysis

3.1.1. Model Fitting and Analysis of Variance

According to the Box-Behnken design principle, 17 groups of tests were carried out. The testing scheme and result of each group are shown in Table 2, including tensile strength, elastic modulus and elongation in the model. The test data were imported into the software Design-Expert 11 for analysis. The regression models of tensile strength Y_1, elastic modulus Y_2 and elongation Y_3 on the independent variables NaOH concentration A, time B and temperature C were established. The alkali treatment results were further investigated using analysis of variance (ANOVA) to determine which factors significantly affect the mechanical properties of coir fiber [24–26]. Tables 3–5 show the ANOVA results.

Table 2. Alkaline treatment conditions and test results of coir fibers.

Std	Run	NaOH Concentration A (%)	Time B (Hours)	Temperature C (°C)	Tensile Strength Y_1 (MPa)	Elastic Modulus Y_2 (GPa)	Elongation Y_3 (%)
12	1	6	22	60	75.56	2.60	23.36
13	2	6	12	40	96.12	2.94	28.87
7	3	2	12	60	84.57	2.44	24.69
9	4	6	2	20	74.16	2.68	23.03
2	5	10	2	40	62.88	2.32	22.40
14	6	6	12	40	93.89	2.94	28.72
5	7	2	12	20	93.71	2.62	28.34
16	8	6	12	40	90.91	3.03	29.13
6	9	10	12	20	78.30	2.45	20.43
10	10	6	22	20	91.01	2.79	25.46
8	11	10	12	60	67.97	2.20	22.04
17	12	6	12	40	95.10	3.02	28.32
1	13	2	2	40	81.71	2.68	25.12
3	14	2	22	40	94.21	2.74	28.82
4	15	10	22	40	71.06	2.59	20.62
11	16	6	2	60	73.51	2.36	24.72
15	17	6	12	40	91.38	2.97	28.50

Table 3. ANOVA for coir fiber tensile strength.

Source	Sum of Squares	df	Mean Square	F-Value	*p*-Value	
Model	1855.98	7	265.14	59.60	<0.0001 **	highly significant
A-NaOH concentration	684.32	1	684.32	153.83	<0.0001 **	
B-Time	195.82	1	195.82	44.02	<0.0001 **	
C-Temperature	158.15	1	158.15	35.55	0.0002 **	
BC	54.76	1	54.76	12.31	0.0066 **	
A^2	190.07	1	190.07	42.73	0.0001 **	
B^2	363.87	1	363.87	81.80	<0.0001 **	
C^2	133.16	1	133.16	29.93	0.0004 **	
Residual	40.04	9	4.45			
Lack of Fit	19.26	5	3.85	0.7416	0.6313	not significant
Pure Error	20.78	4	5.19			
Cor Total	1896.01	16				
$R^2 = 0.9789$ Predicted $R^2 = 0.9365$				Adjusted $R^2 = 0.9625$ Adequate Precision = 21.8151		

Note: *p*-value < 0.01 (highly significant, **).

Table 4. ANOVA for coir fiber elastic modulus.

Source	Sum of Squares	df	Mean Square	F-Value	*p*-Value	
Model	1.05	7	0.1496	90.05	<0.0001 **	highly significant
A-NaOH concentration	0.1058	1	0.1058	63.69	<0.0001 **	
B-Time	0.0578	1	0.0578	34.80	0.0002 **	
C-Temperature	0.1105	1	0.1105	66.49	<0.0001 **	
AB	0.0110	1	0.0110	6.64	0.0299 *	
A^2	0.3511	1	0.3511	211.34	<0.0001 **	
B^2	0.0498	1	0.0498	29.98	0.0004 **	
C^2	0.2929	1	0.2929	176.33	<0.0001 **	
Residual	0.0150	9	0.0017			
Lack of Fit	0.0076	5	0.0015	0.8162	0.5945	not significant
Pure Error	0.0074	4	0.0018			
Cor Total	1.06	16				
$R^2 = 0.9859$ Predicted $R^2 = 0.9533$				Adjusted $R^2 = 0.9750$ Adequate Precision = 28.0769		

Note: *p*-value < 0.01 (highly significant, **); *p*-value < 0.05 (significant, *).

Table 5. ANOVA for coir fiber elongation.

Source	Sum of Squares	df	Mean Square	F-Value	p-Value	
Model	153.25	9	17.03	142.42	<0.0001 **	highly significant
A-NaOH concentration	57.67	1	57.67	482.39	<0.0001 **	
B-Time	1.12	1	1.12	9.35	0.0184 *	
C-Temperature	0.7503	1	0.7503	6.28	0.0407 *	
AB	7.51	1	7.51	62.79	<0.0001 **	
AC	6.92	1	6.92	57.85	0.0001 **	
BC	3.59	1	3.59	30.04	0.0009 **	
A^2	23.61	1	23.61	197.44	<0.0001 **	
B^2	18.57	1	18.57	155.35	<0.0001 **	
C^2	25.59	1	25.59	214.03	<0.0001 **	
Residual	0.8369	7	0.1196			
Lack of Fit	0.4386	3	0.1462	1.47	0.3497	not significant
Pure Error	0.3983	4	0.0996			
Cor Total	154.08	16				
$R^2 = 0.9946$				Adjusted $R^2 = 0.9876$		
Predicted $R^2 = 0.9504$				Adequate Precision = 32.1528		

Note: p-value < 0.01 (highly significant, **); p-value < 0.05 (significant, *).

The significance of variables in the regression model is related to p-value, and it is considered to be significant when the p-value is less than 0.05. Table 3 shows the ANOVA results of tensile strength. It can be seen that variables A, B, C, BC, A^2, B^2 and C^2 exert a significant impact on tensile strength Y_1. To optimize model Y_1, insignificant model terms were eliminated. The regression model is shown in Equation (2). The model of tensile strength Y_1 is highly significant ($p < 0.01$), and the lack-of-fit value is not significant ($p > 0.05$), suggesting a good fitting relationship between the regression model and the actual situation. The regression coefficient R^2 is 0.9789, indicating that the data can properly express the model. The table also shows that the predicted R^2 is 0.9365 and the adjusted R^2 is 0.9625; the difference is less than 0.2. Furthermore, the adequate precision measures the signal to noise ratio. When the ratio is greater than 4, it is desirable. This ratio in the model is 21.8151, indicating that the signal is sufficient and that the model can be used in the navigation design space [24,27].

$$Y_1 = 93.48 - 9.25A + 4.95B - 4.45C - 3.70BC - 6.72A^2 - 9.30B^2 - 5.62C^2 \quad (2)$$

Table 4 shows the ANOVA results of elastic modulus. The p-values corresponding to each variable are compared, and it is shown that the variables A, B, C, AB, A^2, B^2 and C^2 have a significant impact on elastic modulus Y_2. After the insignificant model terms in the model were eliminated, the regression model is obtained, as shown in Equation (3). The model of elastic modulus Y_2 was highly significant, while the lack-of-fit value was not, which shows that the fitting degree between the regression model and the actual situation was good within the testing range. In addition, the regression coefficient R^2 is 0.9859 in the regression model of elastic modulus Y_2, and the predicted R^2 of 0.9533 is in reasonable agreement with the adjusted R^2 of 0.9750. It can be found that they show a significant relationship. Adequate precision measures the signal to noise ratio, which in this model is 28.0769, satisfying the requirements of the navigation design space with an adequate signal [24,27].

$$Y_2 = 2.98 - 0.1150A + 0.0850B - 0.1175C + 0.0525AB - 0.2887A^2 - 0.1087B^2 - 0.2638C^2 \quad (3)$$

Table 5 shows the ANOVA results for elongation. The p-value of the elongation regression equation model is less than 0.0001, indicating that the model is highly significant, while the lack-of-fit value is insignificant. Therefore, it can be concluded that the model is suitable. The regression model is shown in Equation (4). Among them, variables A, B, C, AB, AC, BC, A^2, B^2 and C^2 have a significant impact on elongation Y_3. The same table also shows that the regression coefficient R^2, the predicted R^2 and the adjusted R^2 are 0.9946, 0.9504 and 0.9876, respectively. The predicted value is highly correlated with the actual value, suggesting that the empirical model is significantly reliable. The adequate precision ratio of the model is 32.1528, indicating that the model can be used to navigate the design space [24,27].

$$Y_3 = 28.71 - 2.68A + 0.3738B - 0.3063C - 1.37AB + 1.31AC - 0.9475BC - 2.37A^2 - 2.10B^2 - 2.47C^2 \quad (4)$$

To gain deeper insight into the testing results, the perturbation plots in Figure 3 show the effect of the treatment conditions on tensile strength, elastic modulus and elongation. Conclusions can also be drawn by analyzing the F-value of each treatment condition in Tables 3–5. The influential order for tensile strength is NaOH concentration, time and temperature from the largest to the smallest; that for elastic modulus is temperature, NaOH concentration and time; and that for elongation is NaOH concentration, time and temperature. In addition, NaOH concentration has the greatest effect on the tensile strength, while the time and temperature have relatively little effect on the tensile strength. Compared with the time, NaOH concentration and temperature exert slightly greater impact on elastic modulus. The effect of NaOH concentration on elongation is greater than that of time and temperature. Similar results indicating that the NaOH concentration of alkali treatment has a greater impact on the mechanical properties of natural fibers have been reported elsewhere [26].

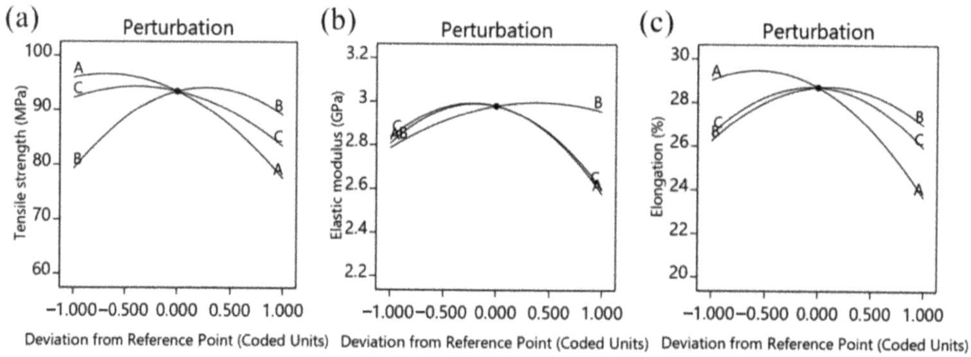

Figure 3. Perturbation plots: (**a**) Effect of factors on tensile strength, (**b**) effect of factors on elastic modulus, (**c**) effect of factors on elongation.

3.1.2. Effect of Interactive Factors on Tensile Strength

The tensile strength of coir fiber should be enhanced to optimize its own properties, as it is significant for coir fiber–reinforced composites. The three-dimensional surface plot and the contour plot shown in Figure 4 illustrate the interactive effect of time and temperature on tensile strength. From the overall three-dimensional surface plot, it can be determined that the tensile strength of coir fiber has changed significantly under the interactive effect of time and temperature. When the NaOH concentration of alkali treatment is fixed at the level of 0 (A = 6%), the temperature rise affects the tensile strength of coir fiber positively when the treatment time remains constant. This effect is clearly seen in the contour plot, where the tensile strength reached the maximum at close to 30 °C and then showed a declining trend as the treatment temperature rose. This is due to the fact that excessively high treatment temperatures can damage the coir fibers, leading to a deterioration of the tensile strength. On the other hand, the tensile strength of coir fiber increases as the treatment time is prolonged and the treatment temperature remains constant. However, excessive soaking time can have a negative effect. Initially, materials attached to the fiber surface will be gradually removed as the alkali treatment proceeds. However, a substantial amount of lignin, whose functions are support and bonding, is dissolved after a long period of time, resulting in the decrease of tensile strength. Jiang et al. also reported that unreasonable alkali treatment conditions would adversely affect the tensile strength of fibers in their research on the effects of alkali treatment on palm fibers [28].

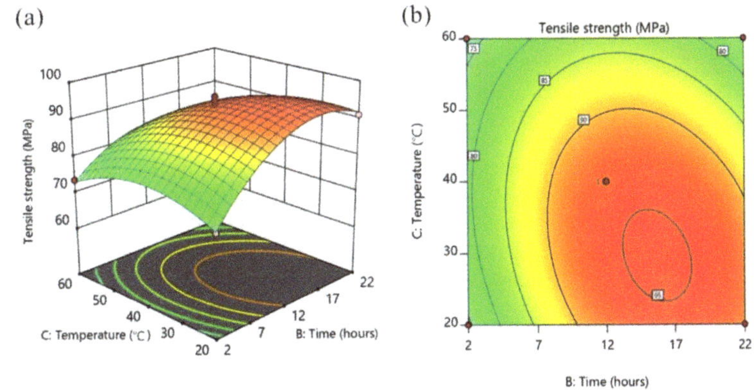

Figure 4. Effect of time and temperature on the tensile strength of coir fiber at a NaOH concentration of 6%: (**a**) three-dimensional surface plot; (**b**) contour plot.

3.1.3. Effect of Interaction Factors on Elastic Modulus

Elastic modulus is one of the most important properties of coir fiber. The interaction of NaOH concentration and time has a significant impact on the elastic modulus of coir fiber. Its three-dimensional surface plot and the contour plot are shown in Figure 5. According to the analysis of the three-dimensional surface plot in Figure 5a, the elastic modulus of coir fiber tends to increase and then decrease as treatment time lengthens in the same NaOH concentration, when the temperature of alkali treatment was fixed at the level of 0 (C = 40 °C), and the trend is the same as NaOH concentration increases when the treatment time remains unchanged. In addition, as shown in the contour plot in Figure 5b, the elastic modulus changes rapidly along the NaOH concentration direction and slowly along the time direction. It is worth noting that excessively high NaOH concentration or excessively long treatment time will cause damage to coir fiber [18]. When the elastic modulus reaches the maximum, any further increase in NaOH concentration and treatment time will reduce the elastic modulus of coir fiber.

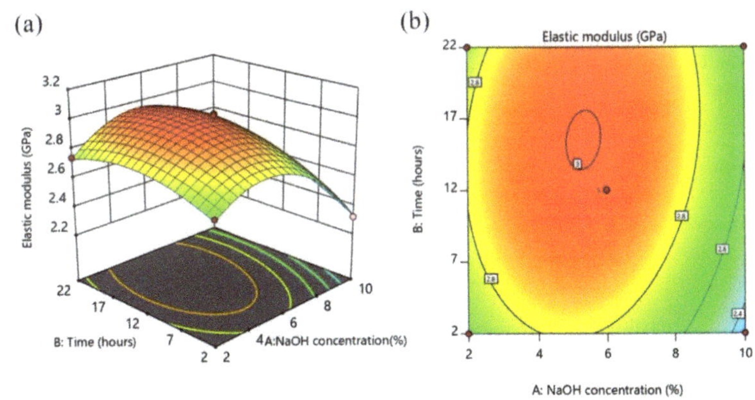

Figure 5. Effect of NaOH concentration and time on elastic modulus of coir fiber at a temperature of 40 °C: (**a**) three-dimensional surface plot; (**b**) contour plot.

3.1.4. Effect of Interaction Factors on Elongation

The elongation of coir fiber is more outstanding compared with most other natural fibers, and it has one of the highest elongations among average natural fibers. In addition, appropriate alkali treatment will further increase the elongation of coir fiber, which will make the fiber more flexible. It can be seen that the bending properties of composites made of coir fiber are also improved. At the same time, it is of great significance for flexible materials.

The interactive effects of NaOH concentration A, time B and temperature C on the elongation of coir fiber are shown in Figure 6. As the pictures show, one factor at the 0 level is fixed, and the influence of the other two factors on the elongation is analyzed. Combined with the F-value of each interactive factor in Table 5, the most significant effect on the elongation of coir fiber is the interaction between NaOH concentration A and time B, followed by that between NaOH concentration A and temperature C, and then between time B and temperature C. The three-dimensional surface plot and the contour plot in Figure 6a,b show the effect of the interaction of NaOH concentration and time on elongation. When the temperature of alkali treatment was fixed at the level of 0 (C = 40 °C), under the same NaOH concentration, the elongation tends to rise with the increase of treatment time and fall slightly after reaching the maximum. In addition, during the same treatment time, the elongation first increases slightly and then decreases as the NaOH concentration rises. The interactive effect of NaOH concentration and temperature on the elongation is shown in Figure 6c,d. When the time of alkali treatment remains fixed at the level of 0 (B = 12 h), the interactive effect of NaOH concentration and temperature on elongation can be clearly seen from the three-dimensional surface plot and contour plot. Among them, the elongation of coir fiber reaches the maximum when the NaOH concentration is about 4% and the treatment temperature is about 35 °C. Then, when the NaOH concentration or temperature exceeds this value, the elongation of coir fiber will decrease. When the NaOH concentration of alkali treatment is fixed at the level of 0 (A = 6%), the interactive effect of time and temperature on elongation is expressed in Figure 6e,f. It is very intuitive to see from the three-dimensional surface plot that the trend of elongation based on temperature during the same treatment time and the trend of elongation based on time at the same treatment temperature both increase and then decrease. At the same time, the maximum elongation can be obtained in the middle of the contour plot.

Jiang et al. and Valášek et al. also studied the effect of alkali treatment on the elongation of natural fibers [18,28]. Reasons for these changes were obtained from the principle analysis. Alkali treatment can remove impurities and some lignin from the surface of coir fiber. Because lignin is easier to degrade than cellulose in alkali solution, the relative content of cellulose increases after alkali treatment. This is conducive to the increase of elongation of coir fiber, whose increase or decrease is inseparable from the main treatment conditions. Excessively small NaOH concentration will not achieve the expected effect of alkali treatment, and excessively short treatment time and low temperature will make the reaction insufficient. On the contrary, excessively large NaOH concentration will damage the fiber structure, and excessively long treatment time and high temperature will dissolve a large amount of lignin, leading to degraded performance. Therefore, large elongation will be obtained by alkali treatment of coir fiber under appropriate NaOH concentration, time and temperature.

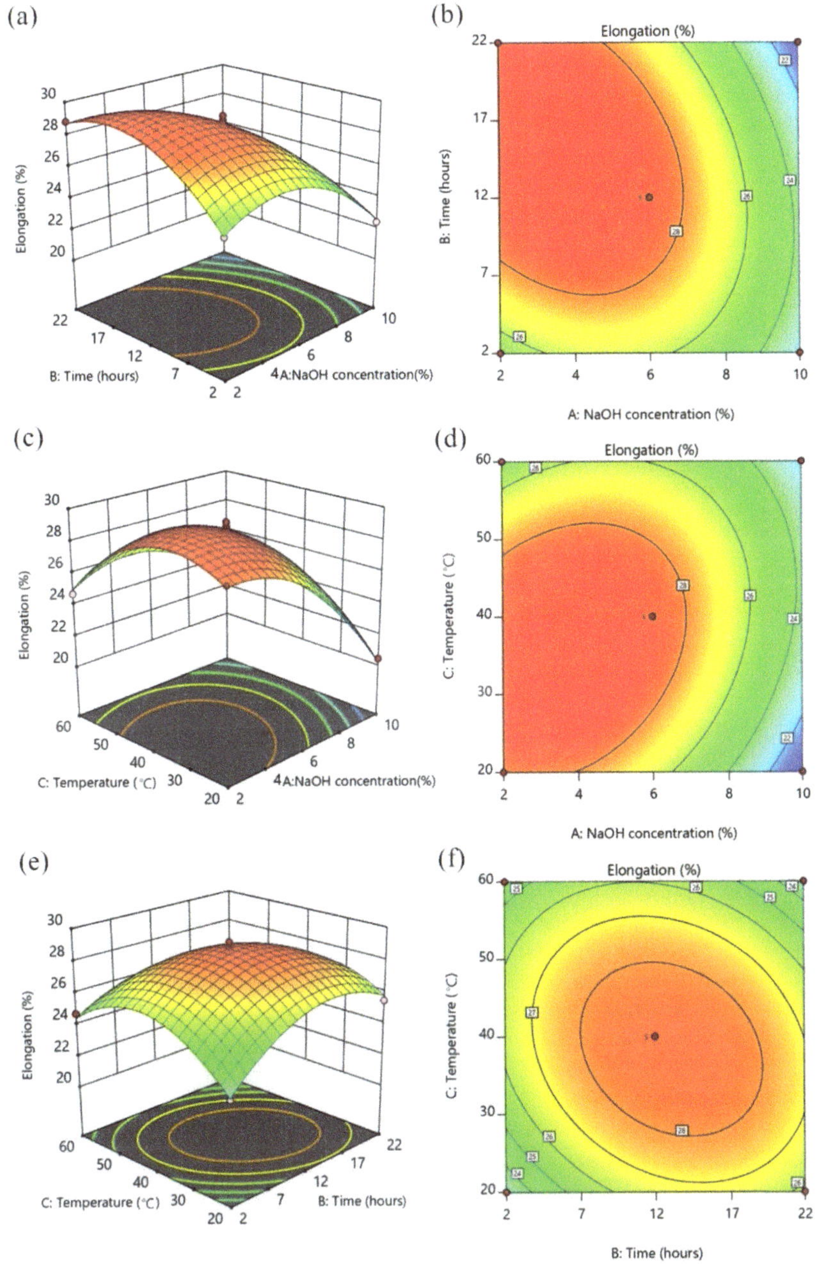

Figure 6. Effect of NaOH concentration and time on elongation of coir fiber at a temperature of 40 °C: (**a**) three-dimensional surface plot; (**b**) contour plot. Effect of NaOH concentration and temperature on elongation of coir fiber at a time of 12 h: (**c**) three-dimensional surface plot; (**d**) contour plot. Effect of time and temperature on elongation of coir fiber at a NaOH concentration of 6%: (**e**) three-dimensional surface plot; (**f**) contour plot.

3.1.5. Optimization of Treatment Conditions

Figure 7a–c show the relationship between the actual and predicted values of tensile strength, elastic modulus and elongation, respectively. All points in the figures are distributed near the diagonal line, indicating a high correlation between the actual and predicted values. They also reflect that the model is highly accurate. It can be seen that these selected models are adequate [24,26].

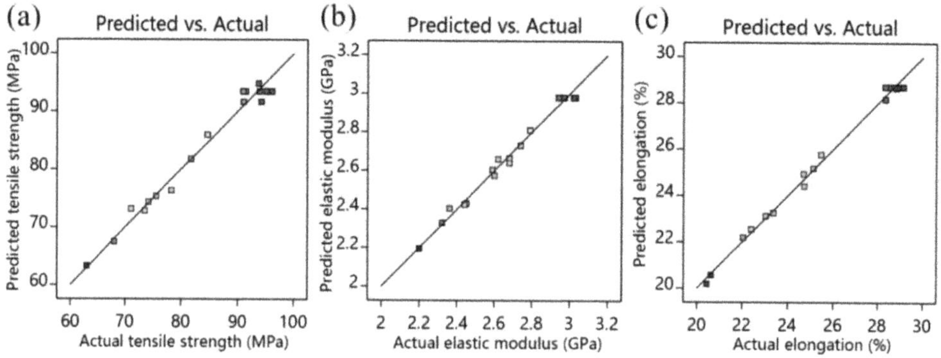

Figure 7. Linear plots: (**a**) predicted and actual values of tensile strength, (**b**) predicted and actual values of elastic modulus and (**c**) predicted and actual values of elongation.

To obtain better properties of coir fibers, suitable alkali treatment conditions are essential. Therefore, the tensile strength, elastic modulus, and elongation exhibited by coir fibers are taken as optimization objectives. In this study, multi-objective optimization is required to optimize the combination of parameters that satisfy several indicators; thus, the NaOH concentration, time, and temperature of the alkali treatment are optimized. The mathematical model is obtained by building a parametric optimization model and analyzing the regression equation as shown in Equation (5).

$$\begin{cases} \max Y_1\ (A, B, C) \\ \max Y_2(A, B, C) \\ \max Y_3(A, B, C) \end{cases} \quad (5)$$

The model is optimized and analyzed by Design-Expert software 11, and a sufficient range is set for the parameters to ensure that the optimization results appear in the given range. As can be observed from Figure 8, the optimal operating conditions for alkali treatment are 4.12% NaOH concentration, 15.08 h of treatment time and a treatment temperature of 34.21 °C. Under these treatment conditions, the tensile strength is 98.13 MPa, elastic modulus is 2.99 GPa and elongation is 29.71%.

3.1.6. Verification Test

To confirm the accuracy of the model prediction, validation tests were performed under the optimized conditions. Five samples were tested, and the average values of tensile strength, elastic modulus and elongation were obtained as 97.14 MPa, 2.98 GPa and 29.35%, respectively. Table 6 shows the comparison between the measured value and the predicted value, from which it can be seen that the errors of tensile strength, elastic modulus and elongation are 1.01%, 0.33% and 1.21%, respectively, indicating that the parameter optimization model is reliable.

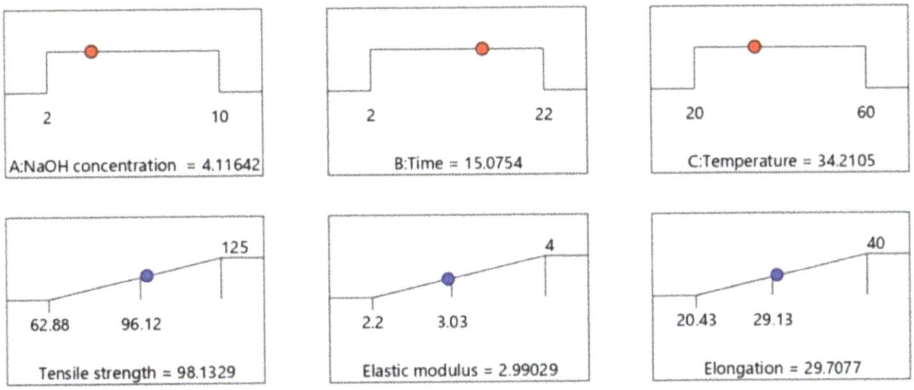

Figure 8. Optimal results.

Table 6. Validation testing results.

Contrast Items	Tensile Strength (MPa)	Elastic Modulus (GPa)	Elongation (%)
Measured value	97.14	2.98	29.35
Predicted value	98.13	2.99	29.71
Error (%)	1.01	0.33	1.21

Figure 9 compares the changes in mechanical properties of untreated coir fibers (UCF) and treated coir fibers under optimal conditions (OCF). The mechanical properties of UCF also come from the average value of five samples. Combined with the pictures, it can be seen that after being treated under optimal conditions, the tensile strength of coir fiber changes from 70.25 MPa to 97.14 MPa, an increase of 38.28%; the elastic modulus from 2.13 GPa to 2.98 GPa, an increase of 39.91%; and the elongation from 23.37% to 29.35%, an increase of 25.59%. The changes exhibited by the test are consistent with the test principle. Silva et al. also reported that the mechanical properties of coir fibers were improved after alkali treatment [21]. However, the conditions of alkali treatment were optimized by targeting tensile strength, elastic modulus and elongation at the same time in this experiment, resulting in a more balanced enhancement of the mechanical properties of coir fibers.

3.2. *Scanning Electron Microscopy Observation and Analysis*

SEM can observe the microscopic changes in the surface morphology of coir fiber before and after the treatment. SEM micrographs of UCF and OCF surfaces are shown in Figure 10. From the overall view, the surface of UCF is tightly wrapped by a large amount of pectin, wax and other impurities, and this feature is reflected in Figure 10a. Due to the wrapping of these substances, the surface of UCF has no obvious bulges and grooves, only impurities attached to the surface of coir fiber. Figure 10b provides SEM micrographs of the UCF surface at higher magnification. It can be observed from the magnified picture that the impurities are flaky and their structure is loose. When UCF is bonded to the matrix material, the existence of these substances between the fibers and the matrix makes it easy for their connection to debond [29,30]. On the contrary, it can be seen from Figure 10c that the surface of OCF is relatively rough, which is due to the removal of some hemicellulose, lignin, pectin and other materials in coir fiber after alkali treatment. As a result, there are many clear bulges and grooves. This structure can be clearly seen in the high magnification SEM micrograph (Figure 10d) of the OCF surface. Other studies also reported the micro changes of the fiber surface after alkali treatment [15,31], which are conducive to improving the interfacial adhesion between coir fiber and the polymer matrix. When the composites are produced, the fluid-like matrix material can enter the bumpy surface structure, and mechanical interlocking between the coir fiber and the polymer matrix can be produced after solidification, thus effectively improving the interfacial bonding performance between them [32,33]. This further proves the effect of alkali treatment under appropriate conditions.

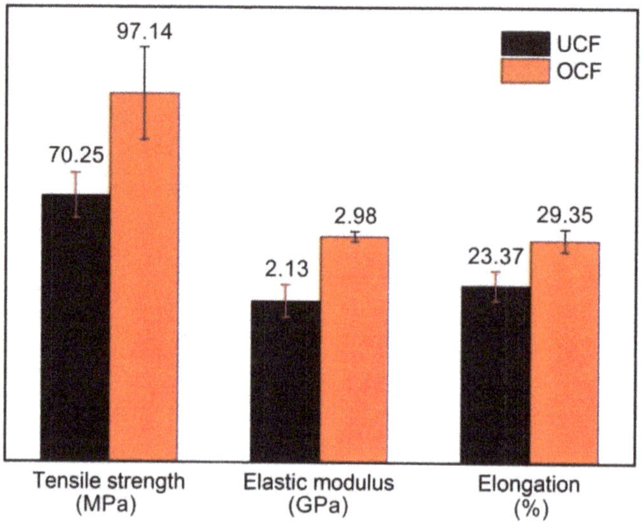

Figure 9. Mechanical properties of untreated and treated coir fibers under optimal conditions.

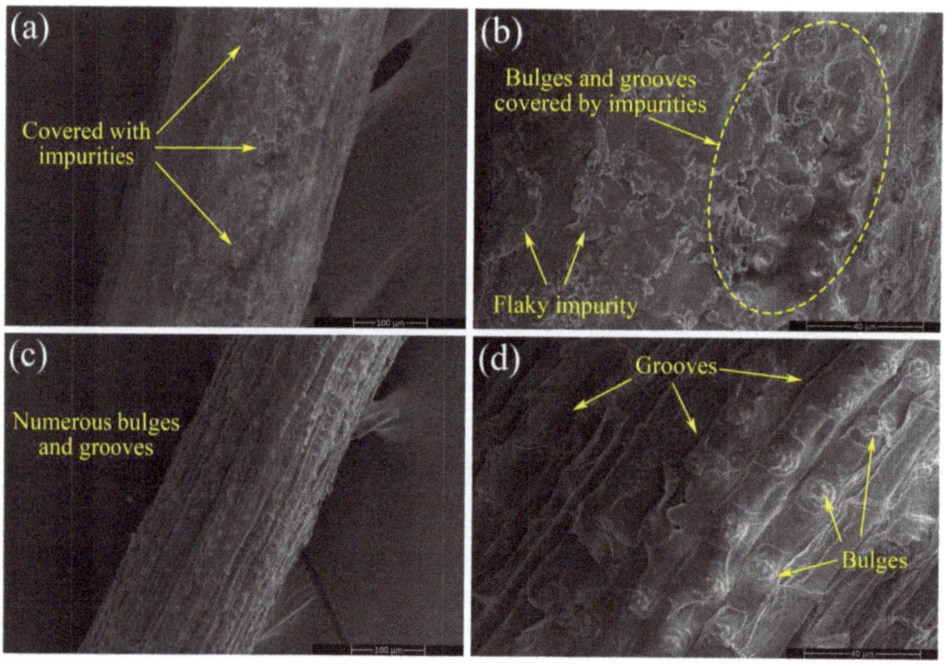

Figure 10. SEM micrographs of coir fiber surfaces: (**a**) untreated fiber (200×), (**b**) untreated fiber (800×), (**c**) treated fiber under optimal conditions (200×) and (**d**) treated fiber under optimal conditions (800×).

Figure 11a,b shows SEM micrographs of UCF cross-sections at different magnifications. Figure 11c,d show SEM micrographs of OCF cross-sections at different magnifications. When the micrographs of UCF and OCF are compared, some differences can be found. It is more obvious that the cross-section of UCF is relatively flat, while that of OCF shows a fine burr-like appearance, due to the erosion by alkali treatment. After the treatment, the non-cellulosic substances in the pores are removed, the coir fiber becomes soft, and a large number of fiber cells contained in them are partially deformed. Additionally, more gaps are observed inside the OCF cross-section compared with UCF, attributable to the network-like association between the deformed fiber cells. These changes improve the elongation of coir fiber [18]. Thus, it can be seen that alkali treatment under appropriate conditions can make the advantageous properties of coir fibers more prominent.

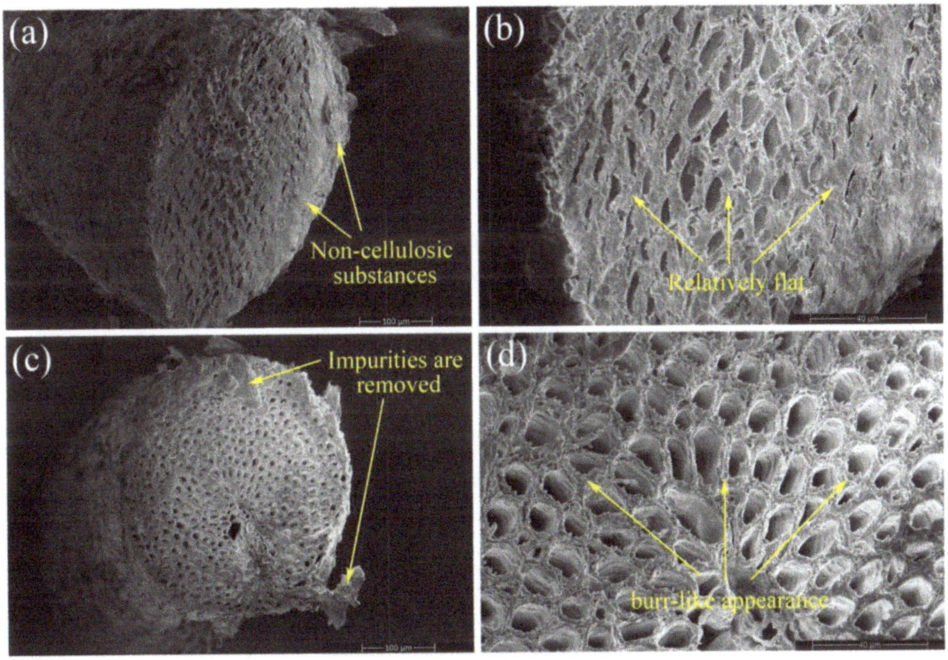

Figure 11. SEM micrographs of coir fiber cross-sections: (**a**) untreated fiber (200×), (**b**) untreated fiber (800×), (**c**) treated fiber under optimal conditions (200×) and (**d**) treated fiber under optimal conditions (800×).

3.3. Thermogravimetric Analysis Results

The thermogravimetric analysis (TGA) curves and the derivative of thermogravimetric (DTG) curves for UCF and OCF are presented in Figure 12. The thermal decomposition of natural fibers is shown in three phases. The first phase is the evaporation of water, the second is the decomposition of hemicellulose, pectin and part of cellulose, and the third is the decomposition of cellulose. It is more difficult for lignin to decompose, and its decomposition is usually considered as part of the whole process. The final residue of the test is ash, and the ash residue of coir fiber after alkali treatment increases from 25.54% to 37.41% at 500 °C. The DTG curve of UCF has three distinct peaks at 47.84 °C, 297.45 °C and 366.68 °C, corresponding to three weight losses in the TGA curve of UCF, which are 6.74%, 24.11% and 34.94%, respectively. The DTG curve of OCF has two distinct peaks at 53.63 °C and 322.58 °C, corresponding to two weight losses in the TGA curve of OCF, which are 4.70% and 45.47%, respectively. The comparison reveals that OCF is missing the second phase of fiber thermal decomposition due to the removal of hemicellulose, lignin and pectin from coir fiber after alkali treatment [28], which is consistent with the results of SEM.

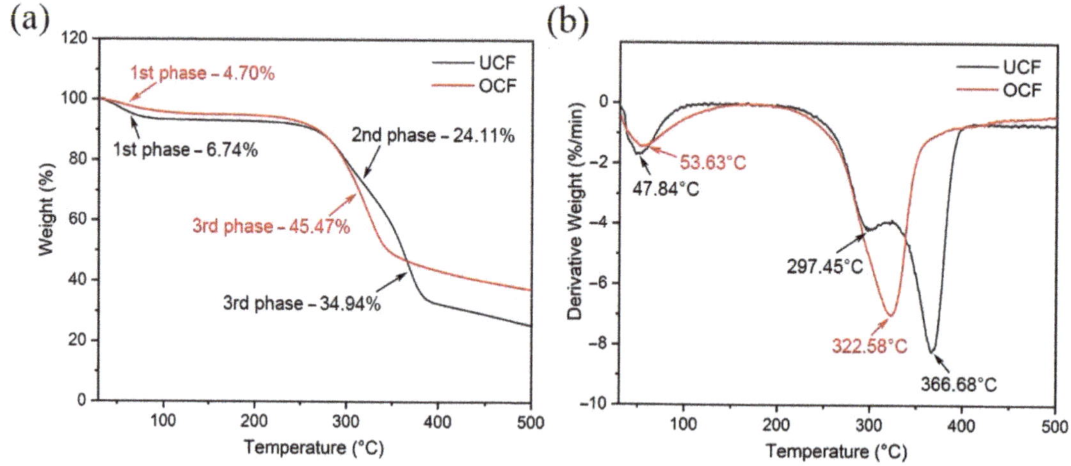

Figure 12. (**a**) The TGA curves of untreated (black) and treated (red) coir fibers under optimal conditions. (**b**) The DTG curves of untreated (black) and treated (red) coir fibers under optimal conditions.

3.4. Fourier-Transform Infrared Spectroscopy Analysis

Both UCF and OCF samples were analyzed using FTIR. Figure 13a shows the FTIR spectra of coir fiber samples from 400 cm^{-1} to 4000 cm^{-1}. It can be seen that the alkali treatment performed under optimal conditions has a significant effect on the FTIR spectra of coir fiber. After these results are analyzed, it is clear that some peaks have changed. These regions are magnified for easy observation and analysis, as shown in Figure 13b. The positions of these absorption peaks and the functional groups to which they belong are listed in Table 7. The intensity of the absorption peak located at 897 cm^{-1} is enhanced, and this is related to the C-H rocking vibration in the cellulose. This indicates that the percentage of cellulose in the samples increases after the removal of some non-cellulosic substances from the OCF [12]. The intensity of the absorption peak at 1248 cm^{-1} decreases obviously, and its shape changes. The absorption peak is related to the C-O-C stretching in lignin, indicating that most of the lignin is removed from the OCF obtained after treatment compared with the UCF [31,34]. The intensity of the absorption peak at 1376 cm^{-1} decreases slightly, and its change is caused by C-H bending vibration, which can be ascribed to the removal of lignin from OCF [35]. The intensity of the absorption peak at 1608 cm^{-1} is slightly weakened, related to the C=C stretching vibration in aromatic lignin. This further explains the decrease of the lignin component content in OCF [7]. In addition, the absorption peak at 1734 cm^{-1} disappears, which is related to the C=O stretching vibration in hemicellulose. This indicates that most of the hemicellulose in coir fibers is degraded after alkali treatment under optimal conditions [36]. From the above changes in FTIR spectra characteristics, it can be concluded that coir fiber alkali treatment has a positive effect on the removal of hemicellulose and lignin.

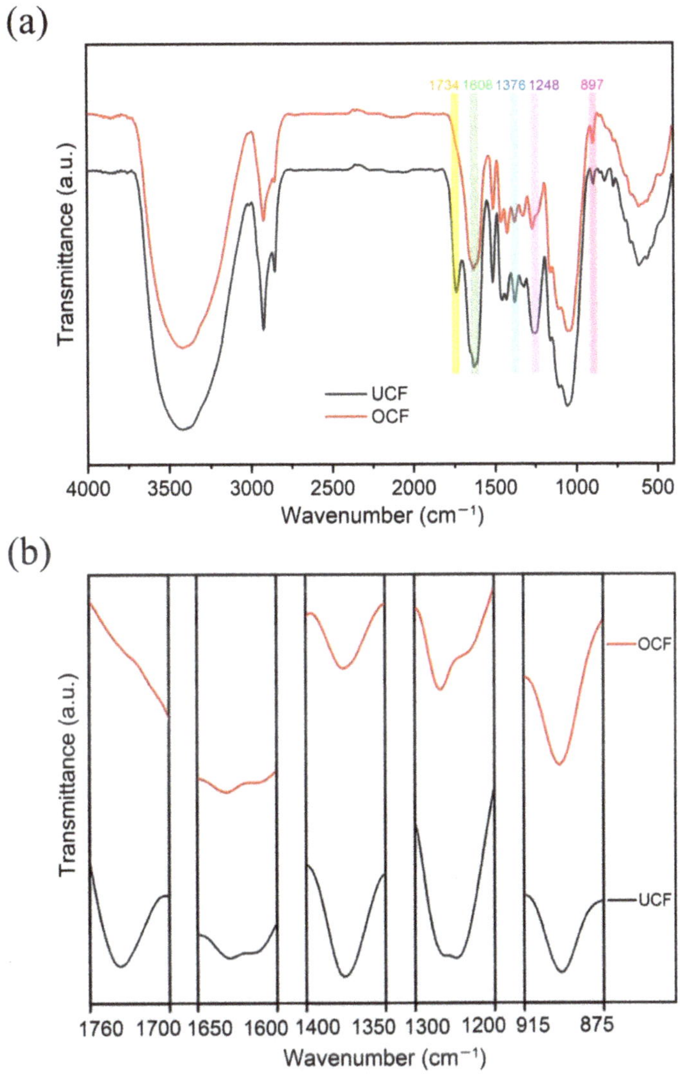

Figure 13. (**a**) FTIR spectra of untreated (black) and treated (red) coir fibers under optimal conditions. (**b**) Enlarged view of labeled regions in FTIR spectra.

Table 7. Peak value of change in FTIR spectrum.

Wavenumbers (cm^{-1})		**Functional Groups**
UCF	**OCF**	
1737	/	C=O
1608	1608	C=C
1376	1376	C-H
1248	1248	C-O-C
897	897	C-H

3.5. X-ray Diffraction Analysis

To investigate changes in coir fiber after treatment, XRD analysis was performed on both samples. Their X-ray diffractograms are shown in Figure 14. It can be seen from the figure that the shapes of the XRD spectra of the two samples are almost the same, with obvious peaks near 16°, 22° and 35°. The peak located around 16° is related to the overlap of the 101 plane and $10\bar{1}$ plane of cellulose, the one located around 22° is mainly related to the 002 plane, and the one located around 35° is related to the 040 plane. This indicates that the cellulose I crystalline structure of coir fiber did not change after alkali treatment. These phenomena were also mentioned by Ma et al. and Wu et al. in the study on alkali treatment of natural fibers [12,17]. In addition, some information can be obtained by comparing the two curves in the diffractograms. The peak of the XRD spectrum of OCF is more obvious, which shows that the crystallinity index of OCF is higher than that of UCF [28]. To verify this conclusion, the crystallinity indexes of the two samples were calculated based on the Segal empirical method. The crystallinity indexes of UCF and OCF are shown in Table 8, which are 35.85% and 40.40%, respectively. This change may be attributed to the large presence of lignin and hemicellulose in UCF and the removal of some amorphous components in OCF, such as hemicellulose and lignin. This increases the relative content of crystalline cellulose, resulting in cellulose crystal to accumulate and cellulose molecules to rearrange [28,37,38]. Therefore, OCF exhibits a higher crystallinity index than UCF, which also suggests that OCF has better mechanical properties than UCF.

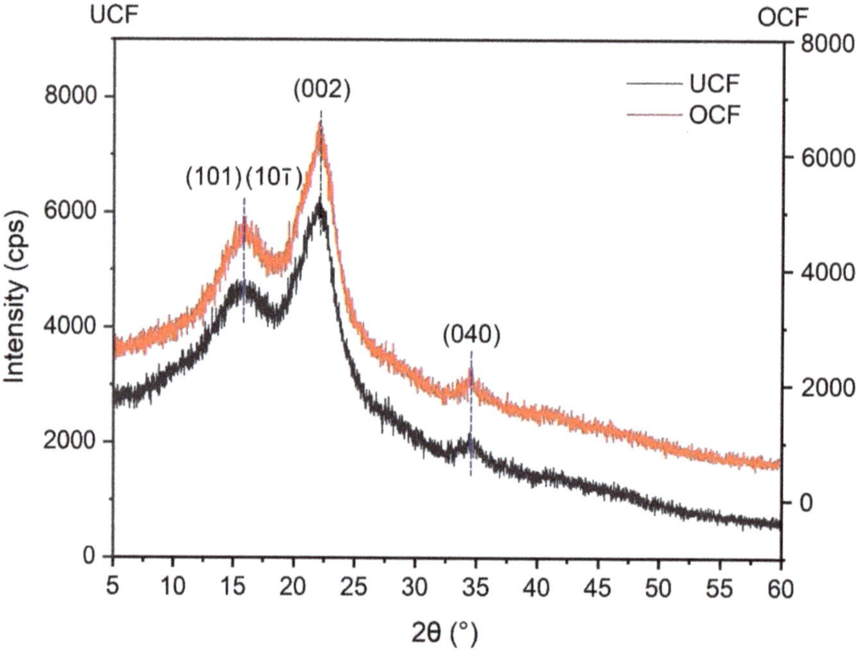

Figure 14. XRD spectra of untreated (black) and treated (red) coir fibers under optimal conditions.

Table 8. Crystallinity index values of coir fibers.

Crystallinity Index (%)	
UCF	OCF
35.85	40.40

4. Conclusions

In this study, the effects of NaOH concentration, time and temperature of alkali treatment on the tensile strength, elastic modulus and elongation of coir fibers were explored based on a research protocol developed by the Box-Behnken design method, and the multi-objective optimization and analysis of the tensile strength, elastic modulus and elongation of coir fiber were completed. As a result, the mechanical properties of coir fiber were comprehensively improved. The necessity of optimization of alkali treatment conditions was proved, which provided a reference for the selection of alkali treatment conditions of coir fiber. The following conclusions are drawn from the study.

Three factors, NaOH concentration, treatment time and treatment temperature, have a significant effect on the mechanical properties of coir fibers. No matter whether the factors are too large or too small, they will be detrimental to the performance of coir fibers.

The optimal alkali treatment conditions are obtained by software analysis at a 4.12% NaOH concentration, 15.08 h of treatment time, and a treatment temperature at 34.21 °C. Under these optimal conditions, the tensile test results of coir fiber show that the tensile strength is 97.14 MPa, the elastic modulus is 2.98 GPa and the elongation is 29.35%, which are consistent with the predicted values.

Compared with the untreated fiber, the coir fiber treated under the best conditions has better tensile strength, elastic modulus and elongation. This is consistent with the analysis results of TGA-DTG, the removal of non-cellulosic substances shown by FTIR and the improvement of crystallinity shown by XRD.

The changes in fiber surface before and after treatment as shown by SEM indicate that appropriate alkali treatment can enhance the bonding between materials.

The optimized combination of this study can be applied to fiber products that require both strength and toughness. In future research, the fibers treated under the optimal conditions are added to the composites instead of other fibers to further improve both strength and toughness of the composites, which is vital to study fiber-reinforced composites.

Author Contributions: Conceptualization, S.R. and C.Z.; methodology, C.Z.; formal analysis, C.Z. and S.Y.; investigation, S.R. and C.Z.; resources, S.R. and S.Y.; writing—original draft preparation, C.Z.; writing—review and editing, S.R. and S.Y.; supervision, S.Y.; project administration, S.R. All authors have read and agreed to the published version of the manuscript.

Funding: This research was funded by the Hainan Provincial Natural Science Foundation of China (521RC494, 521QN0869 and 520RC536).

Data Availability Statement: Not applicable.

Acknowledgments: The authors acknowledge the Hainan Provincial Natural Science Foundation of China for supporting this study.

Conflicts of Interest: The authors declare no conflict of interest.

References

1. Mishra, L.; Basu, G.; Samanta, A.K. Effect of chemical softening of coconut fibres on structure and properties of its blended yarn with jute. *Fibers Polym.* **2017**, *18*, 357–368. [CrossRef]
2. Haque, M.M.; Hasan, M.; Islam, M.S.; Ali, M.E. Physico-mechanical properties of chemically treated palm and coir fiber reinforced polypropylene composites. *Bioresour. Technol.* **2009**, *100*, 4903–4906. [CrossRef] [PubMed]
3. Rocky, B.P.; Thompson, A.J. Production and modification of natural bamboo fibers from four bamboo species, and their prospects in textile manufacturing. *Fibers Polym.* **2020**, *21*, 2740–2752. [CrossRef]
4. Chonsakorn, S.; Srivorradatpaisan, S.; Mongkholrattanasit, R. Effects of different extraction methods on some properties of water hyacinth fiber. *J. Nat. Fibers* **2019**, *16*, 1015–1025. [CrossRef]
5. Moria, S.; Charcaa, S.; Floresa, E.; Savastano, H., Jr. Physical and Thermal Properties of Novel Native Andean Natural Fibers. *J. Nat. Fibers* **2019**, *18*, 475–491. [CrossRef]
6. Santos, J.C.D.; Siqueira, R.L.; Vieira, L.M.G.; Freire, R.T.S.; Mano, V.; Panzera, T.H. Effects of sodium carbonate on the performance of epoxy and polyester coir-reinforced composites. *Polym. Test.* **2018**, *67*, 533–544. [CrossRef]

7. Hasan, K.M.F.; Horváth, P.G.; Kóczán, Z.; Le, D.H.A.; Bak, M.; Bejó, L.; Alpár, T. Novel insulation panels development from multilayered coir short and long fiber reinforced phenol formaldehyde polymeric biocomposites. *J. Polym. Res.* **2021**, *28*, 467. [CrossRef]
8. Saikrishnan, G.; Sumesh, K.R.; Vijayanand, P.; Madhu, S.; Nagarajan, S.; Suganya, P.G. Investigation on the mechanical properties of ramie/kenaf fibers under various parameters using GRA and TOPSIS methods. *Polym. Compos.* **2022**, *43*, 130–143.
9. Sumesh, K.R.; Kavimani, V.; Gopal, P.M.; Petr, S.; Ajithram, A.; Suganya, P.G. The effect of various composite and operating parameters in wear properties of epoxy-based natural fiber composites. *J. Mater. Cycles Waste* **2022**, *24*, 667–679.
10. Ma, Y.; Wu, S.; Zhuang, J.; Tong, J.; Qi, H. Tribological and physio-mechanical characterization of cow dung fibers reinforced friction composites: An effective utilization of cow dung waste. *Tribol. Int.* **2019**, *131*, 200–211. [CrossRef]
11. Yan, L.; Chouw, N.; Huang, L.; Kasal, B. Effect of alkali treatment on microstructure and mechanical properties of coir fibres, coir fibre reinforced-polymer composites and reinforced-cementitious composites. *Constr. Build. Mater.* **2016**, *112*, 168–182. [CrossRef]
12. Wu, J.; Du, X.; Yin, Z.; Xu, S.; Xu, S.; Zhang, Y. Preparation and characterization of cellulose nanofibrils from coconut coir fibers and their reinforcements in biodegradable composite films. *Carbohyd. Polym.* **2019**, *211*, 49–56. [CrossRef]
13. Zaman, H.U.; Beg, M.D.H. Preparation, structure, and properties of the coir fiber/polypropylene composites. *J. Compos. Mater.* **2014**, *48*, 3293–3301. [CrossRef]
14. Wang, F.; Lu, M.; Zhou, S.; Lu, Z.; Ran, S. Effect of fiber surface modification on the interfacial adhesion and thermo-mechanical performance of unidirectional epoxy-based composites reinforced with bamboo fibers. *Molecules* **2019**, *24*, 2682. [CrossRef]
15. Hestiawan, H.; Jamasri; Kusmono. Effect of chemical treatments on tensile properties and interfacial shear strength of unsaturated polyester/fan palm fibers. *J. Nat. Fibers* **2018**, *15*, 762–775. [CrossRef]
16. Loong, M.L.; Cree, D. Enhancement of mechanical properties of bio-resin epoxy/flax fiber composites using acetic anhydride. *J. Polym. Environ.* **2018**, *26*, 224–234. [CrossRef]
17. Ma, Y.; Wu, S.; Zhuang, J.; Tian, Y.; Tong, J. The effect of lignin on the physicomechanical, tribological, and morphological performance indicators of corn stalk fiber-reinforced friction materials. *Mater. Res. Express* **2019**, *6*, 105325. [CrossRef]
18. Valášek, P.; Müller, M.; Šleger, V.; Kolář, V.; Hromasová, M.; D'Amato, R.; Ruggiero, A. Influence of alkali treatment on the microstructure and mechanical properties of coir and abaca fibers. *Materials* **2021**, *14*, 2636. [CrossRef]
19. Frącz, W.; Janowski, G.; Bąk, Ł. Influence of the alkali treatment of flax and hemp fibers on the properties of PHBV based biocomposites. *Polymers* **2021**, *13*, 1965. [CrossRef]
20. Bensalah, H.; Raji, M.; Abdellaoui, H.; Essabir, H.; Bouhfid, R. Thermo-mechanical properties of low-cost "green" phenolic resin composites reinforced with surface modified coir fiber. *Int. J. Adv. Manuf. Technol.* **2021**, *112*, 1917–1930. [CrossRef]
21. Silva, G.G.; Souza, D.A.D.; Machado, J.C.; Hourston, D.J. Mechanical and thermal characterization of native Brazilian coir fiber. *J. Appl. Polym. Sci.* **2000**, *76*, 1197–1206. [CrossRef]
22. Verma, S.; Midha, V.K.; Choudhary, A.K. Multi-objective optimization of process parameters for lignin removal of coir using TOPSIS. *J. Nat. Fibers* **2020**, *19*, 256–268. [CrossRef]
23. Segal, L.; Creely, J.J.; Martin, A.E.; Conrad, C.M. An empirical method for estimating the degree of crystallinity of native cellulose using the X-ray diffractometer. *Text. Res. J.* **1959**, *29*, 786–794. [CrossRef]
24. Hassan, M.Z.; Sapuan, S.M.; Roslan, S.A.; Aziz, S.A.; Sarip, S. Optimization of tensile behavior of banana pseudo-stem (Musa acuminate) fiber reinforced epoxy composites using response surface methodology. *J. Mater. Res. Technol.* **2019**, *8*, 3517–3528. [CrossRef]
25. Pandit, P.; Samanta, K.K.; Teli, M.D. Optimization of atmospheric plasma treatment parameters for hydrophobic finishing of silk using Box Behnken design. *J. Nat. Fibers* **2020**, *19*, 463–474. [CrossRef]
26. Aly, M.; Hashmi, M.S.J.; Olabi, A.G.; Benyounis, K.Y.; Messeiry, M.; Hussain, A.I.; Abadir, E.F. Optimization of alkaline treatment conditions of flax fiber using Box–Behnken method. *J. Nat. Fibers* **2012**, *9*, 256–276. [CrossRef]
27. Manzato, L.; Takeno, M.L.; Pessoa-Junior, W.A.G.; Mariuba, L.A.M.; Simonsen, J. Optimization of Cellulose Extraction from Jute Fiber by Box-behnken Design. *Fiber. Polym.* **2018**, *19*, 289–296. [CrossRef]
28. Jiang, Y.; Deng, P.; Jing, L.; Zhang, T. Tensile properties and structure characterization of palm fibers by alkali treatment. *Fibers. Polym.* **2019**, *20*, 1029–1035. [CrossRef]
29. Shrivastava, R.; Parashar, V. Effect of alkali treatment on tensile strength of epoxy composite reinforced with coir fiber. *Polym. Bull.* **2022**, *1*, 1–13. [CrossRef]
30. Ru, S.; Zhao, C.; Yang, S.; Liang, D. Effect of coir fiber surface treatment on interfacial properties of reinforced epoxy resin composites. *Polymers* **2022**, *14*, 3488. [CrossRef]
31. Pereira, J.F.; Ferreira, D.P.; Bessa, J.; Matos, J.; Cunha, F.; Araújo, I.; Silva, L.F.; Pinho, E.; Fangueiro, R. Mechanical performance of thermoplastic olefin composites reinforced with coir and sisal natural fibers: Influence of surface pretreatment. *Polym. Compos.* **2019**, *40*, 3472–3481. [CrossRef]
32. Manjula, R.; Raju, N.V.; Chakradhar, R.P.S.; Johns, J. Effect of thermal aging and chemical treatment on tensile properties of coir fiber. *J. Nat. Fibers* **2018**, *15*, 112–121. [CrossRef]
33. Freitas, R.R.M.D.; Carmo, K.P.D.; Rodrigues, J.D.S.; Lima, V.H.D.; Silva, J.O.D.; Botaro, V.R. Influence of alkaline treatment on sisal fibre applied as reinforcement agent in composites of corn starch and cellulose acetate matrices. *Plast. Rubber Compos.* **2021**, *50*, 9–17. [CrossRef]

34. Arrakhiz, F.Z.; El Achaby, M.; Kakou, A.C.; Vaudreuil, S.; Benmoussa, K.; Bouhfid, R.; Fassi-Fehri, O.; Qaiss, A. Mechanical properties of high density polyethylene reinforced with chemically modified coir fibers: Impact of chemical treatments. *Mater. Design* **2012**, *37*, 379–383. [CrossRef]
35. Ridzuan, M.J.M.; Majid, M.S.A.; Afendi, M.; Kanafiah, S.N.A.; Zahri, J.M.; Gibson, A.G. Characterisation of natural cellulosic fibre from pennisetum purpureum stem as potential reinforcement of polymer composites. *Mater. Design* **2016**, *89*, 839–847. [CrossRef]
36. Farias, J.G.G.D.; Cavalcante, R.C.; Canabarro, B.R.; Viana, H.M.; Scholz, S.; Simão, R.A. Surface lignin removal on coir fibers by plasma treatment for improved adhesion in thermoplastic starch composites. *Carbohyd. Polym.* **2017**, *165*, 429–436. [CrossRef]
37. Reddy, K.O.; Reddy, K.R.N.; Zhang, J.; Zhang, J.; Rajulu, A.V. Effect of alkali treatment on the properties of century fiber. *J. Nat. Fibers* **2013**, *10*, 282–296. [CrossRef]
38. Sangian, H.F.; Widjaja, A. The effect of alkaline concentration on coconut husk crystallinity and the yield of sugars released. In Proceedings of the IOP Conference Series: Materials Science and Engineering, Manado, Indonesia, 25–26 October 2018; Volume 306, p. 012046.

forests

MDPI

Article

Colour and Chemical Changes of Black Locust Wood during Heat Treatment

František Kačík [1,*], Ivan Kubovský [2], Jiří Bouček [3], Richard Hrčka [4], Milan Gaff [5,6] and Danica Kačíková [7]

1 Department of Chemistry and Chemical Technologies, Faculty of Wood Sciences and Technology, Technical University in Zvolen, T.G. Masaryka 24, 960 01 Zvolen, Slovakia
2 Department of Physics, Electrical Engineering and Applied Mechanics, Faculty of Wood Sciences and Technology, Technical University in Zvolen, T.G. Masaryka 24, 960 01 Zvolen, Slovakia
3 Department of Physics, Faculty of Engineering, Czech University of Life Sciences Prague, Kamýcka 129, 16500 Prague, Czech Republic
4 Department of Wood Science, Faculty of Wood Sciences and Technology, Technical University in Zvolen, T.G. Masaryka 24, 960 01 Zvolen, Slovakia
5 Department of Furniture, Design and Habitat Brno, Mendel University in Brno, 61300 Brno, Czech Republic
6 Experimental Centre, Faculty of Civil Engineering, Czech Technical University in Prague, Thakurova 7, 16629 Prague, Czech Republic
7 Department of Fire Protection, Faculty of Wood Sciences and Technology, Technical University in Zvolen, T.G. Masaryka 24, 960 01 Zvolen, Slovakia
* Correspondence: kacik@tuzvo.sk

Abstract: Black locust is a fast-growing deciduous tree species with multiple industrial purposes due to its valuable traits. However, the heterogeneity of colour distribution between sapwood and heartwood limits its application. Thermal modification is an environment-friendly technology for improving various wood properties, especially dimensional stability, decay resistance, and colour homogeneity. In this work, black locust (*Robinia pseudoacacia* L.) wood samples were thermally modified at temperatures of 160, 180, and 210 °C. Extractives and main wood components were analysed by wet chemical methods, colour was measured by spectrometry, and structural changes by Fourier transform infrared spectroscopy. The obtained results show that the darkening of black locust wood, unlike other wood species of the temperate zone, is mainly caused by changes in extractives. Their content decreases during thermal treatment, but new chromophores are formed, especially in quinones. Degradation of hemicelluloses and the partial degradation of cellulose also contribute to colour changes. At higher temperatures, condensation reactions can occur in lignin, leading to the formation of some chromophores. Statistical analysis confirmed that temperature can be considered a very significant factor affecting the colour of the wood surface.

Keywords: black locust wood; ThermoWood process; colour; extractives; chemical changes; FTIR spectroscopy

Citation: Kačík, F.; Kubovský, I.; Bouček, J.; Hrčka, R.; Gaff, M.; Kačíková, D. Colour and Chemical Changes of Black Locust Wood during Heat Treatment. *Forests* **2023**, *14*, 73. https://doi.org/10.3390/f14010073

Academic Editor: Mathieu Petrissans

Received: 2 December 2022
Revised: 26 December 2022
Accepted: 29 December 2022
Published: 30 December 2022

1. Introduction

Nowadays, wood modification is defined as a process implemented to improve the physical, mechanical, or aesthetic properties of sawn timber, veneer or wood particles used in the fabrication of wood composites, and other wood-based materials. This process produces a material that can be disposed of at the end of the life cycle of a product cycle without presenting any environmental hazards greater than those associated with the disposal or combustion of unmodified wood [1,2]. Thermal modification is an environment-friendly technology for improving various wood properties, especially dimensional stability, decay resistance, and colour homogeneity. Colour is one of the most important properties for consumers, and the probability of changing the natural colour of wood without chemicals is important for some markets. An attractive darker colour is an important advantage of heat-treated wood. The variation in colour of the heat-treated wood is influenced by the heterogeneous structure of wood and the different technologies that can be applied [3,4]. Moreover, colour can serve as a quality indicator for industrially manufactured ThermoWood [5].

Thermally modified wood becomes a darker brown colour; however, due to UV light, it turns silvery grey. In some cases, colour is a key factor in choosing a particular wood, as the aesthetic aspect often prevails in some end uses. Darkening can be an essential advantage of its heating, which gives the wood a more beneficial aspect in some territories where exotic wood is not found. The darker shade imparted to wood by heat treatment is due to the production of coloured degradation products from hemicelluloses [6] and extractives [7]. Patzelt et al. [8] suggested that colour change could also be used as a categorization method for thermally treated wood, as it has a considerable relationship with treatment intensity [9,10], both modulus of elasticity and bending strength [11], as well as the thermal process used. In addition, chemical, mechanical, and colour traits of thermally modified spruce wood showed close relationships, and therefore they can be used for their mutual predictions [12].

The essence of the colour change is complicated, as all the main wood components, including extractives, can contribute to the change. The colour of wood can be changed by modifying the wood structure's main components (cellulose, hemicellulose, and lignin), mainly due to heat, humidity, light, or UV radiation [13]. Matsuo et al. [14] reported that the darker colour of heat-treated wood was attributed to the formation of degradation products from hemicelluloses, changes in extractives, and the formation of oxidation products such as quinones. The colour of thermally modified wood also depends on its origin, and heating conditions including temperature, moisture content, and presence or absence of oxygen.

Black locust (*Robinia pseudoacacia* L.), a fast-growing deciduous tree species, is a kind of quality wood of multiple industrial purposes due to its valuable traits of stiffness, wear resisting, and high basic density. However, the heterogeneity of colour distribution between sapwood and heartwood limits its application [15,16].

Black locust is an economically important species as a fast-growing tree producing valuable timber. Today, its wood is used to make furniture, gardens, children's equipment, energy fuel, and as a source for biorefineries based on the production of soluble sugars and lignin [17,18]. This wood species is aboriginal to eastern North America. Its eastern range is centred on the Appalachian Mountains and extends from central Pennsylvania and southern Ohio to northeastern Alabama, northern Georgia, and northwest South Carolina. The western section of its native range includes parts of Missouri, Arkansas, and Oklahoma, and populations also exist in Indiana and Kentucky. Black locust was first introduced to Europe in the early 17th century, and since then, it has been widely introduced to temperate Asia, Australia, and New Zealand, northern and southern Africa, and temperate South America [19]. Black locust is widespread across Europe, occurring from Sicily in Italy to South Norway and longitudinally from the Portugal littoral regions up to the Caucasus. Currently, it occurs in 42 European countries, for example, 400,000 ha in Hungary, 200,000 ha in France, 250,000 ha in Romania, and 230,000 ha in Italy [20].

As the use of thermally treated wood continues to increase, many research works are focused on its physical, mechanical, and chemical changes, including colour ones. Unlike other trees of the temperate zone (spruce, pine, birch, aspen, ash, etc.), the changes in thermal treatment of black locust wood are not sufficiently investigated.

This study therefore aimed to investigate the effect of temperature on black locust chemical and colour changes during heat treatment for making products with desirable surface colour. Since this tree species is very easily available in the territory of Central Europe, thanks to this treatment, it is possible to replace exotic ones with it. Due to the fact that the heat treatment of wood leads to a deterioration of its strength, it is not recommended to use it for load-bearing and heavily mechanically stressed structural elements.

2. Materials and Methods

2.1. Samples Preparation

Black locust (*Robinia pseudoacacia* L.) wood specimens (200 mm × 100 mm × 20 mm; longitudinal × tangential × radial) were air conditioned (RH 65 ± 3% and temperature 20 ± 2 °C) to an equilibrium moisture content (EMC) of 12%. Forty samples were divided into four groups. Three groups were thermally modified according to the ThermoWood process using three peak temperatures for 3 h (160, 180, and 210 °C); the fourth group without modification was left as a reference and denoted as "20 °C" [21,22]. All samples were then mechanically disintegrated and milled using a POLYMIX PX-MFC 90D laboratory mill (Kinematica, Luzern, Switzerland) and dried (4 h at 103 ± 2 °C).

2.2. Chemical Analyses

The fraction with a particle size of 0.5–1.0 mm was extracted in the Soxhlet apparatus (Sigma-Aldrich, Munich, Germany) according to ASTM D1107-21 [23]. The lignin content was determined according to the National Renewable Energy Laboratory (NREL) procedure [24], cellulose according to the Seifert's method [25], and the holocellulose according to Wise et al. [26]. The hemicelluloses were determined by subtracting cellulose content from holocellulose.

2.3. Colour Measurement

The colour was measured using a benchtop-type Spectrophotometer CM-5 (Konica Minolta, Japan) with a wavelength resolution of 10 nm. The measurement spot diameter of 3 mm was selected on the top of the spectrophotometer according to dimensions of specimen's surfaces prior to colour measurement. The spectrophotometer was calibrated prior each measurement according to the procedure recommended by the producer. The white standard was built in the spectrophotometer. The whole calibration was guided by a computer using the Colour Data Software CM-S100W SpectraMagicTNNX (Konica Minolta, Osaka, Japan). The reflectance spectrum was measured under illuminant D65 with specular component included. The 2°-degree standard observer was set on prior to computation of colour coordinates. The spectrophotometer provided the three parameters of CIE $L^*a^*b^*$ colour space and two parameters of chromaticity diagram among others. The provided lightness L^*, parameter a^* and b^*, also served for computing polar coordinates of CIE $L^*a^*b^*$ colour space: saturation C_{ab} and hue h_{ab}. The change of colour was computed according to total colour difference formula:

$$\Delta E = \sqrt{\left(L_2^* - L_1^*\right)^2 + \left(a_2^* - a_1^*\right)^2 + \left(b_2^* - b_1^*\right)^2}.$$

where index of 1 denotes target reference (unmodified sample) and of 2 denotes sample (modified sample).

The reflectance spectra were converted into K/S spectra using the Kubelka–Munk equation according to [27]. The eight samples free of visual defects (cracks) were included in the measurement of each set. The colour measurement consisted of the 64 recordings for each modified and unmodified sample on both sides of the tangential surfaces. The sets were distinguished with selected modifying temperatures. Overall, the 256 measurements were performed.

2.4. FTIR Analysis

FTIR spectroscopy measurements of wood samples were performed on a Nicolet iS10 FT-IR spectrometer (Thermo Fisher Scientific, Waltham, MA, USA). The ATR crystal plate was from Diamond, and solid materials could be placed into intimate physical contact with the sampling area. The spectra were acquired by accumulating 32 scans at a spectral resolution of 4 cm^{-1} in a range from 4000 to 650 cm^{-1}. Results were evaluated by applying OMNIC 9.0 software (Thermo Fisher Scientific, Waltham, MA, USA). Each specimen was measured on the tangential face, and four measurements were performed for each sample.

2.5. Statistical Analysis

For evaluating the results, ANOVA analysis was used. Based on the P-level value and Fisher F tests, it was determined whether a factor affected the values of the monitored characteristics. Diagrams were constructed for the 95% confidence interval, reflecting the significance level of 0.05 ($p < 0.05$), and the results were verified with Duncan's tests.

3. Results and Discussion

3.1. Chemical Changes

The chemical composition of untreated black locust wood (Table 1) is like the results reported previously for different *Robinia* species, e.g., extractives 2.5%–8.3%, cellulose 40.6%–43.1%, hemicelluloses 16.18%–33.16%, lignin 19.73%–22.7% [28–31].

Table 1. Chemical composition of sound and thermally modified black locust wood, oven-dry weight percentages (mean ± SD).

Trait	Temperature			
	20 °C	160 °C	180 °C	210 °C
Extractives	9.29 ± 0.08	8.71 ± 0.29	8.38 ± 0.17	7.64 ± 0.15
Lignin	24.11 ± 0.06	23.94 ± 0.51	25.23 ± 0.14	28.82 ± 0.07
Holocellulose	68.46 ± 0.41	67.35 ± 0.20	63.97 ± 0.16	59.30 ± 0.17
Cellulose	42.50 ± 0.21	44.88 ± 0.08	47.63 ± 0.20	54.45 ± 0.28
Hemicelluloses	25.96 ± 0.20	22.47 ± 0.25	16.33 ± 0.34	4.85 ± 0.22

Some extractives decompose at a higher temperature, but new ones are created due to the decomposition of main wood components. During the heating of eucalypt wood, nearly all the original extractives decomposed, and new ones were created, such as anhydrosugars, mannosan, galactosan, and levoglucosan [32]. Only a few works investigated the effect of extractives on the colour changes of black locust wood, but even those did not report changes in the content of extractives during heat treatment. [15,27,33]. Generally, extractives' content changes depend on the wood species and heating conditions, and various trends have been published. A decrease in extractives was observed in our experiments (Table 1) and their influence on colour changes will be discussed below.

Thermal modification causes the decrease of polysaccharide content, primarily by the degradation of hemicelluloses. The amount of other wood biopolymers (lignin, cellulose) in modified wood increases (Table 1). Lignin yield generally increases at wood treatment at a low pH and high temperature. The content of acid-insoluble lignin is higher in the modified material than in the untreated ones. This trend is due to the formation of pseudo-lignin following the condensation reactions of degradation products of lignin and polysaccharides [34–36].

Cellulose begins to degrade at higher temperatures compared with hemicelluloses and lignin. However, its relative amount in heat-treated wood depends on the method used for the determination. In this work, the increase in cellulose content was observed during thermal modification, as determined by Seifert's method (Table 1). A similar trend in cellulose content was observed at the ThermoWood process of poplar wood by differential scanning calorimetry [37].

In this work, the most vulnerable main wood component during thermal treatment was hemicelluloses. Their amount decreases by 81.32% in modified wood at 210 °C, accompanied by a relative increase in cellulose (28.12%) and lignin (19.54%), respectively (Table 1).

3.2. Colour Changes with Statistical Evaluation

The colour of wood is a property with probabilistic character. Heat treatment of black locust wood caused colour changes. The lightness of the surface (L^*) decreased with increasing the treatment temperature, while the highest decrease was recorded on samples treated at 160 °C. The chromaticity values (a^* and b^*) recorded changes, which, together with the change in light, resulted in an increase in the total colour difference ΔE^* up to a value of approximately 42. These colour changes can be evaluated as very significant, since the colour changed from originally pale yellowish brown to dark brown.

The largest colour changes were recorded on samples treated at 160 °C (Figure 1, Table 2). The most significant change was shown by the lightness value L^*, where a decrease, compared with the reference sample (20 °C), of approximately 45% (from 69.5 to 38.0) was evident. Such a decrease in lightness represents a significant darkening of the surface of treated black locust wood. Chromaticity values a^* and b^* also changed significantly (Table 1). The a^* value increased by 100%, while the b^* value decreased by more than 58%. Thermal modification at 180 and 210 °C affected the colour of the surface less significantly. The decrease in the value of L^* compared with the samples treated at 160 °C was already much milder. The a^* and b^* values had a similar trend. On the samples treated at 180 °C, the further decrease in L^* was only 10%; the value of a^* decreased by 17% and b^* by 20%. Compared with the values measured on the samples treated at 180 °C, the samples treated at 210 °C showed a decrease in L^* by another 5%, the value of a^* decreased by 8%, and the value of b^* decreased by 10.5%. The listed changes in L^*, a^*, and b^* values represent a colour change of the wood surface from the original cream to dark brown (Figure 1). Our findings are analogous to those obtained for other heat-treated wood species [38–40].

Figure 1. Samples of the original and thermally modified black locust wood.

Table 2. The averages of black locust colour properties in CIE $L^*a^*b^*$ colour space (mean ± SD).

Temperature	L^*	a^*	b^*	ΔE
20 °C	69.5 ± 1.96	3.7 ± 0.63	28.0 ±0.98	0.0 ± 0.00
160 °C	38.0 ± 1.87	7.4 ± 0.51	11.6 ± 1.41	35.7 ± 3.27
180 °C	34.0 ± 1.16	6.1 ± 0.58	9.2 ± 1.43	40.3 ± 2.09
210 °C	32.2 ± 0.95	5.6 ± 0.46	8.2 ± 1.25	42.3 ± 4.57

The measured colour data were visualized with the proposed method of Hrčka [41] (Figure 2). The brown crosses represent the measured data in CIE $L^*a^*b^*$ colour space. The black and RGB ellipses are projections of data clouds to CIE $L^*a^*b^*$ planes. The black ellipses represent the colour of unmodified samples (20 °C). The red ellipses represent the colour of modified samples at 160 °C. The green ellipses represent the colour of modified samples at 180 °C. The blue ellipses represent the colour of modified samples at 210 °C. The alpha level of 0.1% was selected, because of the large variability of wood colour data. Despite the alpha-level low value, the observed lightness of modified wood decreased dramatically. The average L^* value of 70 at 20 °C decreased to 38 on longitudinal surfaces modified with a temperature of 160 °C. The significant drop of the b^* coordinate average was also evident on longitudinal surfaces of unmodified and modified samples. The average b^*

value of 28 at 20 °C decreased to 12 on longitudinal surfaces modified with a temperature of 160 °C. The *a** value increased with increasing temperature between 20 °C and 160 °C. The additional increasing of temperature resulted in significant decreasing of the *a** coordinate. The modifying of the *a** coordinate with temperature followed the characteristic behaviour observed also in the research of Tolvaj et al. [42] and Banadics et al. [43]. The hue was a less variable property than saturation. Both properties decreased with the increase in the modifying temperature (Table 2). The significant difference of wood colours indicates the possibility to prepare the interesting changes of black locust colour from almost-saturated yellow green to less-saturated dark brown.

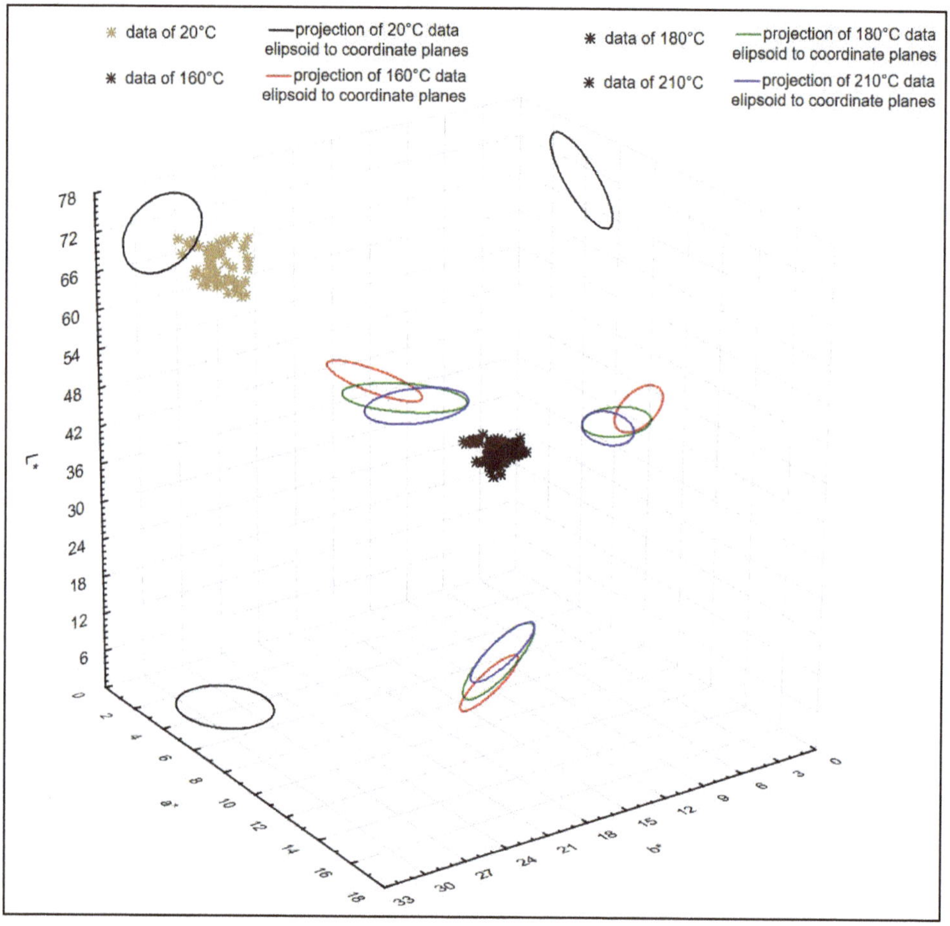

Figure 2. The measured colour data of black locust and their projections to coordinate planes of CIE *L*a*b**.

The statistical analysis of the variance evaluating the impact of different temperature values on the parameters *L**, *a**, *b**, and Δ*E* (Tables 3–6) show that in all observed cases temperature can be considered a very significant factor affecting colour parameters.

Table 3. Statistical evaluation of the factors affecting the change in the L^* value.

Effect	Sum of Squares	Degree of Freedom	Variance	Fisher's F-Test	Significance Level p
Intercept	482,613.3	1	482,613.3	200,991.5	***
Thermal modification	59,316.7	3	19,772.2	8234.4	***
Error	605.1	252	2.4		

NS—not significant, ***—significant, $p < 0.005$.

Table 4. Statistical evaluation of the factors affecting the change in the a^* value.

Effect	Sum of Squares	Degree of Freedom	Variance	Fisher's F-Test	Significance Level p
Intercept	8251.33786	1	8251.33786	27,279.9329	***
Thermal modification	435.647217	3	145.215739	480.101007	***
Error	76.2222234	252	0.302469141		

NS—not significant, ***—significant, $p < 0.005$.

Table 5. Statistical evaluation of the factors affecting the change in the b^* value.

Effect	Sum of Squares	Degree of Freedom	Variance	Fisher's F-Test	Significance Level p
Intercept	51,978.5851	1	51,978.5851	31,701.9997	***
Thermal modification	16,479.5435	3	5493.18116	3350.31873	***
Error	413.179092	252	1.63959957		

NS—not significant, ***—significant, $p < 0.005$.

Table 6. Statistical evaluation of the factors affecting the change in the ΔE value.

Effect	Sum of Squares	Degree of Freedom	Variance	Fisher's F-Test	Significance Level p
Intercept	224,333.835	1	224,333.835	47,625.563	***
Thermal modification	76,244.1131	3	25,414.7044	5395.48395	***
Error	1187.01224	252	4.71036604		

NS—not significant, ***—significant, $p < 0.005$.

A graphic presentation of the results describing the interaction of different temperature values on the colour parameter is shown in Figure 1, Table 2. A significant decrease in L^* values caused by the effect of a temperature of 160 °C compared with the untreated samples can be observed. A further increase in temperature (180 and 210 °C) caused only a slight decrease in the lightness value.

In the case of parameter a^*, we observed the opposite trend (Table 2). Because of the temperature of 160 °C, a statistically significant increase in the values of the parameter a^* can be observed in comparison with the untreated test specimens (20 °C). Due to the influence of higher temperatures (180 °C and 210 °C), a statistically significant decrease in parameter values can be observed compared with the values measured in bodies treated with a temperature of 160 °C.

The parameter b^*, as well as the whole curve, have an identical course with the curve describing the values of parameter L^* (Table 2). Even in this case, we can observe a significant drop in values caused by the effect of the temperature of 160 °C compared with untreated ones. As a result of the influence of external temperatures (180 °C and 210 °C), this decrease increased statistically significantly, but the difference between the values was already significantly lower.

When looking at the ΔE parameters, its significant increase caused by the effect of increasing temperature values is evident in all observed cases (Table 2).

Chemical reactions during heat treatment often cause the formation of coloured oxidation and degradation products. These changes are complex and require detailed study.

Chen et al. [27,33] reported that the effect of degradation of hemicellulose on colour changes at thermal treatment of black locust wood should be relatively small compared with lignin. Furthermore, other authors discovered a good correlation between lightness decrease changes and chemical changes in glucose, hemicelluloses, and lignin [32]. In this work, no good correlation between colour changes and contents of main wood components were observed. Black locust wood contains a large content of extractives, and we assume that their changes during thermal treatment can significantly affect colour changes, as observed by several authors [15,27,33] during the thermal treatment of extracted and non-extracted black locust wood. This is supported by the decrease in the yellow index b^* (Table 2); a similar trend was discovered at the heat treatment of black locust wood [15]. On the other hand, the b^* parameter indicated increased values compared with untreated timber at the thermal treatment of ailanthus, spruce, and pine wood [6,39].

The colour of any material depends on its chemical composition, namely, a combination of chromophores that absorb certain wavelengths of the visible light, in such a way, altering the spectral composition of the reflected radiation, which, when entering the human eye, creates a sense of definite colour. Therefore, the reflection curve in the visible light region is an accurate representation of the colour of a material [44,45].

It is obvious that there was an absorption change in visible region (Figure 3) and the difference in absorption spectra (Figure 4) can represent changes due to heat treatment. These changes indicated the generation of some types of chromophores in the degradation, condensation, and oxidation processes. They can be assigned to the formation of coloured quinoid compounds, arising from the degradation and oxidation of the aromatic hydroxyls of lignin and aromatic extractives upon heat treatment. Regions between 420 and 160 nm can be caused mainly by *p*-quinone, and the region between 500 and 580 nm can be caused mainly by *o*-quinone [27,46–48].

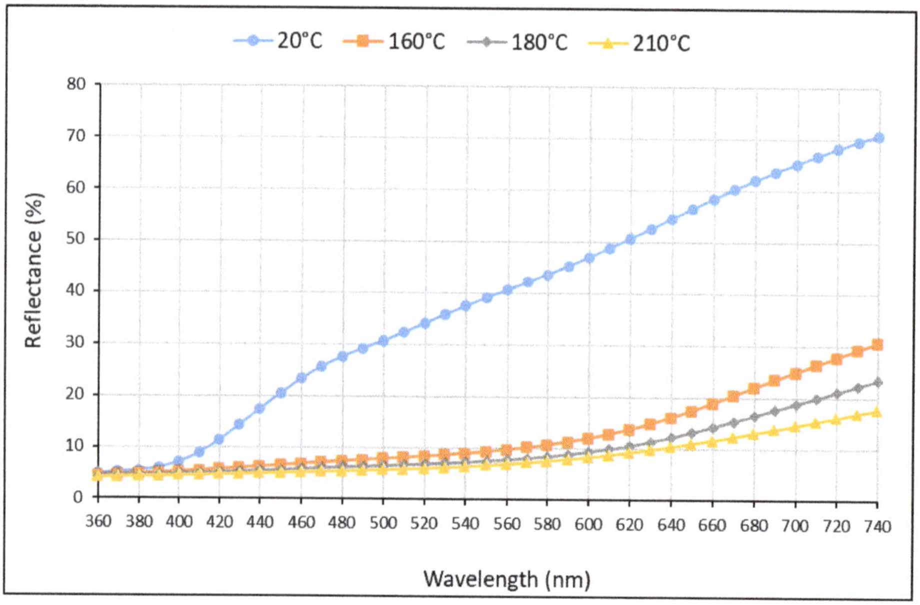

Figure 3. VIS reflectance spectra of untreated and heat-treated black locust wood.

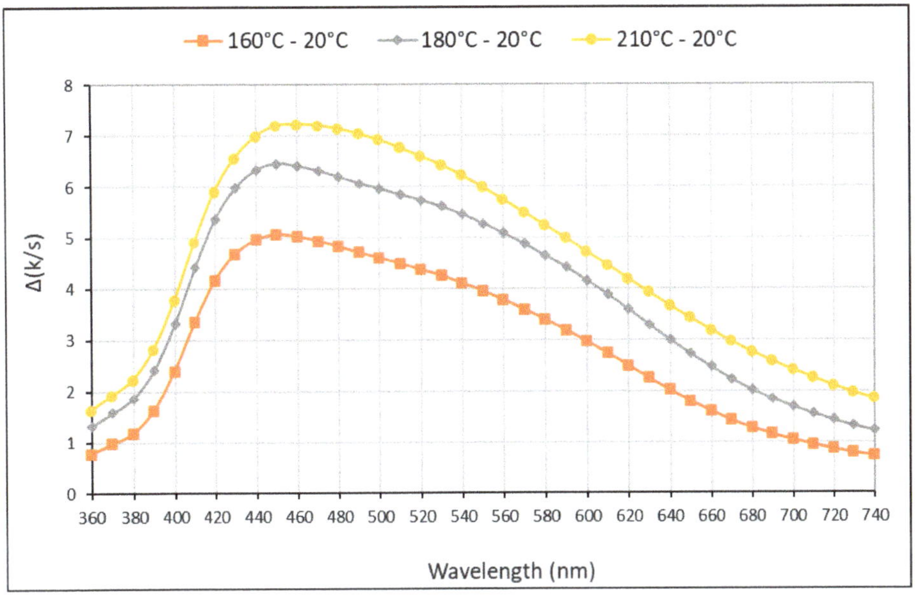

Figure 4. Absorption difference spectra Δ(k/s) of heat-treated black locust wood.

3.3. Changes in FTIR Spectra

FTIR spectra (Figure 5) were measured on the surface of untreated wood (20 °C, considered as a reference) and subsequently after heat treatment (at 160, 180, and 210 °C).

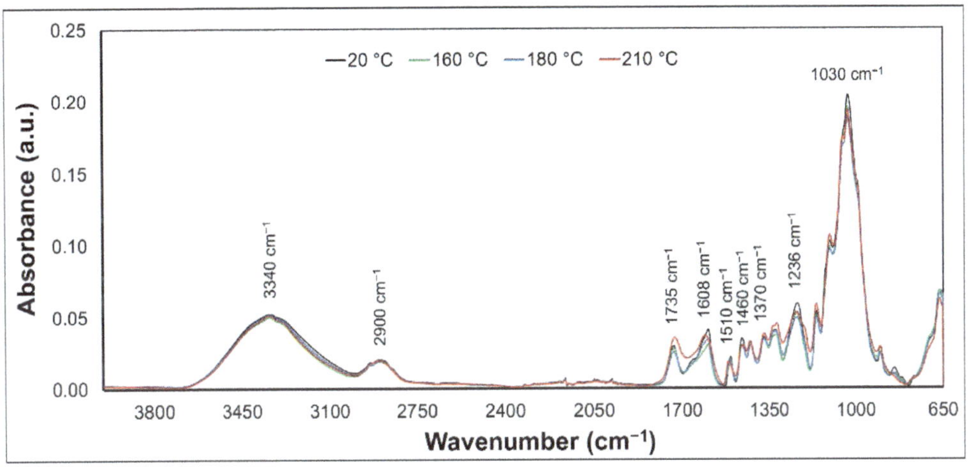

Figure 5. FTIR spectra of untreated and thermally treated black locust wood.

The height of each bar in differential FTIR spectra (Figure 6) represents the difference between the absorbance of a given band of the thermally treated sample and that of the reference sample for a given wavenumber. Positive values mean an increase, and negative values mean a decrease in absorbance.

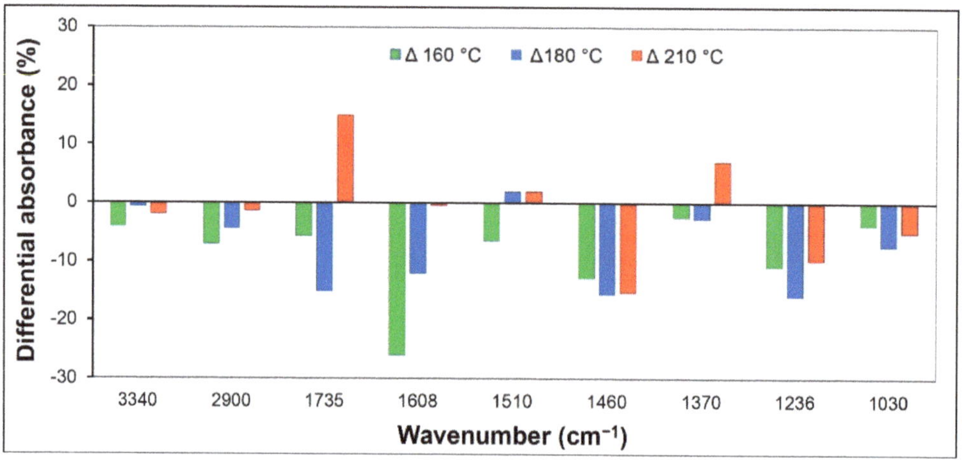

Figure 6. Differential FTIR spectra of untreated and thermally treated black locust wood.

It follows from the obtained spectral dependences that the band changes between 3000 and 3600 cm^{-1} show the decrease, which is in line with the observations of Chen et al. [27]. This widest band is due to O—H vibrations in lignin and hemicelluloses [49]. The band with a maximum absorbance at 2900 cm^{-1} is due to symmetric and asymmetric C—H stretching vibrations in aliphatic compounds. Inari et al. [50] show a similar behaviour. More evident changes were observed in bands in the region from 1800 to 800 cm^{-1} (bands are assigned to tensile and deformational vibrations of all wood components). This area is characteristic for wooden materials. At the band 1735 cm^{-1} (C=O stretching in unconjugated carbonyl groups) [51], a decrease in absorbance was initially recorded (samples treated at 160 and 180 °C). The increase was observed only at the highest temperature (210 °C). Analogous statements have been reported by Lagaňa et al. [52] about the thermal modification of beech wood. The drop at the early stage of the heat treatment may be due to the breaking of acetyl or acetoxy groups in xylan due to the lowest thermal stability of hemicelluloses compared with cellulose and lignin.

The increased intensity of these bands with longer thermal treatment may be due to an increase in carbonyl or carboxyl groups in lignin or in carbohydrates by oxidation [53]. In our case, the decrease in the band can be mainly caused by the deacetylation of hemicelluloses during the thermal treatment of the wood surface components [54]. Thermally caused processes are related with the degradation of carbonyl groups in lignin as well as in hemicelluloses, where the cleavage of C=O bonds affects the change in the content of chromophores. These structures are particularly responsible for the colouring of the wood surface. The band in the vicinity of 1600 cm^{-1} (C=C stretching vibration conjugated with aromatic ring in lignin) decreased continuously, while the lowest decrease occurred at 210 °C. Slight changes were noted by the band at approximately 1500 cm^{-1} (C=C stretching vibration and aromatic skeleton vibration). A decrease, practically independent of temperature, was characteristic of the band at 1460 cm^{-1} (asymmetric CH_3 bending in methoxyl groups in lignin). At 1370 cm^{-1} (symmetric and asymmetric CH_3 bending) [55], the absorbance also decreased (except for a temperature of 210 °C, where the opposite trend was observed). The band at 1236 cm^{-1} is assigned to lignin and hemicelluloses [56,57]. Its decrease was the lowest at the highest temperature (210 °C). The decrease in the absorbance of the analysed wavenumber areas supports the assumption of degradation processes in lignin, caused by the splitting of bonds and the subsequent disintegration of its structure. At higher temperatures, condensation reactions can occur in lignin [58]. The permanent decrease in the band at 1030 cm^{-1} (C—O deformation vibrations in cellulose) points to the degradation changes of cellulose [59].

4. Conclusions

This work presents the results of a study focused on the colour and chemical changes of black locust wood caused by thermal degradation. Wood samples were heat treated, during the ThermoWood process, at temperatures of 160, 180, and 210 °C. The temperature of 160 °C caused a significant darkening of the wood surface (a decrease in lightness L^*), while a further increase in temperature had a less significant effect on lightness. Changes in chromatic values (a^*, b^*) indicate a gradual shift from the original pale-yellow colour to dark brown. Thermal degradation of wood leads to the formation of chromophore structures, which is reflected in a change in its colour. Because black locust contains a large content of extractives, it can be assumed that thermal oxidation processes in these substances have a significant role in changing wood colour. Statistical analysis confirmed that temperature can be considered a statistically very significant factor affecting the wood surface colour.

Author Contributions: F.K., I.K. and D.K. conceived and designed the experiments; J.B., M.G. and D.K. prepared samples; I.K. measured and evaluated FTIR spectra; R.H. performed and evaluated colour measurements; D.K., I.K., J.B., M.G. and F.K. wrote the paper. All authors have read and agreed to the published version of the manuscript.

Funding: This work was supported by the Slovak Research and Development Agency under the Contract No. APVV-16-0326 (20%), APVV-20-0159 (20%), and by the Scientific Grant Agency of the Ministry of Education SR and the Slovak Academy of Sciences, Grant No. 1/0117/22 (50%) and 1/0577/22 (10%).

Data Availability Statement: Not applicable.

Conflicts of Interest: The authors declare no conflict of interest.

References

1. Jones, D.; Kržišnik, D.; Hočevar, M.; Zagar, A.; Humar, M.; Popescu, C.-M.; Popescu, M.-C.; Brischke, C.; Nunes, L.; Curling, S.F.; et al. Evaluation of the Effect of a Combined Chemical and Thermal Modification of Wood though the Use of Bicine and Tricine. *Forests* **2022**, *13*, 834. [CrossRef]
2. Bekhta, P.; Krystofiak, T.; Lis, B.; Bekhta, N. The Impact of Sanding and Thermal Compression of Wood, Varnish Type and Artificial Aging in Indoor Conditions on the Varnished Surface Color. *Forests* **2022**, *13*, 300. [CrossRef]
3. Esteves, B.; Ayata, U.; Gurleyen, L. Effect of heat treatment on the colour and glossiness of black locust, wild pear, linden, alder and willow wood. *Drewno* **2019**, *62*, 39–52. [CrossRef]
4. Zhan, T.; Zhu, J.; Liu, Z.; Li., T.; Peng, H.; Jiang, J.; Lyu, J. Meta-analysis of chromatic properties of heat-treated wood. *Eur. J. Wood Prod.* **2022**, *80*, 851–858. [CrossRef]
5. Torniainen, P.; Jones, D.; Sandberg, D. Colour as a quality indicator for industrially manufactured ThermoWood®. *Wood Mater. Sci. Eng.* **2021**, *16*, 287–289. [CrossRef]
6. Liu, X.Y.; Tu, X.W.; Liu, U. Effects of light thermal treatments on the color, hygroscopity and dimensional stability of wood. *Wood Res.* **2021**, *66*, 95–104. [CrossRef]
7. Sundqvist., B.; Morén, T. The influence of wood polymers and extractives on wood colour induced by hydrothermal treatment. *Eur. J. Wood Wood Prod.* **2002**, *60*, 375–376. [CrossRef]
8. Patzelt, M.; Emsenhuber, G.; Stingl, R. Colour Measurement as means of quality control of thermally treated wood. In Proceedings of the Abstracts of the first European conference on wood modification, Ghent, Belgium, 3–4 April 2003.
9. Mitsui, K.; Takada, H.; Sugiyama, M.; Hasegawa, R. Changes in the properties of light-irradiated wood with heat treatment: Part 1 Effect of treatment conditions on the change in colour. *Holzforschung* **2001**, *55*, 601–605. [CrossRef]
10. Jiang, H.C.; Lu, Q.J.; Li, G.J.; Li, M.; Li, J.N. Effect of heat treatment on the surface color of rubber wood (*Hevea brasiliensis*). *Wood Res* **2020**, *65*, 633–644. [CrossRef]
11. Bekhta, P.; Niemz, P. Effect of high temperature on the change in color, dimensional stability and mechanical properties of spruce wood. *Holzforschung* **2003**, *57*. [CrossRef]
12. Kačíková, D.; Kačík, F.; Čabalová, I.; Durkovič, J. Effects of thermal treatment on chemical, mechanical and colour traits in Norway spruce wood. *Bioresour. Technol.* **2013**, *144*, 669–674. [CrossRef]
13. Kubovský, I.; Kačík, F. Changes of the wood surface colour induced by CO_2 laser and its durability after the xenon lamp exposure. *Wood Res.* **2013**, *58*, 581–590.
14. Matsuo, M.; Yokoyama, M.; Umemura, K.; Gril, J.; Yano, K.; Kawai, S. Color changes in wood during heating: Kinetic analysis by applying a time temperature superposition method. *Appl. Phys. A* **2010**, *99*, 47–52. [CrossRef]
15. Fan, Y.; Gao, J.; Chen, Y. Colour responses of black locust (*Robinia pseudoacacia* L.) to solvent extraction and heat treatment. *Wood Sci. Technol.* **2010**, *44*, 667–678. [CrossRef]

16. Moya, R.; Fallas, R.S.; Bonilla, P.J.; Tenorio, C. Relationship Between Wood Color Parameters Measured by the CIELab System and Extractive and Phenol Content in *Acacia mangium* and *Vochysia guatemalensis* from Fast-Growth Plantations. *Molecules* **2012**, *17*, 3639–3652. [CrossRef]
17. Rédei, K.; Veperdi, I. The role of black locust (*Robinia pseudoacacia* L.) in establishment of short-rotation energy plantations in Hungary. *Int. J. Hortic. Sci.* **2009**, *15*, 41–44. [CrossRef]
18. Jiménez-López, L.; Martín-Sampedro, R.; Eugenio, M.E.; Santos, J.I.; Sixto, H.; Cañellas, I.; Ibarra, D. Co-production of soluble sugars and lignin from short rotation white poplar and black locust crops. *Wood Sci. Technol.* **2020**, *54*, 1617–1643. [CrossRef]
19. Li, G.; Xu, G.; Guo, K.; Du, S. Mapping the Global Potential Geographical Distribution of Black Locust (*Robinia Pseudoacacia* L.) Using Herbarium Data and a Maximum Entropy Model. *Forests* **2014**, *5*, 2773–2792. [CrossRef]
20. Sitzia, T.; Cierjacks, A.; de Rigo, D.; Caudullo, G. Robinia pseudoacacia in Europe: Distribution, Habitat, Usage and Threats. In *European Atlas of Forest Tree Species*; San-Miguel-Ayanz, J., de Rigo, D., Caudullo, G., Houston Durrant, T., Mauri, A., Eds.; Publications Office of the EU: Luxembourg, 2016; p. e014e79+.
21. Sikora, A.; Kačík, F.; Gaff, M.; Vondrová, V.; Bubeníková, T.; Kubovský, I. Impact of thermal modification on color and chemical changes of spruce and oak wood. *J. Wood Sci.* **2018**, *64*, 406–416. [CrossRef]
22. Sikora, A.; Hájková, K.; Jurczyková, T. Degradation of Chemical Components of Thermally Modified *Robinia pseudoacacia* L. Wood and Its Effect on the Change in Mechanical Properties. *Int. J. Mol. Sci.* **2022**, *23*, 15652. [CrossRef] [PubMed]
23. *ASTM D1107-21*; Standard Test Method for Ethanol-Toluene Solubility of Wood. ASTM International: West Conshohocken, PA, USA, 2021.
24. Sluiter, A.; Hames, B.; Ruiz, R.; Scarlata, C.; Sluiter, J.; Templeton, D.; Crocker, D. *Determination of Structural Carbohydrates and Lignin in biomass: Laboratory Analytical Procedure (LAP)*; NREL/TP-510-42618. Version 08-03-2012; National Renewable Energy Laboratory: Golden, CO, USA.
25. Seifert, V.K. Über ein neues Verfahren zur Schnellbestimmung der Rein—Cellulose. *Papier* **1956**, *10*, 301–306.
26. Wise, L.E.; Maxine, M.; D'Addieco, A.A. Chlorite holocellulose, its fractionation and bearing on summative wood analysis and on studies on the hemicelluloses. *Paper Trade J.* **1946**, *122*, 35–43.
27. Chen, Y.; Fan, Y.; Gao, J.; Stark, N.M. The effect of heat treatment on the chemical and color change of black locust (*Robinia pseudoacacia*) wood flour. *BioResources* **2012**, *7*, 1157–1170. [CrossRef]
28. Ferreira, S.; Gil, N.; Queiroz, J.A.; Duarte, A.P.; Domingues, F.C. An evaluation of the potential of Acacia dealbata as raw material for bioethanol production. *Bioresour. Technol.* **2011**, *102*, 4766–4773. [CrossRef]
29. Yáñez, R.; Romaní, A.; Garrote, G.; Alonso, J.L.; Parajó, J.C. Experimental evaluation of alkaline treatment as a method for enhancing the enzymatic digestibility of autohydrolysed *Acacia dealbata*. *J. Chem. Technol. Biotechnol.* **2009**, *84*, 1070–1077. [CrossRef]
30. Hoa, D.T.; Man, T.D.; Hau, N.V. Pretreatment of lignocellulosic biomass for enzymatic hydrolysis. *Asian J. Appl. Sci. Technol.* **2008**, *25*, 341–346. [CrossRef]
31. Kačík, F.; Ďurkovič, J.; Kačíková, D.; Zenková, E. Changes in the chemical composition of black locust wood after hot-water pretreatment before bioethanol production. *Acta Facultatis Xylologiae Zvolen.* **2016**, *58*, 15–23. [CrossRef]
32. Esteves, B.; Graca, J.; Pereira, H. Extractive composition and summative chemical analysis of thermally treated eucalypt wood. *Holzforschung* **2008**, *62*, 344–351. [CrossRef]
33. Chen, Y.; Gao, J.; Fan, Y.; Tshabalala, M.A.; Stark, N.M. Heat-induced chemical and color changes of extractive-free black locust (*Robinia pseudoacacia*) wood. *BioResources* **2012**, *7*, 2236–2248. [CrossRef]
34. Hu, F.; Jung, S.; Ragauskas, A. Pseudo-lignin formation and its impact on enzymatic hydrolysis. *Bioresour. Technol.* **2012**, *117*, 7–12. [CrossRef]
35. Shinde, S.D.; Meng, X.; Kumar, R.; Ragauskas, A.J. Recent advances in understanding the pseudo-lignin formation in a lignocellulosic biorefinery. *Green Chem.* **2018**, *20*, 2192–2205. [CrossRef]
36. Kučerová, V.; Lagaňa, R.; Výbohová, E.; Hýrošová, T. The Effect of Chemical Changes during Heat Treatment on the Color and Mechanical Properties of Fir Wood. *Bioresources* **2016**, *11*, 9079–9094. [CrossRef]
37. Bayani, S.; Bazyar, B.; Mirshokraie, S.A.; Taghiyari, H.R. Effects of heat treatment on the relative amounts of cellulose in nanosilver-impregnated and untreated poplar wood (*Populus alba*). *Floresta Ambiente* **2019**, *26*, e20160398. [CrossRef]
38. Nguyen, T.H.V.; Nguyen, T.T.; Ji, X.; Minghui Guo, M. Predicting Color Change in Wood During Heat Treatment Using an Artificial Neural Network Model. *BioResources* **2018**, *13*, 6250–6264. [CrossRef]
39. Torniainen, P.; Popescu, C.-M.; Jones, D.; Scharf, A.; Sandberg, D. Correlation of Studies between Colour, Structure and Mechanical Properties of Commercially Produced ThermoWood® Treated Norway Spruce and Scots Pine. *Forests* **2021**, *12*, 1165. [CrossRef]
40. Piernik, M.; Woźniak, M.; Pinkowski, G.; Szentner, K.; Ratajczak, I.; Krauss, A. Color as an Indicator of Properties in Thermally Modified Scots Pine Sapwood. *Materials* **2022**, *15*, 5776. [CrossRef]
41. Hrčka, R. Identification of discoloration of beech wood in CIELAB space. *Wood Res.* **2008**, *53*, 119–124.
42. Tolvaj, L.; Molnár, S.; Németh, R.; Varga, D. Color modification of black locust depending on the steaming parameters. *Wood Res.* **2010**, *55*, 81–88.
43. Banadics, E.A.; Tolvaj, L.; Varga, D. Steaming of Poplar, Black Locust and Beech Timbers Simultaneously to Investigate Colour Modification Effect of Extractive Transport. *Drewno* **2022**, *65*. [CrossRef]

44. Hon, D.N.S. Weathering and photochemistry. In *Wood and Cellulosic Chemistry*; Hon, D.N., Shiraishi, N., Eds.; Marcel Dekker: New York, NY, USA, 2001; pp. 513–546.
45. Cirule, D.; Kuka, E.; Kevers, M.; Andersone, I.; Andersons, B. Photodegradation of Unmodified and Thermally Modified Wood Due to Indoor Lighting. *Forests* **2021**, *12*, 1060. [CrossRef]
46. Polčin, J.; Rapson, W.H. Effect of Bleaching Agents on the Absorption of Lignin in Groundwood Pulps. Part II. Oxidative-Reductive Bleaching. *Pulp Paper Mag. Can.* **1971**, *72*, 80–91.
47. Tribulová, T.; Kačík, F.; Evtuguin, D.; Čabalová, I. Assessment of chromophores in chemically treated and aged wood by UV-VIS diffuse reflectance spectroscopy. *Cellul. Chem. Technol.* **2016**, *50*, 659–667.
48. Wei, Y.X.; Wang, M.J.; Zhang, P.; Chen, Y.; Gao, J.M.; Fan, Y.M. The role of phenolic extractives in color changes of locust wood (*Robinia pseudoacacia*) during heat treatment. *Bioresources* **2017**, *12*, 7041–7055.
49. Kubovský, I.; Kačíková, D.; Kačík, F. Structural Changes of Oak Wood Main Components Caused by Thermal Modification. *Polymers* **2020**, *12*, 485. [CrossRef]
50. Inari, G.N.; Petrissans, M.; Gerardin, P. Chemical reactivity of heat-treated wood. *Wood Sci. Technol.* **2007**, *41*, 157–168. [CrossRef]
51. Tolvaj, L. Traditions, anomalies, mistakes and recommendations in infrared spectrum measurement for wood. *Wood. Sci. Technol.* **2022**. [CrossRef]
52. Lagaňa, R.; Csiha, C.; Horváth, N.; Tolvaj, L.; Andor, T.; Kúdela, J.; Németh, R.; Kačík, F.; Ďurkovič, J. Surface properties of thermally treated European beech wood studied by PeakForce Tapping atomic force microscopy and Fourier-transform infrared spectroscopy. *Holzforschung* **2021**, *75*, 56–64. [CrossRef]
53. Esteves, B.; Velez Marques, A.; Domingos, I.; Pereira, H. Chemical changes of heat treated pine and eucalypt wood monitored by FTIR. *Maderas Cienc. Tecnol.* **2013**, *15*, 245–258. [CrossRef]
54. Kubovský, I.; Kačík, F. Colour and chemical changes of the lime wood surface due to CO_2 laser thermal modification. *Appl. Surf. Sci.* **2014**, *321*, 261–267. [CrossRef]
55. Müller, G.; Schöpper, C.; Vos, H.; Kharazipour, A.; Polle, A. FTIR-ATR spectroscopic analysis of changes in wood properties during particle and fibreboard production of hard and softwood trees. *BioResources* **2009**, *4*, 49–71. [CrossRef]
56. Kubovský, I.; Oberhofnerová, E.; Kačík, F.; Pánek, M. Surface changes of selected hardwoods due to weather conditions. *Forests* **2018**, *9*, 557. [CrossRef]
57. Bhagia, S.; Ďurkovič, J.; Lagaňa, R.; Kardošová, M.; Kačík, F.; Cernescu, A.; Schäfer, P.; Yoo, C.G.; Ragauskas, A.J. Nanoscale FTIR and Mechanical Mapping of Plant Cell Walls for Understanding Biomass Deconstruction. *ACS Sustainable Chem. Eng.* **2022**, *10*, 3016–3026. [CrossRef]
58. Popescu, M.C.; Froidevaux, J.; Navi, P.; Popescu, C.M. Structural modifications of *Tilia cordata* wood during heat treatment investigated by FT-IR and 2D IR correlation spectroscopy. *J. Mol. Struct.* **2013**, *1033*, 176–186. [CrossRef]
59. Ciolacu, D.; Ciolacu, F.; Popa, V.I. Amorphous cellulose-structure and characterization. *Cellul. Chem. Technol.* **2011**, *45*, 13–21.

forests

MDPI

Article

Properties of Plywood Panels Composed of Thermally Densified and Non-Densified Alder and Birch Veneers

Pavlo Bekhta [1,2,*], Tomáš Pipíška [2], Vladimír Gryc [2], Ján Sedliačik [3], Pavel Král [2], Jozef Ráheľ [2] and Jan Vaněrek [4]

[1] Department of Wood-Based Composites, Cellulose and Paper, Ukrainian National Forestry University, 79057 Lviv, Ukraine

[2] Department of Wood Science and Technology, Faculty of Forestry and Wood Technology, Mendel University in Brno, Zemědělská 3, 613 00 Brno, Czech Republic

[3] Department of Furniture and Wood Products, Technical University in Zvolen, T.G. Masaryka 24, 960 01 Zvolen, Slovakia

[4] Faculty of Civil Engineering, Institute of Technology of Building Materials and Components, Brno University of Technology, 602 00 Brno, Czech Republic

* Correspondence: bekhta@nltu.edu.ua

Abstract: Ukrainian companies mainly use birch in the manufacture of plywood, but species, such as black alder, are not yet widely used in the manufacture of plywood due to their poorer properties. It is well known that thermal compression is often used to improve the properties of solid wood. Good lay-up schemes of veneer can maximize the advantages and minimize the disadvantages of these wood species, and generally improve the utility value of the plywood. This research aimed to develop plywood panels with two wood species and two types of veneer treatments in order to evaluate the influences of different lay-up schemes on the properties of the plywood. Five-layer plywood panels were formed with 16 different lay-up schemes using birch (*Betula verrucosa* Ehrh.) (B) and black alder (*Alnus glutinosa* L.) (A) veneers, which were non-densified (N) and thermally densified (D). The different lay-up schemes were used to identify opportunities to improve the mechanical and physical properties of the plywood by replacing the birch veneer in the plywood structure with an alternative alder veneer. The veneer sheets were thermally densified in a laboratory hot press at a temperature of 180 °C and pressure of 2 MPa for 3 min. The conducted study showed that the bending strength, modulus of elasticity and shear strength of mixed-species plywood (B^D–A^N–A^N–A^N–B^D) increased by up to 31.5%, 34.4% and 16.8%, respectively, in comparison to those properties of alder plywood from non-densified veneer (A^N–A^N–A^N–A^N–A^N), by positioning alder non-densified veneers in the core layers and birch densified veneers in the outer layers. Moreover, the surface roughness of plywood panels with outer layers of birch veneer was lower than that of panels with outer layers of alder veneer. It was shown that non-treated alder veneer, despite exhibiting somewhat lower strength properties than birch veneer, could be successfully used with proper lay-up schemes in the veneer-based products industry.

Keywords: birch; black alder; thermal densification; plywood properties; wood veneers

Citation: Bekhta, P.; Pipíška, T.; Gryc, V.; Sedliačik, J.; Král, P.; Ráheľ, J.; Vaněrek, J. Properties of Plywood Panels Composed of Thermally Densified and Non-Densified Alder and Birch Veneers. *Forests* **2023**, *14*, 96. https://doi.org/10.3390/f14010096

Academic Editor: Antonios Papadopoulos

Received: 13 December 2022
Revised: 2 January 2023
Accepted: 3 January 2023
Published: 4 January 2023

1. Introduction

Nowadays, veneer-based products, including plywood and laminated veneer lumber (LVL), are highly effective engineering wood products for construction. The plywood manufacturing process creates products with several dimensions to suit architectural and structural purposes, and mechanical properties equivalent or superior to those of the initially solid wood [1]. However, various process variables should be considered to obtain panels with satisfactory properties, such as the wood species used, type of adhesive, density, moisture content, lay-up scheme of the veneers, etc. [2]. The choice of wood species is very important for the properties and price of the panel [3]. Ukrainian companies mainly use birch (*Betula verrucosa* Ehrh.) in the manufacture of plywood, but species such as black alder (*Alnus glutinosa* L.) are not yet widely used in the manufacture of plywood due to their poorer mechanical properties. However, mixing veneers from different species can be a good strategy for obtaining panels with a good cost–benefit relationship [4].

Various modification methods can be used to improve the properties of low-grade wood species, one of which is the thermal densification of wood at elevated temperatures and pressure [5]. The combination of heat and compression is an environmentally friendly modification method that does not require any chemicals. Previous studies have shown that thermal densification not only improves the aesthetic properties and reduces surface roughness, but also improves the mechanical properties and bonding quality [6–11]. Moreover, the thermal densification of the veneer used for the manufacture of plywood or lamination of wood-based panels reduces the consumption of adhesive and pressing pressure by 40%, the consumption of lacquer materials and the bonding strength improves [12,13].

Various studies on veneer-based products indicated that the lay-up scheme is a key factor in improving the mechanical properties of products and beneficial use of available low-grade and lower cost wood species [14–24]. LVL panels, made by combining veneers from durable and non-durable wood species, showed improved durability when two faces and one core veneer were from decay-resistant species [14]. In other studies, it was shown that the most suitable lay-up schemes for poplar and beech veneers [15] or poplar and birch veneers [17] were eight- and ten-ply LVL, respectively, with high bending properties. Several authors [18,20] studied the effect of the thermal modification process on the properties of plywood panels manufactured by combining thermally treated and untreated poplar veneers. It was found that the addition of thermally treated veneers caused a decrease in the mechanical properties and improved the dimensional stability of plywood panels. On the contrary, other results revealed that the mechanical properties of plywood composed of fully and partially heat-treated veneers were not significantly different from those of untreated plywood [24].

One of the lay-up schemes (construction strategy) of veneers is to place the high-density wood veneers in the outer layers and low-density veneers in the inner layers of veneer-based panels [19,25]. H'ng et al. [25] manufactured 11-ply and 15-ply LVL panels by combining high-density keruing veneers (as outside veneers) with low-density wood veneers such as pulai, sesendok and kekabu hutan. They found that combining the keruing (as surface layers) and low-density wood veneers gave a greater bending strength and more stable material.

Another lay-up strategy is to combine softwood and hardwood veneers to manufacture mixed-species veneer-based products [21,22,26]. Combining even a small amount of spotted gum veneer with plantation hoop pine veneer resulted in an improved mechanical performance, especially in flatwise bending [22]. These authors showed that the substitution of only two spotted gum veneers on the faces of the hoop pine 12-ply LVL panels yielded an increase of up to 34.5% (MOE) and 38.5% (MOR) compared with the all hoop pine 12-ply LVL panels. Another study [19] on manufacturing 7-ply LVL panels using five different lay-up strategies that combined ash and Turkish red pine veneers showed that the best results in terms of the density, MOR and MOE were achieved in the panels with outer layers comprising ash veneers. Ozarska [4], when comparing softwoods with hardwoods, noted that the softwoods were more susceptible to bonding, but the hardwoods provides increased strength and structural rigidity to the finished product.

In our previous work [27], we presented the possibility of creating plywood panels with different strength properties when combining beech veneers of different thicknesses, as well as thermally densified and non-densified veneers. It was found that the use of densified veneers increased the mechanical properties of plywood but worsened the thickness of swelling and water absorption of panels.

However, there is no information in the literature about the manufacture of plywood panels by combining thermally densified and non-densified veneers of different high-grade and low-grade hardwood species in one plywood panel. It is hypothesized that mixing veneers from high- and low-grade wood species may improve the properties of the resulting mixed-species plywood panels by maximizing the advantages and overcoming the disadvantages of the two resources. Therefore, the aim of this work was to evaluate the effect of combining different high-grade (birch, which has relatively better structural properties) and low-grade (black alder, with poorer properties but a lower cost) wood species and different lay-up schemes on the mixed-species plywood properties made from alternate layers of densified and non-densified veneers.

2. Materials and Methods

2.1. Materials

Commercially produced rotary-cut veneers of birch (*Betula verrucosa* Ehrh.) and black alder (*Alnus glutinosa* L.) were used in the study. Birch wood is commonly used to manufacture veneer and plywood in Ukraine. Alder wood was chosen based on its economic advantages because it is a cheap species and is commonly used for the manufacture of plywood in Ukraine, as a substitution for traditionally used birch wood. Birch and alder veneers had an equal thickness of 1.6 mm, moisture content of 5.7% and 7.0%, respectively, and a density of 677 kg/m^3 and 487 kg/m^3, respectively. A total of 240 veneer samples with a dimension of 300 × 300 mm were prepared for the production of plywood.

Commercial urea-formaldehyde (UF) resin (density = 1.28 g/m^3, solids content = 65 ± 2%, pH = 7.5 ± 0.5, viscosity 1000–2500 mPa·s at 20 °C), kaolin (as a filler), ammonium nitrate NH_4NO_3 (as a hardener) and distilled water were used to prepare the adhesive. For the preparation of UF adhesive, 5% hardener and 2% filler were used.

2.2. Thermal Compression of Veneer

Veneer samples without visible defects were selected for thermal compression. The veneer sheets were densified between smooth metal sheets, which were placed on the hot press. Each veneer sheet was densified at a temperature of 180 °C, pressure of 2 MPa and for a duration of 3 min (during the last minute of compression, the pressure was gradually released).

The changes of the densified veneer properties were evaluated using veneer thickness and density measurements. To achieve this, the weight and dimensions of the veneer specimens were measured before and after thermal compression. Based on these data, the compression ratio (*CR*) and densification ratio (*DR*) of the veneer sheets after the thermal compression process were calculated as follows:

$$CR = (T_n - T_d)/T_n \times 100 \quad (1)$$

where T_n is the thickness of the veneer before compression (mm) and T_d is the thickness of the veneer after compression (mm).

$$DR = (D_d - D_n)/D_n \times 100 \quad (2)$$

where D_n is the density of the veneer before compression (kg/m^3) and D_d is the density of the veneer after compression (kg/m^3).

2.3. Manufacturing and Testing of Plywood Samples

Five-layer plywood panels with dimensions of 300 × 300 mm were manufactured. The veneer assembly for the production of plywood samples was formed from either birch or black alder veneers only, or from birch veneer combined with black alder veneer, with non-densified and densified veneers in one package. There was a total of 16 different veneer panel set options (Table 1). The adhesive spread was 110 g/m^2 and adhesive was applied by hand to one side of each veneer using a roller. Conditions for pressing plywood samples were as follows: temperature 130 °C, pressure 1.8 MPa, time 5.5 min (during the last minute, the pressure was smoothly released).

Table 1. Configuration of five-layer plywood panels.

Code of Panels	Type of Panels	Veneer Assembly Pattern by Wood Species	Veneer Assembly Pattern by Type of Veneer Treatment
I	B^N–B^N–B^N–B^N–B^N	B–B–B–B–B	N–N–N–N–N
II	B^N–A^N–B^N–A^N–B^N	B–A–B–A–B	N–N–N–N–N
III	B^N–A^N–A^N–A^N–B^N	B–A–A–A–B	N–N–N–N–N
IV	A^N–B^N–B^N–B^N–A^N	A–B–B–B–A	N–N–N–N–N
V	A^N–B^N–A^N–B^N–A^N	A–B–A–B–A	N–N–N–N–N
VI	A^N–A^N–A^N–A^N–A^N	A–A–A–A–A	N–N–N–N–N
VII	B^D–A^N–A^N–A^N–B^D	B–A–A–A–B	D–N–N–N–D
VIII	A^D–B^N–B^N–B^N–A^D	A–B–B–B–A	D–N–N–N–D
IX	B^D–A^N–B^D–A^N–B^D	B–A–B–A–B	D–N–D–N–D
X	A^D–B^N–A^D–B^N–A^D	A–B–A–B–A	D–N–D–N–D
XI	B^D–B^D–B^D–B^D–B^D	B–B–B–B–B	D–D–D–D–D
XII	B^D–A^D–B^D–A^D–B^D	B–A–B–A–B	D–D–D–D–D
XIII	B^D–A^D–A^D–A^D–B^D	B–A–A–A–B	D–D–D–D–D
XIV	A^D–B^D–B^D–B^D–A^D	A–B–B–B–A	D–D–D–D–D
XV	A^D–B^D–A^D–B^D–A^D	A–B–A–B–A	D–D–D–D–D
XVI	A^D–A^D–A^D–A^D–A^D	A–A–A–A–A	D–D–D–D–D

B—birch veneer; A—alder veneer; N—non-densified veneer; D—densified veneer.

After pressing, the plywood panels were conditioned for one week in an environmental chamber maintained at 65 ± 5% relative humidity and 20 ± 2 °C before their properties were evaluated. For each variant in the experimental plan (Table 1), three panels were made, that is, there was a total of 48 panels. After conditioning, samples were cut from the manufactured panels into required test sizes according to relevant standards. In accordance with European standards, the following number of samples was evaluated for each panel variant: three samples for moisture content (EN 322) [28], ten samples for density (EN 323) [29], three samples for bending strength (MOR) (EN 310) [30] and modulus of elasticity (MOE) (EN 310) [30], ten samples for shear strength (EN 314) [31,32] and ten samples for thickness swelling (*TS*) (EN 317) [33] and water absorption (*WA*). The samples for shear strength testing were immersed in water at 20 ± 3 °C for 24 h for plywood bonding class 1 (dry conditions).

The samples for *TS* and *WA* testing were immersed in distilled water for 24 h. After this time, the test pieces were removed from the water, weighed, and the thickness was measured. The percent change from the original thickness represents the *TS*, and the percent weight change from the original weight represents the *WA*. The *WA* and *TS* of plywood panels samples were calculated according to the following formulae:

$$WA = (W_2 - W_1)/W_1 \times 100 \quad (3)$$

$$TS = (T_2 - T_1)/T_1 \times 100 \quad (4)$$

where W_1 and T_1 are the initial weight (g) and thickness (mm) before soaking, and W_2 and T_2 are the final weight (g) and thickness (mm) after soaking.

Surface roughness of plywood panels was measured using a digital microscope VHX-5000 (Keyence, Itasca, IL, USA) with a VH-Z100R wide depth of field objective. 3D images of the surface were taken, from which the overall profile and height parameters of the surface were measured. Three surface roughness characteristics were determined, i.e., the arithmetic average height (R_a), average peak to valley roughness (R_z) and root-mean-square deviation of the profile (R_q). The measurement was performed for two samples from the measured group (two random repetitions on each sample) in the transverse direction. The resulting values were reported as mean values per group.

Samples (5 × 5 mm) were cut from the panel to analyze microscopic structure of bond line and effect of veneer densification on adhesive penetration of plywood panels using an electron microscope (MIRA3 LMU, Tescan, a. s., Brno, Czech Republic). The bond line was softened with water and then immediately cleaned up with a razor blade. An accelerating voltage of 7 keV and a beam current of approximately 30 pA were used for result visualizations.

2.4. Statistical Analysis

The effects of combining veneers of different wood species and different types of veneer treatment on the properties of the laboratory-fabricated plywood panels were analyzed using analysis of variance (ANOVA), with a 0.05 level of significance using statistical software program STATISTICA 12.0 package (StatSoft Inc., Tulsa, OK, USA). Duncan's range tests were conducted to determine significant differences between mean values.

3. Results and Discussion

3.1. Compression and Densification Ratio of Veneers

The average values of the physical characteristics of thermally densified veneers are given in Table 2. During thermal compression, some changes in the physical properties of the thermally densified veneers could be observed. Thermal compression reduced the thickness of the veneer, and, therefore, the volume, which increased the density. In particular, the density of alder and birch veneers increased after the thermal compression to 528 and 708 kg/m^3, respectively. This result was in agreement with those reported by other researchers [34], who found that veneer density increased by 14% under thermal compression at a pressure of 2.7 MPa and temperature of 180 °C for 2 min. Considering the lower density of the alder veneer compared to the birch veneer, it is natural that the *CR* and *DR* of the alder veneer were higher (11.5% and 8.3%, respectively) than similar indicators for the birch veneer (8.2% and 4.6%, respectively). The values of *CR* were higher than the values of *DR*. This difference was explained by the fact that the *CR* takes into account changes in the thickness of the veneer only, and the *DR* takes into account both changes in the thickness and mass of the veneer. The mass of the veneer also decreased in the process of thermal compression. The decrease in mass could be explained by the loss (evaporation) of moisture under the influence of pressure and temperature [12]. In previous studies [35], no chemical changes were observed in thermally densified wood veneers in the case of a similar modification. Similar results were also reported by Rautkari et al. [36], who, using FTIR spectroscopic analysis, showed that no significant chemical changes occurred during frictional compression. The difference between the absolute values of *CR* and *DR* was 3.2% and 3.6%, respectively, for the alder and birch veneers. This was explained by the slightly higher moisture content of the alder veneer (7.0%) than the birch veneer (5.7%). It should also be noted that during thermal compression of the veneer the surface roughness of the veneer decreased and its aesthetic properties improved [8–10].

Table 2. The mean values of physical properties of thermally densified veneers.

Wood Species	Density of Veneer (kg/m^3)		Compression Ratio (%)	Densification Ratio (%)
	Non-Densified	Densified		
Alder	487 (21)	528 (20)	11.5 (2.3)	8.3 (1.9)
Birch	677 (41)	708 (51)	8.2 (1.4)	4.6 (0.8)

Values in parentheses are standard deviations.

3.2. Thickness and Density of Plywood Samples

The physical properties of plywood samples, which were made by combining veneers of different wood species and different types of treatment in one panel, are summarized in Table 3. The thinnest and thickest plywood samples were 6.7 mm and 7.4 mm and were made of only densified alder veneer (A^D–A^D–A^D–A^D–A^D) and only non-densified birch veneer (B^N–B^N–B^N–B^N–B^N), respectively. Between all other samples, the thickness of the panels differed insignificantly and was in the range of 6.9–7.2 mm. The thickness of the plywood panels in this study did not exceed the tolerances for unsanded panels according to European standard EN 315 [37]. The lowest density of 583 kg/m^3 was found in plywood (A^N–A^N–A^N–A^N–A^N) from non-densified alder veneer, as expected. The greatest densities of 779 and 832 kg/m^3 were found in plywood made of non-densified birch veneer (B^N–B^N–B^N–B^N–B^N) and densified birch veneer (B^D–B^D–B^D–B^D–B^D), respectively. The lowest *WA* values of 29.3% and 30.4% were found in plywood samples made of densified (B^D–B^D–B^D–B^D–B^D) and non-densified birch veneer (B^N–B^N–B^N–B^N–B^N), respectively. The highest *WA* values of 41.9% and 43.0% were found in plywood samples made of densified (A^D–A^D–A^D–A^D–A^D) and non-densified (A^N–A^N–A^N–A^N–A^N) alder veneer, respectively. The smallest and greatest values of *TS* of 7.2% and 12.1% were observed in plywood made from non-densified (A^N–A^N–A^N–A^N–A^N) and densified (A^D–A^D–A^D–A^D–A^D) alder veneer, respectively.

Table 3. Values of physical properties of plywood samples.

Code of Panel	Type of Panel	Thickness (mm)	Density (kg/m^3)	*WA* (%)	*TS* (%)
I	B^N–B^N–B^N–B^N–B^N	7.4 ^{g} (0.1)	779 ^{h} (31)	30.4 ^{a} (2.4)	7.9 ^{abc} (0.9)
II	B^N–A^N–B^N–A^N–B^N	7.2 ^{def} (0.2)	685 ^{cde} (24)	36.3 ^{c} (2.0)	9.4 ^{cdef} (1.7)
III	B^N–A^N–A^N–A^N–B^N	7.2 ^{ef} (0.2)	684 ^{cde} (20)	36.3 ^{c} (2.2)	8.0 ^{abc} (1.3)
IV	A^N–B^N–B^N–B^N–A^N	7.0 ^{bcdef} (0.3)	709 ^{ef} (27)	33.2 ^{b} (2.7)	8.5 ^{abcd} (1.4)
V	A^N–B^N–A^N–B^N–A^N	7.0 ^{bcdef} (0.3)	661 ^{bc} (27)	35.6 ^{bc} (1.3)	7.6 ^{ab} (2.2)
VI	A^N–A^N–A^N–A^N–A^N	7.2 ^{cdef} (0.1)	583 ^{a} (18)	43.0 ^{e} (3.9)	7.2 ^{a} (1.3)
VII	B^D–A^N–A^N–A^N–B^D	7.1 ^{cdef} (0.1)	673 ^{cd} (15)	39.0 ^{d} (2.5)	9.0 ^{bcde} (1.1)
VIII	A^D–B^N–B^N–B^N–A^D	7.1 ^{bcdef} (0.1)	726 ^{fg} (52)	36.2 ^{c} (2.8)	10.3 ^{ef} (1.4)
IX	B^D–A^N–B^D–A^N–B^D	7.1 ^{bcdef} (0.3)	712 ^{ef} (18)	35.0 ^{bc} (1.4)	10.0 ^{def} (1.9)
X	A^D–B^N–A^D–B^N–A^D	7.0 ^{bc} (0.1)	703 ^{ef} (29)	35.3 ^{bc} (2.2)	9.3 ^{cdef} (1.0)
XI	B^D–B^D–B^D–B^D–B^D	7.2 ^{fg} (0.2)	832 ^{i} (44)	29.3 ^{a} (1.4)	10.9 ^{fg} (2.3)
XII	B^D–A^D–B^D–A^D–B^D	7.0 ^{bcd} (0.3)	744 ^{g} (47)	35.5 ^{bc} (1.9)	10.9 ^{fg} (1.5)
XIII	B^D–A^D–A^D–A^D–B^D	7.0 ^{bcde} (0.1)	675 ^{cd} (6)	37.3 ^{cd} (3.3)	10.3 ^{ef} (1.0)
XIV	A^D–B^D–B^D–B^D–A^D	6.9 ^{b} (0.1)	708 ^{ef} (14)	34.6 ^{bc} (2.2)	10.1 ^{ef} (1.5)
XV	A^D–B^D–A^D–B^D–A^D	7.0 ^{bcd} (0.1)	695 ^{de} (18)	37.3 ^{cd} (2.5)	10.9 ^{fg} (2.1)
XVI	A^D–A^D–A^D–A^D–A^D	6.7 ^{a} (0.2)	638 ^{b} (24)	41.9 ^{e} (4.6)	12.1 ^{g} (1.9)

Values in parentheses are standard deviations; means followed by the same letter were not significantly different at $p \leq 0.05$.

The ANOVA analysis (Table 4) showed that combinations of wood species and having different types of veneer treatment in one panel significantly affected the thickness, density and *WA* of plywood samples. In addition, the combination of a densified and non-densified veneer in one panel affected the thickness and density of plywood samples to almost the same extent, taking into account approximately the same influencing factors (F = 10.355 and F = 13.334). Meanwhile, the combination of veneers of different wood species in one panel had a stronger effect on the density (F = 101.404) and a weaker effect on the thickness (F = 9.464) of plywood samples. *TS* did not depend on the mixing of wood species (F = 1.527) in one panel, but significantly depended on the mixing of densified and non-densified veneers (F = 31.045) (Table 4).

Table 4. The influence of the variable factors on the physical properties of plywood.

Source of Variation	Thickness		Density		*WA*		*TS*	
	F	Sig.	*F*	Sig.	*F*	Sig.	*F*	Sig.
W	9.464	0.000	101.404	0.000	51.493	0.000	1.527	0.185
T	10.355	0.000	13.334	0.000	4.148	0.007	31.045	0.000
W × T	1.837	0.084	4.625	0.000	0.966	0.458	2.526	0.018

W—combination of veneers of different wood species; T—combination of veneers with different treatment.

A graphic illustration of the effect of combining veneers of different wood species and different treatments in one panel on the thickness, density, *WA* and *TS* of plywood panels is presented in Figure 1. The smallest average thickness (6.7 mm) was in plywood samples made from a densified alder veneer, and the largest (7.4 mm) was in samples made from a non-densified birch veneer (Table 3, Figure 1). Decreasing the share of alder veneer in the panel increased the thickness of the plywood samples, although this increase was insignificant. For the lower proportion (40%) of alder veneer in one panel, the thickness of samples with outer layers of non-densified (A^N–B^N–B^N–B^N–A^N) or densified (A^D–B^D–B^D–B^D–A^D) alder veneer was lower (7.0 mm and 6.9 mm, respectively) than the thickness of samples with outer layers of non-densified (B^N–A^N–B^N–A^N–B^N) or densified (B^D–A^D–B^D–A^D–B^D) birch veneer (7.2 mm and 7.0 mm, respectively); however, the difference between them was insignificant ($p > 0.05$). For the higher proportion (60%) of alder veneer in one panel, the thickness of samples (A^N–B^N–A^N–B^N–A^N) with outer layers of non-densified alder veneer was lower (7.0 mm) than the thickness of samples (B^N–A^N–A^N–A^N–B^N) with outer layers of non-densified birch veneer (7.2 mm). For the same conditions, the thickness of samples (A^D–B^D–A^D–B^D–A^D) and (B^D–A^D–A^D–A^D–B^D) with outer layers of densified veneers was equal (7.0 mm). This effect of veneer wood species on the thickness of plywood samples was mainly explained by the difference in the density of alder and birch wood. Birch veneer, having a higher density, is compressed less (8.2%) than alder veneer (11.5%), and, accordingly, formed thicker plywood samples. Moreover, the effect of the birch veneer on the formation of panel thickness was more pronounced when a denser birch veneer was placed in the outer layers of the panel. The smallest thicknesses (6.7–7.2 mm) of plywood samples were in samples made from densified veneer, and the greatest thicknesses (7.0–7.4 mm) were in the samples made from non-densified veneer (Table 3, Figure 1); however, the difference between them was significant. Plywood, in which the share of densified veneer (D) was greater, had a smaller thickness than plywood with a greater share of non-densified veneer (N), although the difference between them was insignificant.

The combination of wood species in one panel had a much stronger effect ($F = 101.404$) on the density of plywood samples than the combination of the type of treated veneer ($F = 13.334$) (Table 4). Plywood samples (A^N–A^N–A^N–A^N–A^N) made from non-densified alder veneer had the lowest density (583 kg/m^3), while samples (B^D–B^D–B^D–B^D–B^D) made from densified birch veneer had the highest density (832 kg/m^3). The predominance of non-densified alder veneer in one panel (60%) provided a lower density of 684 kg/m^3 and 661 kg/m^3 for plywood samples (B^N–A^N–A^N–A^N–B^N) and (A^N–B^N–A^N–B^N–A^N), respectively, in comparison to the density of 779 kg/m^3 for a birch-only plywood (B^N–B^N–B^N–B^N–B^N). However, these densities were greater than the density of 583 kg/m^3 obtained for the alder-only plywood (A^N–A^N–A^N–A^N–A^N). A similar trend was observed for the densified veneer. It is obvious that the density of the birch veneer was greater than the density of the alder veneer. The effect of the type of veneer treatment (N and D) and its combination on the density of plywood samples was not as strong as the effect of the wood species of the veneer (Table 4). With the same schemes of veneer assembly of different species in one panel, plywood samples with a densified veneer in the outer layers provided a greater density of plywood samples compared to samples comprising non-densified veneers. With the same assembly scheme of veneer and different wood species in one panel, increasing the proportion of densified veneer in the panel, in particular in the outer layers, reduced the density of sample (B^D–A^N–A^N–A^N–B^D) to 673 kg/m^3, compared to the density of 684 kg/m^3 obtained for sample (B^N–A^N–A^N–A^N–B^N), which was made from non-densified veneer only. However, the difference between the densities was insignificant. Meanwhile, the addition of densified veneer in the outer layers of sample (A^D–B^N–B^N–B^N–A^D) led to an increase in density to 726 kg/m^3 compared to the density of 709 kg/m^3 obtained for sample (A^N–B^N–B^N–B^N–A^N), which was made from non-densified veneer only. The higher *CR* and *DR* of alder veneer than birch veneer explained this (Table 2).

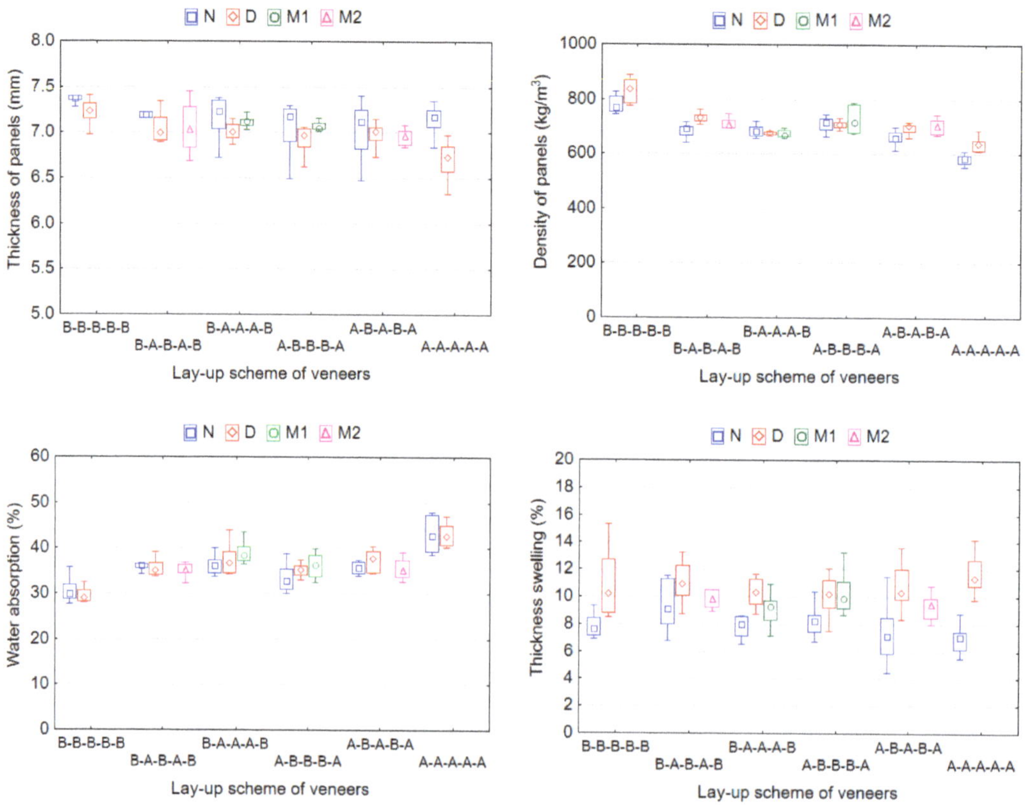

Figure 1. Effect of wood veneer species and type of veneer treatment on the physical properties of plywood panels. N—non-densified veneers in one panel; D—densified veneer in one panel; M1—mix of veneers (D–N–N–N–D) in one panel; M2—mix of veneers (D–N–D–N–D) in one panel; Statistical features are median, 25% minimum and 75% maximum.

The combination of different wood species and different types of veneer treatment in one plywood panel significantly affected the *WA* of plywood samples. In addition, the combination of veneers from different wood species in one panel had a much stronger effect ($F = 51.493$) on *WA* than the type of veneer treatment used (a combination of densified and non-densified veneer in one panel) ($F = 4.148$) (Table 4). The lowest and highest *WA* values of 29.3% and 43.0% were in plywood samples made from densified birch veneers (B^{D}–B^{D}–B^{D}–B^{D}–B^{D}) and non-densified alder veneers (A^{N}–A^{N}–A^{N}–A^{N}–A^{N}), respectively. With an increase in the proportion of alder veneer in the inner layers of birch plywood made from non-densified veneer (B^{N}–A^{N}–B^{N}–A^{N}–B^{N} and B^{N}–A^{N}–A^{N}–A^{N}–B^{N}), the *WA* increased and its value was significantly higher (36.3% for B^{N}–A^{N}–B^{N}–A^{N}–B^{N} and B^{N}–A^{N}–A^{N}–A^{N}–B^{N}) than the value 30.4% for the *WA* of a birch-only plywood made from non-densified veneer (B^{N}–B^{N}–B^{N}–B^{N}–B^{N}). However, the addition of birch veneers to the inner layers of an alder plywood resulted in a reduction in the *WA* of plywood made from alder only. A similar trend was observed for plywood panels made from densified veneer (Table 3, Figure 1). An increase in the proportion of densified veneer in one panel led to a decrease in the *WA* of plywood samples. This was explained by the density of plywood samples. It is well known that there is a relationship between *WA* and panel density [38]. The dependence of *WA* on the density of plywood panels is shown in Figure 2a. As the density of plywood samples increases, their *WA* decreases, as the number and size of available cavities through which water can enter the sample decreases. Several authors [27] also showed that *WA* is related to panel density, with a higher density resulting in a lower number of pores and, consequently, a lower *WA*.

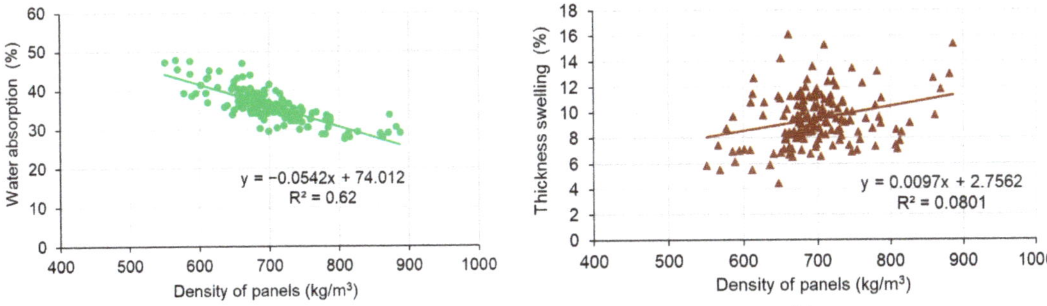

Figure 2. Influence of density on water absorption (**a**) and thickness swelling (**b**) of plywood samples.

The combination of veneers of different wood species in one panel had no effect on the *TS*, while the type of veneer treatment had a significant effect on the *TS* (Table 4). The lowest (7.2%) and highest (12.1%) values of *TS* were found in alder plywood samples made from non-densified (A^N–A^N–A^N–A^N–A^N) and densified (A^D–A^D–A^D–A^D–A^D) veneers, respectively. The mixed-species plywood samples using only densified veneers had a higher *TS* than samples using only non-densified veneers (Table 3, Figure 1). Increasing the share of densified veneer in one panel led to an increase in the *TS* of the mixed-species plywood samples. In this case, this was explained by the increase in the density of plywood samples. This is well known from sources in the literature [12,27,38]. However, in this study, no clear dependence of *TS* on sample density was found (Figure 2b).

The surface roughness of plywood panels with outer layers of non-densified and densified birch and alder veneer was also compared (Figure 3). In addition to the fact that plywood with outer layers of alder veneer had lower values of MOR and MOE (Table 5), the surface of such plywood also had a greater surface roughness compared to the birch plywood (Figure 3). It was found that for plywood made of non-densified alder veneer (A^N–A^N–A^N–A^N–A^N), the roughness parameters R_a and R_q were 19.4% and 5.4% greater, respectively, than the similar parameters for plywood made of non-densified birch veneer (B^N–B^N–B^N–B^N–B^N). On the contrary, the roughness parameter R_z for plywood made of non-densified alder veneer was 18.3% lower than for plywood made of non-densified birch veneer. For plywood made of densified alder veneer, the roughness parameters R_a, R_z and R_q were 41.6%, 0.4% and 14.2% higher than similar parameters for plywood made of densified birch veneer, respectively. This was logical, as birch wood has less porosity than alder wood. However, the differences in R_a, R_z and R_q found between non-densified/densified birch and alder veneers were statistically insignificant. Nevertheless, the thermal compression led to a decrease in the surface roughness values for both the birch and alder veneers. In our previous work [10], we also found that the surface of wood veneers became smoother and roughness values decreased significantly due to the thermal densification process.

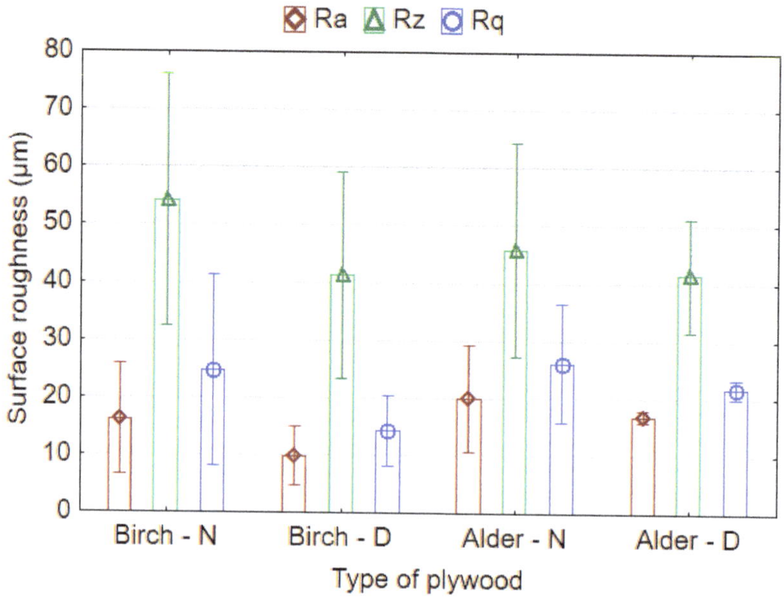

Figure 3. Surface roughness of plywood panels. N—non-densified veneers in surface layers; D—densified veneer in surface layers.

Table 5. Values of mechanical properties for plywood samples.

Code of Panels	Type of Panels	MOR (MPa)	MOE (MPa)	Shear Strength (MPa)
I	$B^N–B^N–B^N–B^N–B^N$	110 ^{d} (10)	12405 ^{f} (405)	2.8 ^{bc} (0.6)
II	$B^N–A^N–B^N–A^N–B^N$	97^{c} (17)	9765 ^{cd} (1644)	3.2 ^{cde} (0.5)
III	$B^N–A^N–A^N–A^N–B^N$	101 ^{cd} (10)	11,033 ^{e} (996)	2.9 ^{cd} (0.6)
IV	$A^N–B^N–B^N–B^N–A^N$	75 ^{a} (8)	7934 ^{a} (1363)	3.7 ^{fg} (0.5)
V	$A^N–B^N–A^N–B^N–A^N$	88 ^{b} (14)	9392 ^{bc} (1583)	3.0 ^{cde} (0.8)
VI	$A^N–A^N–A^N–A^N–A^N$	75 ^{a} (5)	8103 ^{a} (648)	2.9 ^{cd} (0.6)
VII	$B^D–A^N–A^N–A^N–B^D$	99 ^{c} (8)	10,893 ^{e} (798)	3.4 ^{ef} (0.6)
VIII	$A^D–B^N–B^N–B^N–A^D$	73 ^{a} (6)	8644 ^{ab} (1221)	3.7 ^{fg} (0.6)
IX	$B^D–A^N–B^D–A^N–B^D$	105 ^{cd} (8)	10,970 ^{e} (785)	3.3 ^{def} (0.8)
X	$A^D–B^N–A^D–B^N–A^D$	79 ^{a} (7)	8796 ^{abc} (1170)	3.8 ^{g} (0.7)
XI	$B^D–B^D–B^D–B^D–B^D$	119 ^{e} (14)	12,290 ^{f} (1409)	3.2 ^{cde} (0.6)
XII	$B^D–A^D–B^D–A^D–B^D$	102 ^{cd} (11)	10,758 ^{de} (557)	2.2 ^{a} (0.2)
XIII	$B^D–A^D–A^D–A^D–B^D$	101 ^{cd} (5)	10,670 ^{de} (995)	3.2 ^{cde} (0.3)
XIV	$A^D–B^D–B^D–B^D–A^D$	79 ^{a} (5)	8913 ^{abc} (717)	3.2 ^{cde} (0.6)
XV	$A^D–B^D–A^D–B^D–A^D$	80 ^{ab} (7)	9361 ^{bc} (691)	3.7 ^{fg} (0.5)
XVI	$A^D–A^D–A^D–A^D–A^D$	78 ^{a} (9)	8345 ^{ab} (1111)	2.5 ^{ab} (0.5)

Values in parentheses are standard deviations; means followed by the same letter are not significantly different at $p \leq 0.05$.

3.3. Mechanical Properties of Plywood Samples

The mechanical properties of plywood samples, which were made by combining veneers of different wood species and different types of treatment in one panel, are summarized in Table 5. The highest (119 MPa) and lowest (73 MPa) values of MOR were observed in plywood samples ($B^D–B^D–B^D–B^D–B^D$), a panel made of densified birch veneer, and ($A^D–B^N–B^N–B^N–A^D$), a panel with outer layers of densified alder veneer and inner layers of non-densified birch veneer, respectively. The highest (12,405 MPa) and lowest (7934 MPa) values of MOE were found in plywood samples ($B^N–B^N–B^N–B^N–B^N$), a panel made of non-densified birch veneer, and ($A^N–B^N–B^N–B^N–A^N$), a panel with external layers of non-densified alder veneer and internal layers of non-densified birch veneer, respectively. The highest (3.8 MPa) and lowest (2.2 MPa) shear strength values were observed in plywood samples (A^D–B^N–A^D–B^N–A^D) and (B^D–A^D–B^D–A^D–B^D), respectively. Thus, the highest shear strength was observed between densified alder veneer and non-densified birch veneer (A^D–B^N–A^D–B^N–A^D). The lowest shear strength was found between a densified birch and densified alder veneer (B^D–A^D–B^D–A^D–B^D).

The ANOVA analysis showed (Table 6) that combining veneers of different wood species in one panel had a significant effect on the MOR and MOE, while combining different types of veneer treatment had an insignificant effect on the MOR and MOE. The mixed-species plywood panels manufactured from densified veneers had a higher MOR and MOE than panels made from non-densified veneers (Table 5, Figure 4). The mixed-species plywood panels made with outer layers comprising birch veneers (non-densified or densified) had higher MOR and MOE than panels made with outer layers comprising alder veneers (non-densified or densified). At the same content of birch and alder veneers in one panel (samples that included non-densified veneers B^N–A^N–B^N–A^N–B^N and A^N–B^N–B^N–B^N–A^N, or samples that included densified veneers B^D–A^D–B^D–A^D–B^D and A^D–B^D–B^D–B^D–A^D), plywood with outer layers of birch veneer had about 22% and 18% higher MOR and MOE, respectively, when compared to plywood with outer layers of alder veneer. The birch plywood panels made from non-densified or densified veneers had higher MOR and MOE than alder panels made from non-densified or densified veneers. Birch plywood samples made from non-densified (B^N–B^N–B^N–B^N–B^N) or densified (B^D–B^D–B^D–B^D–B^D) veneer had 31.5% and 34.8% higher MOR, and 34.7% and 32.1% higher MOE than alder plywood samples made from non-densified (A^N–A^N–A^N–A^N–A^N) or densified (A^D–A^D–A^D–A^D–A^D) veneers, respectively. This was explained by the higher strength and density of birch wood compared to alder wood. However, the difference between the MOR values for mixed-species plywood samples (B^N–A^N–A^N–A^N–B^N, B^N–A^N–B^N–A^N–B^N, B^D–A^D–A^D–A^D–B^D and B^D–A^D–B^D–A^D–B^D) made from non-densified or densified veneers was insignificant, so in practice, during the manufacture of birch plywood, alder veneers can be used to form the inner layers without deteriorating the MOR of the plywood. Similar results were obtained in the research conducted by Bal [19], in which seven-ply LVL panels were manufactured from fast-growing poplar veneers and used as the inner layers, while eucalyptus veneers were used as the outer layers. The author demonstrated that using two eucalyptus veneers on the faces significantly increased the MOE (30%) and MOR (12%) in comparison to poplar-only LVL panels. In another work [26] where seven-ply LVL panels were manufactured by combining higher-density Austrian pine veneers and lower-density Lombardy poplar veneers, it was found that as the share of pine veneers increased in the mixed-species panels, the MOR and MOE increased by up to 40% and 69% on average, compared to panels manufactured only with poplar. Xue and Hu [17] also found that the strength of the LVL made of birch veneers on the outer surface was much greater than the LVL made of poplar veneers.

Table 6. The influence of the variable factors on the mechanical properties of plywood.

Source of Variation	MOR		MOE		Shear Strength	
	F	Sig.	*F*	Sig.	*F*	Sig.
W	62.765	0.000	48.591	0.000	12.048	0.000
T	0.998	0.396	0.782	0.506	9.562	0.000
W × T	1.881	0.077	1.825	0.086	7.326	0.000

W—combination of veneers of different wood species; T—combination of veneers with different treatments used.

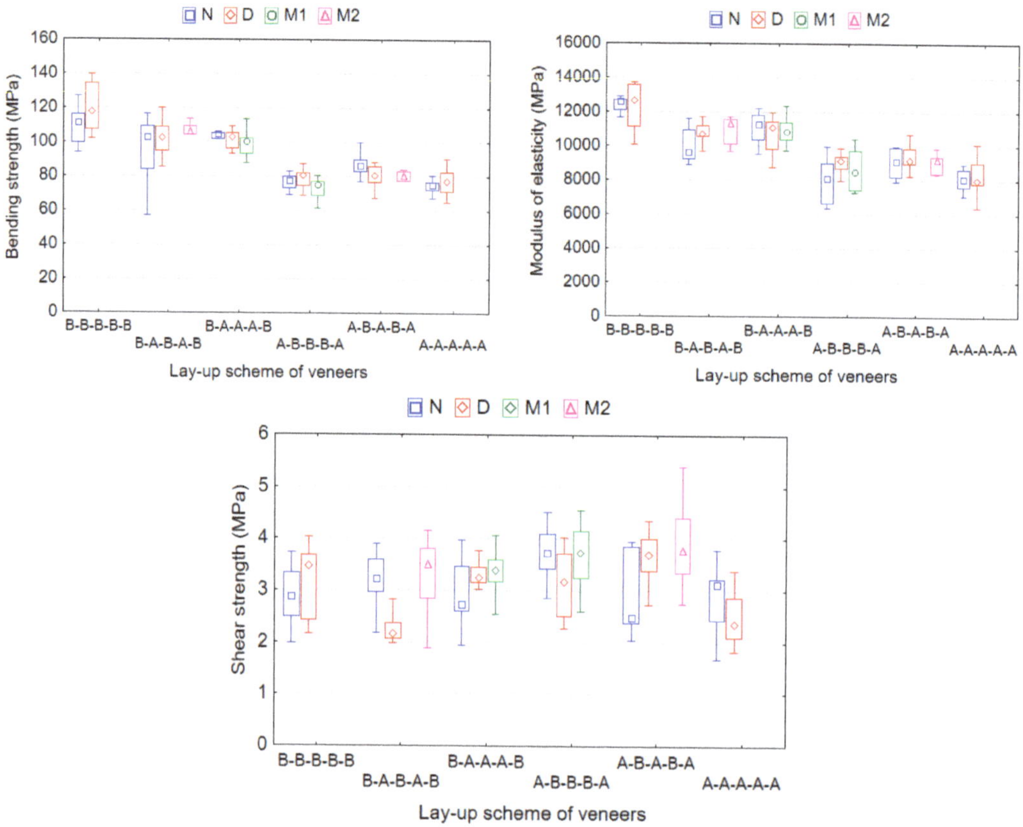

Figure 4. Effect of wood veneer species and type of veneer treatment on the MOR, MOE and shear strength of plywood panels. N—non-densified veneers in one panel; D—densified veneer in one panel; M1—mix of veneers (D–N–N–N–D) in one panel; M2—mix of veneers (D–N–D–N–D) in one panel. Statistical features are median, 25% minimum and 75% maximum.

Figure 5 shows the linear dependence of the MOR and MOE of plywood samples on their density. These dependences were not strong, which indicates that, in addition to density, other factors, such as the veneer thickness, species and type and amount of adhesive used, affected the MOR and MOE. In another work [39], it was also observed that the MOR and MOE of plywood panels made from densified and non-densified veneers increased when the density increased.

With an increase in the proportion of alder veneer in the inner layers of birch plywood made from non-densified veneers (B^N–A^N–B^N–A^N–B^N and B^N–A^N–A^N–A^N–B^N), the MOR and MOE increased, although its values were lower (97 MPa and 9765 MPa for B^N–A^N–B^N–A^N–B^N, respectively; 101 MPa and 11033 MPa for B^N–A^N–A^N–A^N–B^N, respectively) than the MOR and MOE values of 110 MPa and 12405 MPa, respectively, obtained for plywood made from birch alone (B^N–B^N–B^N–B^N–B^N). The addition of birch veneer to the inner layers of an alder plywood also led to an increased MOR and MOE. A similar trend was observed for plywood panels made from densified veneers. An increase in the proportion of densified veneer in one panel contributed to an increase in the MOR and MOE. Densified veneer had a higher density than non-densified veneer. It is known that MOR and MOE increases with increasing density [39]. This was in good agreement with the linear relationship between density and both MOR and MOE obtained in this study (Figure 5).

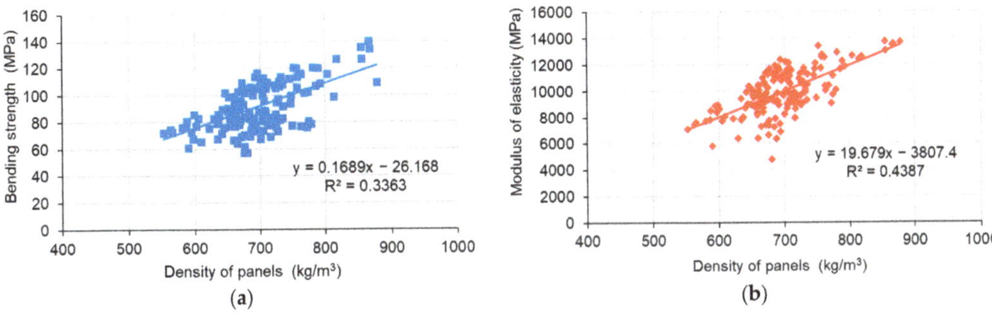

Figure 5. Dependence of bending strength (**a**) and modulus of elasticity (**b**) of plywood samples on their density.

Combinations of different wood species and types of veneer treatment in one panel significantly affected the shear strength of plywood samples (Table 6). In addition, the effect of wood species was stronger ($F = 12.048$) than the effect of the type of veneer treatment used ($F = 9.562$). It was found that there was no significant difference ($p > 0.05$) in shear strength values (2.8 MPa and 2.9 MPa, respectively) between birch (B^N–B^N–B^N–B^N–B^N) and alder (A^N–A^N–A^N–A^N–A^N) plywood panels made from non-densified veneers. However, a significant difference ($p \leq 0.05$) in shear strength values (3.2 MPa and 2.5 MPa, respectively) was observed between birch (B^D–B^D–B^D–B^D–B^D) and alder (A^D–A^D–A^D–A^D–A^D) plywood panels made from densified veneers. When the cut was along the birch (the second sheet-layer of veneer in one panel), the strength was higher, while the strength was lower when the cut was along the alder (the second sheet-layer of veneer in one panel). However, in this study, all plywood composed of densified and non-densified veneers still met the requirements (>1.0 MPa) of EN 314-2 [32]. This showed that all the combinations with high-density birch veneer and low-density alder veneer bonded well.

From the obtained experimental results, it can be observed that the addition of alder veneer to the inner layers of birch plywood had a positive effect on the shear strength, increasing it compared to the strength of a birch-only plywood made from non-densified veneer. Similarly, the addition of birch veneer to the inner layers of an alder plywood made from both non-densified or densified veneers led to an increased shear strength.

Higher values of shear strength were observed in mixed-species plywood panels with alternating sheets of densified and non-densified veneers in adjacent layers (panels B^D–A^N–A^N–A^N–B^D, A^D–B^N–B^N–B^N–A^D, B^D–A^N–B^D–A^N–B^D and A^D–B^N–A^D–B^N–A^D). In our previous work [27], it was also observed that using densified veneers increased the mechanical performance of plywood panels, but worsened the *TS* and *WA* of panels.

Some authors [1,40] have affirmed that the density of wood is an important factor that affects the formation of an adhesive bond between veneers. In their opinion, low-density woods will absorb a larger quantity of adhesive, due to its higher porosity. Based on this, it can be assumed that when non-densified veneer is in contact with another non-densified veneer (N–N), more adhesive is absorbed by the two surfaces to be bonded and less adhesive remains on the surfaces to be bonded. Starvation bonding may occur in this case. When densified veneer is in contact with non-densified veneer (D–N), it can be expected that less adhesive will be absorbed by the densified surface (due to its lower porosity) and more adhesive will remain between the bonded surfaces. This will ensure a good bonding strength and prevent "hungry" bonding. In addition, it should be taken into account that birch and alder veneers have a different density and porosity.

Microscopic images of a bond line in birch and alder plywood are presented in Figures 6 and 7.

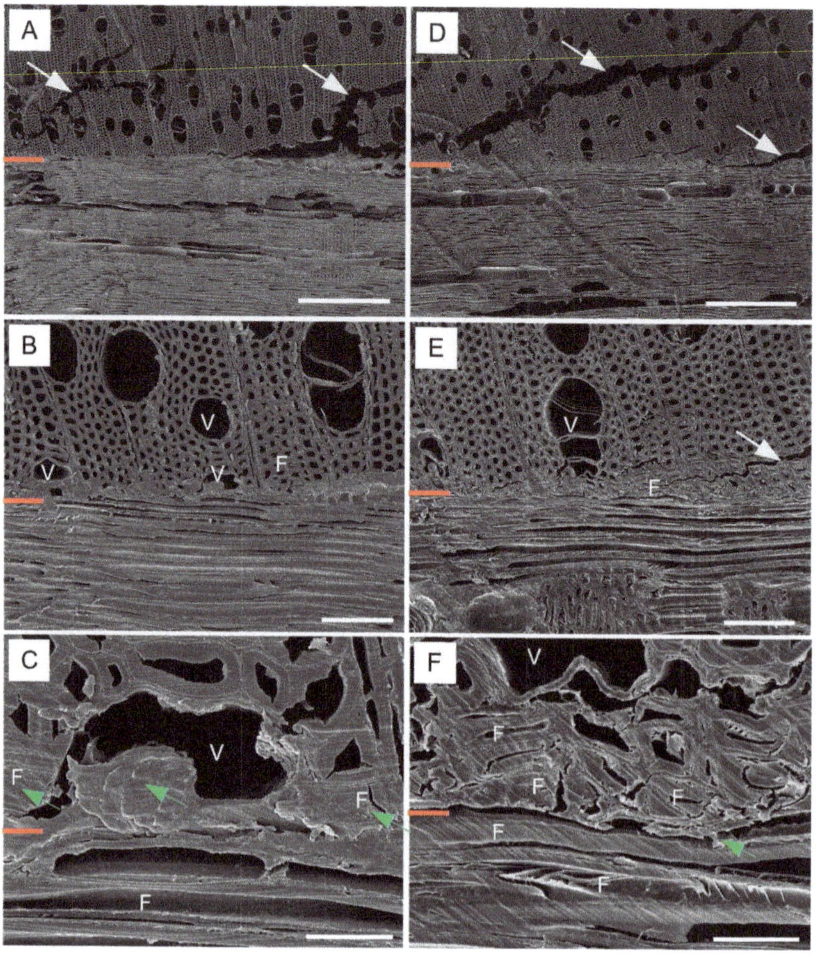

Figure 6. SEM observation of bond line in birch plywood. (**A–C**)—non-densified veneers (panel I); (**D–F**)—densified veneers (panel XI); V—vessel, F—fiber, red line—bond line, white arrow—crack, green arrow—glue. Scale bar: (**A**,**D**) = 500 μm, (**B**,**E**) = 100 μm, (**C**,**F**) = 20 μm.

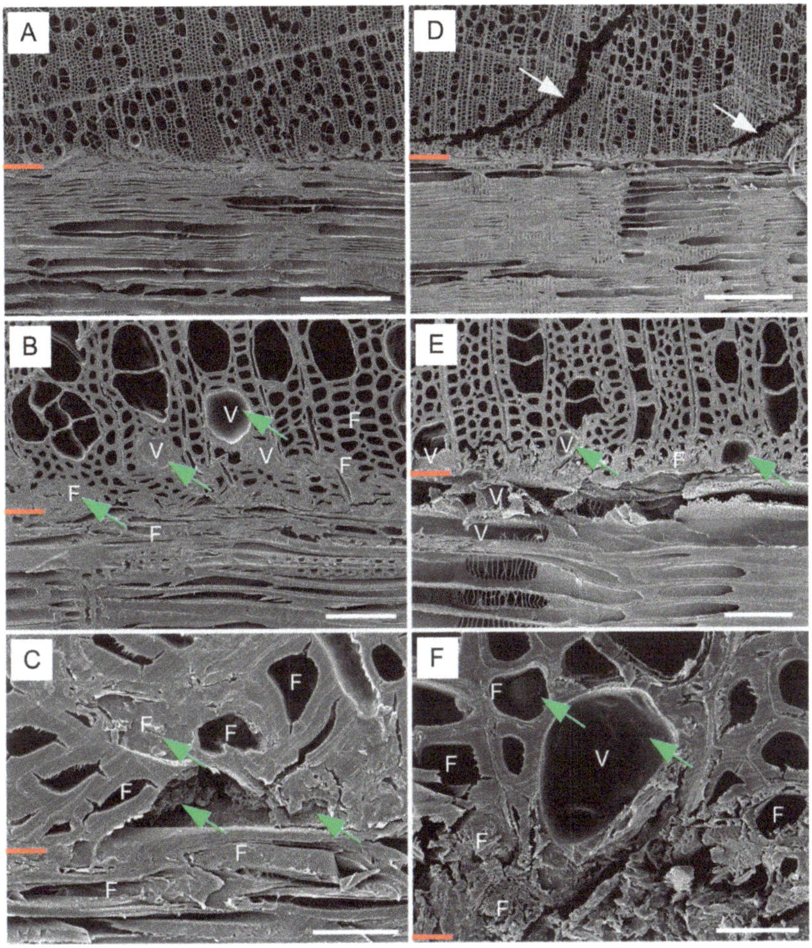

Figure 7. SEM observation of bond line in alder plywood. (**A–C**)—non-densified veneers (panel VI); (**D–F**)—densified veneers (panel XVI); V—vessel, F—fiber, red line—bond line, white arrow—crack, green arrow—glue. Scale bar: (**A,D**) = 500 µm, (**B,E**) = 100 µm, (**C,F**) = 20 µm.

Figure 6 shows that the birch wood had a smaller percentage of vessels compared to the alder (Figures 6A and 7A). Wagenführ [41] stated that the number of vessels in birch wood ranges from 40 to 60 per 1 mm^2, while in the case of alder wood the number of vessels is 75–145 per 1 mm^2; there was also a difference in the porosity of the woods, with 59% in birch compared to 71% in alder. The arrangement of the anatomical elements and their morphometric parameters influenced the resulting penetration of the glue from the bond line into the deeper layers of the veneer. The bond line in non-densified birch veneers was thin and the glue did not penetrate into the deeper layers of the glued veneers (Figure 6B,C). The penetration of glue into the deeper layers of the densified veneer was additionally prevented by a layer of compressed fibers and vessels (Figure 6F). The densification process of the veneer surface layers resulted in deformed anatomical elements and eliminated lumens. The densification of the surface layers of birch veneers was also manifested by a reduction in the surface roughness (Figure 3).

Figure 7 shows a bond line in alder plywood made from non-densified and densified veneers. The alder wood had a higher number of vessels than the birch, and the wood fibers were thin-walled (Figure 7A,B). The higher porosity of the alder wood, thus, enabled an easier penetration of the glue into the deeper layers of the veneer (Figure 7B,C). In the case of densified alder veneers, there was also a distinct layer of deformed fibers and vessels. The densification of the surface layers, as in the case of birch, prevented the penetration of the glue into the deeper layers of the veneer, and the bond line was, therefore, thinner. In both types of wood panels, it was possible to observe distinct microscopic cracks that occurred when the veneer was peeled off (Figure 6A,B and Figure 7D).

4. Conclusions

The results showed that the type of construction, wood species and applied thermal densification of the veneer affected the examined physical and mechanical properties. According to the ANOVA analysis, the *WA*, MOR, MOE and shear strength of plywood were more sensitive to the mixing of wood species in one panel than the mixing of densified and non-densified veneers. Moreover, the results indicated the great potential of black alder wood in plywood manufacturing. Alder veneers can be used to form the inner layers of plywood panels without reducing the shear strength. The B^D–A^N–B^D–A^N–B^D and B^D–A^N–A^N–A^N–B^D panels were determined to be the most reasonable lay-up schemes when the shear strength, MOR and MOE values of mixed-species plywood panels manufactured were examined.

The plywood thicknesses for all types of panel were in the range 6.7–7.4 mm and they did not go beyond tolerances for unsanded panels in accordance with standard EN 315.

Increasing the proportion of thermally densified veneer in one panel led to a lower thickness and *WA*, but higher density, MOR, MOE, shear strength and *TS* of plywood panels. It is important to note that positive effects can be achieved not only by increasing the proportion of densified veneers in one panel, but also with a change of construction by mixing the wood species used.

Plywood with outer layers of alder veneer had a lower bending strength than plywood with outer layers of birch veneer. Alder plywood with inner layers of birch veneer or adjacent birch and alder veneers in the core had a lower MOR and MOE, but higher shear strength than birch plywood with inner layers of alder veneer or adjacent alder and birch veneers in the core. It was found that plywood panels manufactured from a mixture of species offered higher bending properties when compared to panels manufactured from alder veneers only.

For single-species plywood panels, the birch plywood performed best in terms of MOR, MOE, shear strength and *WA*, while the alder plywood performed the worst. The negative influence of the alder veneer was attenuated by combining the densified and non-densified veneer, and having alder and birch veneers in one panel. The mixed-species plywood panels allowed an increased use of the lower cost, low-grade, and low-density alder wood veneers as core veneers in panels to reduce production costs and increase the mechanical properties of predominately low-density alder wood plywood.

Author Contributions: Conceptualization, methodology, software, validation, formal analysis, P.B., T.P., P.K. and V.G.; investigation, resources, and data curation, P.B., T.P., P.K., J.S., J.R. and J.V.; writing—original draft preparation, P.B.; writing—review and editing, P.B., T.P., P.K., V.G., J.S., J.R. and J.V.; visualization and supervision, P.B.; supervision, P.B. All authors have read and agreed to the published version of the manuscript.

Funding: This work was supported by the European Union's Horizon 2020 research and innovation program under grant agreement no. 952314; the Slovak Research and Development Agency under contracts APVV-18-0378, APVV-19-0269 and APVV-21-0051; project VEGA 1/0264/22; and the project entitled "Innovative construction and technology of massive solid wooden panel" (Technology Agency of the Czech Republic, reg. no. FW01010579).

Acknowledgments: The authors thank Radim Rousek for his help with the electron microscope.

Conflicts of Interest: The authors declare no conflict of interest.

References

1. Baldwin, R.F. *Plywood and Veneer-Based Products: Manufacturing Practices*; Miller Freeman: San Francisco, CA, USA, 1995; 388p.
2. Aydin, I.; Colak, S.; Colakoglu, G.; Salih, E. A comparative study on some physical and mechanical properties of Laminated Veneer Lumber (LVL) produced from Beech (*Fagus orientalis* Lipsky) and Eucalyptus (*Eucalyptus camaldulensis* Dehn.) veneers. *Eur. J. Wood Prod.* **2004**, *62*, 218–220. [CrossRef]
3. Ido, H.; Nagao, H.; Kato, H.; Miyatake, A.; Hiramatsu, Y. Strength properties of laminated veneer lumber in compression perpendicular to its grain. *J. Wood Sci.* **2010**, *56*, 422–428. [CrossRef]
4. Ozarska, B. A review of the utilization of hardwoods for LVL. *Wood Sci. Technol.* **1999**, *33*, 341–351. [CrossRef]
5. Navi, P.; Sandberg, D. *Thermo-Hydro-Mechanical Wood Processing*; EPFL Press: Lausanne, Switzerland; CRC Press, Taylor & Francis Group, LLC.: Boca Raton, FL, USA, 2012; 360p.
6. Candan, Z.; Hiziroglu, S.; Mcdonald, A.G. Surface quality of thermally compressed Douglas fir veneer. *Mater. Des.* **2010**, *31*, 3574–3577. [CrossRef]
7. Buyuksari, U. Surface characteristics and hardness of MDF panels laminated with thermally compressed veneer. *Compos. B Eng.* **2013**, *44*, 675–678. [CrossRef]
8. Bekhta, P.; Proszyk, S.; Krystofiak, T. Colour in short-term thermo-mechanically densified veneer of various wood species. *Eur. J. Wood Prod.* **2014**, *72*, 785–797. [CrossRef]
9. Bekhta, P.; Proszyk, S.; Lis, B.; Krystofiak, T. Gloss of thermally densified alder (*Alnus glutinosa* Goertn.), beech (*Fagus sylvatica* L.), birch (*Betula verrucosa* Ehrh.), and pine (*Pinus sylvestris* L.) wood veneers. *Eur. J. Wood Prod.* **2014**, *72*, 799–808. [CrossRef]
10. Bekhta, P.; Proszyk, S.; Krystofiak, T.; Mamonova, M.; Pinkowski, G.; Lis, B. Effect of thermomechanical densification on surface roughness of wood veneers. *Wood Mater. Sci. Eng.* **2014**, *9*, 233–245. [CrossRef]
11. Pelit, H.; Sönmez, A.; Budakçi, M. Effects of thermomechanical densification and heat treatment on density and Brinell hardness of Scots pine (*Pinus sylvestris* L.) and Eastern beech (*Fagus orientalis* L.). *BioResources* **2015**, *10*, 3097–3111. [CrossRef]
12. Bekhta, P.; Sedliačik, J.; Jones, D. Effect of short-term thermomechanical densification of wood veneers on the properties of birch plywood. *Eur. J. Wood Prod.* **2018**, *76*, 549–562. [CrossRef]
13. Bekhta, P.; Krystofiak, T.; Proszyk, S.; Lis, B. Adhesion strength of thermally compressed and varnished wood (TCW) substrate. *Prog. Org. Coat.* **2018**, *125*, 331–338. [CrossRef]
14. Nzokou, P.; Zyskowski, J.; Boury, S.; Kamdem, D.P. Natural decay resistance of LVL made of veneers from durable and non-durable wood species. *Holz Roh Werkst.* **2005**, *63*, 173–178. [CrossRef]
15. Burdurlu, E.; Kilic, M.; Ilce, A.C.; Uzunkavak, O. The effects of ply organization and loading direction on bending strength and modulus of elasticity in laminated veneer lumber (LVL) obtained from beech (*Fagus orientalis* L.) and lombardy poplar (*Populus nigra* L.). *Constr. Build. Mater.* **2007**, *21*, 1720–1725. [CrossRef]
16. Nazerian, M.; Ghalehno, M.D.; Kashkooli, A.B. Effect of wood species, amount of juvenile wood and heat treatment on mechanical and physical properties of laminated veneer lumber. *J. Appl. Sci.* **2011**, *11*, 980–987. [CrossRef]
17. Xue, B.; Hu, Y. Mechanical Properties Analysis and reliability Assessment of Laminated Veneer Lumber (LVL) Having Different Patterns of Assembly. *BioResources* **2012**, *7*, 1617–1632. [CrossRef]
18. Zdravković, V.; Lovrić, A.; Stanković, B. Dimensional Stability of Plywood Panels Made from Thermally Modified Poplar Veneers in the Conditions of Variable Air Humidity. *Drv. Ind.* **2013**, *64*, 175–181. [CrossRef]
19. Bal, B.C. Some technological properties of laminated veneer lumber produced with fast-growing poplar and eucalyptus. *Maderas Cienc. Y Tecnol.* **2016**, *18*, 24–39. [CrossRef]
20. Lovrić, A.; Zdravković, V.; Popadić, R.; Milić, G. Properties of Plywood Boards Composed of Thermally Modified and Non-modified Poplar Veneer. *BioResources* **2017**, *12*, 8581–8594.
21. Ilce, A. Mechanical properties of laminated veneer lumber made from ash and red pine woods. *BioResources* **2018**, *13*, 8653–8661. [CrossRef]
22. McGavin, R.L.; Nguyen, H.H.; Gilbert, B.P.; Dakin, T.; Faircloth, A. A comparative Study on the Mechanical Properties of Laminated Veneer Lumber (LVL) Produced from Blending Various Wood Veneers. *BioResources* **2019**, *14*, 9064–9081.
23. Kallakas, H.; Rohumaa, A.; Vahermets, H.; Kers, J. Effect of Different Hardwood Species and Lay-Up Schemes on the Mechanical Properties of Plywood. *Forests* **2020**, *11*, 649. [CrossRef]
24. Hsu, F.Y.; Hung, K.C.; Xu, J.W.; Wu, T.L.; Wu, J.H. Physicomechanical properties and creep behavior of plywood composed of fully and partially heat-treated veneers. *Wood Sci. Technol.* **2021**, *55*, 445–460. [CrossRef]
25. H'ng, P.S.; Paridah, M.; Chin, K. Bending properties of laminated veneer lumber produced from Keruing (*Dipterocarpus* sp.) reinforced with low density wood species. *Asian J. Sci. Res.* **2010**, *3*, 118–125. [CrossRef]
26. Kilic, Y.; Burdurlu, E.; Elibol, G.C.; Ulupinar, M. Effect of layer arrangement on expansion, bending strength and modulus of elasticity of solid wood and laminated veneer lumber (LVL) produced from pine and poplar. *Gazi Univ. J. Sci.* **2010**, *23*, 89–96.
27. Bekhta, P.; Salca, E.A.; Lunguleasa, A. Some properties of plywood panels manufactured from combinations of thermally densified and non-densified veneers of different thickness in one structure. *J. Build. Eng.* **2020**, *29*, 101116. [CrossRef]
28. *EN 322*; Wood-Based Panels—Determination of Moisture Content. European Committee for Standardization: Brussels, Belgium, 1993.
29. *EN 323*; Wood-Based Panels—Determination of Density. European Committee for Standardization: Brussels, Belgium, 1993.

30. *EN 310*; Wood-Based Panels—Determination of Modulus of Elasticity in Bending and of Bending Strength. European Committee for Standardization: Brussels, Belgium, 1993.
31. *EN 314-1*; Plywood—Bonding Quality—Part 1: Test Methods. European Committee for Standardization: Brussels, Belgium, 2004.
32. *EN 314-2*; Plywood—Bonding Quality—Part 2: Requirements. European Committee for Standardization: Brussels, Belgium, 1993.
33. *EN 317*; Particleboards and Fibreboards: Determination of Swelling in Thickness after Immersion in Water. European Committee for Standardization: Brussels, Belgium, 1993.
34. Arruda, L.; Del Menezzi, C.H.S. Effect of thermomechanical treatment on physical properties of wood veneers. *Int. Wood Prod. J.* **2013**, *4*, 217–224. [CrossRef]
35. Bekhta, P.; Proszyk, S.; Krystofiak, T.; Sedliačik, J.; Novák, I.; Mamonova, M. Effect of short-term thermomechanical densification on the structure and properties of wood veneers. *Wood Mater. Sci. Eng.* **2017**, *12*, 40–54. [CrossRef]
36. Rautkari, L.; Hughes, M.; Properzi, M.; Pichelin, F. Properties and chemical changes of surface densified wood. In Proceedings of the 11th World Conference on Timber Engineering, Trentino, Italy, 20–24 June 2010; Trees and Timber Institute, National Research Council: Trentino, Italy, 2010; pp. 235–237.
37. *EN 315*; Plywood. Tolerances for Dimensions. European Committee for Standardization: Brussels, Belgium, 2000.
38. Sheldon, S.; Walker, J. Wood-based composites: Plywood and veneer-based products. In *Primary Wood Processing: Principles and Practice*, 2nd ed.; Walker, J., Ed.; Springer: New York, NY, USA, 2006; pp. 391–426.
39. Salca, E.-A.; Bekhta, P.; Seblii, Y. The effect of veneer densification temperature and wood species on the plywood properties made from alternate layers of densified and non-densified veneers. *Forests* **2020**, *11*, 700. [CrossRef]
40. Marra, A.A. *Technology of Wood Bonding: Principles in Practice*; Van Nostrand Reinhold: New York, NY, USA, 1992; 453p.
41. Wagenführ, R. *Holzatlas*; Fachbuchverlag Leipzig im Carl Hanser: München, Germany, 2000; 707p.

forests

Article

Influence of *Chrysoporthe deuterocubensis* Canker Disease on the Chemical Properties and Durability of *Eucalyptus urograndis* against Wood Rotting Fungi and Termite Infestation

Rasdianah Dahali [1], Seng Hua Lee [2,*], Paridah Md Tahir [1,3,*], Sabiha Salim [3], Muhammad Syahmi Hishamuddin [3], Atikah Che Ismail [3], Pui San Khoo [4], Tomasz Krystofiak [5,*] and Petar Antov [6]

[1] Institute of Tropical Forestry and Forest Products (INTROP), Universiti Putra Malaysia, Serdang 43400, Selangor, Malaysia
[2] Department of Wood Industry, Faculty of Applied Sciences, Universiti Teknologi MARA (UiTM) Cawangan Pahang Kampus Jengka, Bandar Tun Razak 26400, Pahang, Malaysia
[3] Faculty of Forestry and Environment, Universiti Putra Malaysia, Serdang 43400, Selangor, Malaysia
[4] Centre for Advanced Composite Materials (CACM), Faculty of Mechanical Engineering, Universiti Teknologi Malaysia, Johor Bahru 81310, Johor, Malaysia
[5] Faculty of Forestry and Wood Technology, Poznan University of Life Sciences, 60-637 Poznan, Poland
[6] Faculty of Forest Industry, University of Forestry, 1797 Sofia, Bulgaria
* Correspondence: leesenghua@hotmail.com (S.H.L.); parida.introp@gmail.com (P.M.T.); tomasz.krystofiak@up.poznan.pl (T.K.)

Abstract: In this study, the effects of stem canker disease caused by *Chrysoporthe deuterocubensis* on the chemical properties and durability of a Eucalyptus hybrid (*E. urophylla* x *E. grandis*) were investigated. Eleven-year-old healthy and infected trees were collected. The samples were grouped into four different classes based on the infection severity: healthy (class 1), moderately infected (class 2), severely infected (class 3), and very severely infected (class 4). The changes in chemical properties were evaluated via chemical analysis and Fourier transform infrared spectroscopy (FTIR) analysis. A resistance test against fungal decay (*Pcynoporus sanguineus* and *Caniophora puteana*) and termite (*Coptotermes curvignathus*) was also performed. The results showed that reductions in cellulose and hemicellulose content from 53.2% to 45.4% and 14.1% to 13.9%, respectively, were observed in the infected samples. Meanwhile, the percentages of lignin and extractives increased from 18.1% to 20.5% and 14.6% to 20.2%, respectively. The resistance against fungi and termites varied between severity classes. Generally, infected wood behaved better than healthy wood in terms of durability against fungi and termites. The durability classes for both tests were significantly improved, from resistant to highly resistant and poor to moderately resistant, respectively. These results suggest that *E. urograndis* that is infected by *C. deuterocubensis* might have a better potential use in lumber production with regard to its durability and processing cost compared to pulp and paper products.

Keywords: *Eucalyptus urograndis*; *Chrysoporthe deuterocubensis*; infection classes; chemical analysis; FTIR analysis; fungal decay; termite attack

Citation: Dahali, R.; Lee, S.H.; Md Tahir, P.; Salim, S.; Hishamuddin, M.S.; Che Ismail, A.; Khoo, P.S.; Krystofiak, T.; Antov, P. Influence of *Chrysoporthe deuterocubensis* Canker Disease on the Chemical Properties and Durability of *Eucalyptus urograndis* against Wood Rotting Fungi and Termite Infestation. *Forests* **2023**, *14*, 350. https://doi.org/10.3390/f14020350

Academic Editor: Miha Humar

Received: 17 January 2023
Revised: 6 February 2023
Accepted: 6 February 2023
Published: 9 February 2023

1. Introduction

Eucalyptus spp. is commonly used in indigenous reforestation programs in Australia, Brazil, South Africa, and China to meet the demands of various forest-based industries, with the majority used for veneer in plywood, pulp and paper, or charcoal [1]. It is popular due to its rapid growth, with trees being harvested between the ages of 6 and 7 years. Numerous cloning programs have begun in these countries to improve the quality of this material, primarily through the development of hybrid species [2–5]. In particular, hybrid trees of *Eucalyptus urophylla* × *E. grandis*, also known as *E. urograndis*, represent the most commonly used hybrid for producing pulp, paper, and cellulose for industrial use [6,7]. The crossing of these two Eucalyptus species permits fast growth, a characteristic of *E. grandis*, as well as improved physical properties of the wood, i.e., increased wood density from 460 to 650 kg/m^3), which is a characteristic of *E. urophylla*. It is also thought that this hybrid is well adapted to various ecological conditions [8]. According to Luo et al. [1], the majority of Eucalyptus plantations in China produce logs that are excellent for solid wood applications, such as veneer production, due to characteristics such as low taper, good straightness, desirable wood density, rigidity, and wood surface texture [9,10]. Therefore, hybrid clones of *E. urophylla* × *E. grandis* possess ideal characteristics in terms of both growth and density, with numerous industries continuing to invest in research on the genetic development of this hybrid [11–13]. Recently, fluctuations in the price of pulp and paper have caused many of the major forestry companies to begin considering alternative uses for Eucalyptus wood [14]. Factors such as scarcity and high cost of native woods, as well as ecological pressure from overharvesting, have contributed to the rise in the utilization of Eucalyptus wood materials, particularly in the flooring and lumber industry [15]. Although the major properties of wood from *E. urograndis* have been reported in terms of pulp and paper, studies specifically concerning the potential use of these materials in lumber products are lacking. Meanwhile, the utilization of Eucalyptus species, particularly Eucalyptus hybrids, in Malaysia is still new [16–18] and requires much attention and care, especially concerning the susceptibility of this hybrid to biological deterioration. According to research from other countries [18], *Eucalyptus* spp. is susceptible and vulnerable to disease and infection [19,20]; for example, disease had a negative impact on Eucalyptus plantations in many African countries [21].

Nonetheless, despite their promising future, Eucalyptus trees are vulnerable to fungal disease infection, particularly stem canker (*Chrysoporthe deuterocubensis*). Stem canker is one of the most important pathogens that infected and killed Eucalyptus trees in several Malaysian plantation locations [22]. The cankers cause broken limbs and trunks, stunted and distorted growth, and, frequently, mortality. This disease is known to kill a large number of trees, particularly in young plantations, and as a result, it is a major impediment to the successful establishment of Eucalyptus plantations [23]. The infected trees are normally left to rot on site without any intervention. This, however, raises two concerns. The first is that if it is left untreated, it may infect other healthy trees nearby. The second concern is whether the infected tree can be used in another way to avoid total loss. In this regard, we conducted a number of studies to investigate the effects of stem canker disease (*Chrysoporthe deuterocubensis*) infection on the physical, mechanical, and machining properties of *E. urograndis* in the Sabah plantation [24,25]. The results revealed that the strength properties of the infected trees were reduced. However, the reduction was minor, as the most severely damaged trees lost 21.3% of their modulus of rupture when compared to healthy trees, while lightly damaged trees lost only 11.6%. These infected trees are said to be suitable for non-structural applications such as furniture, interior finishing, window frames, and doors [24]. Infected trees, on the other hand, were found to have acceptable machining characteristics, justifying their use in furniture production [25].

Another important criterion for the efficient utilization of wood is its natural durability, particularly when applied externally. Despite the fact that the infected trees are still able to be used as furniture or door frames, their durability against fungi and termites still remains unknown. In addition to related research, the chemical properties and durability of infected *E. urograndis* against wood-degrading organisms such as fungi and termites were investigated. The primary goal of this study was to assess the impact of *C. deuterocubensis* infection on the chemical properties and durability of *E. urograndis* against wood rotting fungi and termite infestation. The relationship between severity classes and the extent of fungi and termite damage was also investigated.

2. Materials and Methods

Trees of 11-year-old *Eucalyptus urophylla* x *E. grandis* hybrid clones originating from China were collected in this study. The plantation was planted and managed by the company Sabah Softwoods Berhad, Tawau, Malaysia, with geographical coordinates 4°33′16.3″ N, 117°42′58.7″ E. The trees were planted in an entirely randomized block design with three types of planting spaces (1.5 m × 3 m, 1.8 m × 3 m, and 3 m × 3 m). The region where the experiment was deployed is about 11.7 ha and has a moist tropical climate with two well-defined seasons (dry and wet) [26]. The dry season lasts from May to September and the rainy season lasts from October to April. The infected trees were identified, and a forest pathologist confirmed by visual examination that the pathogen that attacked the trees was *C. deuterocubensis* [20]. Pictures were taken, the characteristics of the infection were recorded, and the extent of the infection was determined based on the severity of the symptoms that occurred. Based on careful examinations of the pictures, as well as the physical symptoms/characteristics of the disease and its percentage of occurrence, the infected trees were grouped into four classes according to the severity of the infection, as depicted in Table 1. Meanwhile, Table 2 displays the physical properties of healthy and infected *E. urograndis* wood that have been observed from our previous study [24] such as equilibrium moisture content (EMC), density, and volumetric shrinkage (Vol_{sh}).

Table 1. Classes of severity for *Eucalyptus urograndis* infected with *Chrysoporthe deuterocubensis* [a].

Class	Category	Symptom
1	Healthy	Stem appears normal without any symptom of being infected
2	Moderate	Swollen bark (callus) Cracking Fruiting structure Fresh kino pocket Canker
3	Severe	Swollen bark (callus) Cracking Fruiting structure Fresh kino pocket and fresh kino/gummosis Canker Sunken Rotten
4	Very severe	Swollen bark (callus) Cracking Fruiting structure Dried kino pocket & dried kino/gummosis Canker Sunken Rotten Shoot

[a] Adapted with permission from Rasdianah et al. [24].

Table 2. Physical properties of *Eucalyptus urograndis* [a].

Infection Classes	EMC (%)	Density (Kg/m^3)	Vol_{sh} (%)
1	10.5	670.8	15.1
2	10.1	618.9	14
3	10.2	706.8	14.5
4	9.7	542.3	12.9

[a] Adapted with permission from Rasdianah et al. [24].

2.1. Evaluation of Chemical Properties

2.1.1. Wet Chemical Analysis

Small splints (1 to 2 cm long) were taken from each infection class after conditioning for evaluation of their chemical properties. To produce sawdust, the splints were ground using a Wiley cutting mill (Wiley, Model 4, Swedesboro, NJ, USA); to obtain homogeneous sizes, sawdust was screened using a 40 to 60 mesh (0.4 to 0.6 mm) sieve according to the Technical Association of the Pulp and Paper Industry (TAPPI) standard method, T 257 cm-02 [27]. The chemical composition analysis of *E. urograndis* particles was conducted at the Malaysian Agricultural Research and Development Institute (MARDI) lab, Serdang, by using neutral detergent fiber (NDF) and acid detergent fiber (ADF). This is a common way to evaluate the main fiber constituents, cellulose, hemicelluloses, and lignin. All chemical constituent percentages for each class of severity samples were determined as mean values of triplicate experiments.

2.1.2. Fourier Transform Infrared (FTIR) Analysis

The FTIR analysis was performed at the Biocomposite Laboratory of the Institute of Tropical Forestry and Forest Product (INTROP), UPM, using FTIR-attenuated total reflection (ATR) spectrometer instrument (iS10 FTIR, Thermo Nicolet, Durham, NH, USA) with spectrometer resolution of 0.4 cm^{-1}. The FTIR analysis was performed to qualitatively determine organic components, including chemical bond, functional group, and organic content (e.g., protein, carbohydrate, and lipid), and to correlate the result of chemical composition change from wet chemical analysis between different infection severity classes of the samples. The study included healthy samples as well as moderately, severely, and very severely (class 1, 2, 3, and 4) infected samples. As a control for comparison, a healthy sample was prepared. FTIR-ATR with a spectral range cutoff of 525 cm^{-1} was used to analyze the components of all samples. The particles were measured in the absorption mode using a mid-IR spectrum with wavenumbers ranging from 400 to 4000 cm^{-1}, and all spectra were plotted on the transmittance axis using the same scale [28].

2.2. Evaluation of Biological Properties

2.2.1. Decay Resistance

The decay resistance test was determined by exposing the samples (20 mm × 20 mm × 10 mm) to white rot and brown rot fungi (namely *Pycnoporus sanguineus* (WML 006) and *Coniophora puteana* (WML 004)) in a soil block test for 12 weeks according to procedures outlined in ASTM D2017 standard [29]. These fungi (Figure 1a,b) were collected from the Mycology and Pathology Laboratory, Forest Research Institute Malaysia (FRIM) cultured collection. These fungi were chosen because they are widely distributed in tropical regions. Even in allegedly durable timber, they cause significant damage [30]. Twenty replicates of each class of severity (class 1, 2, 3, and 4) were tested against each fungus. The test blocks were labeled and oven-dried at 103 ± 2 °C until they reached a constant weight. The blocks initial weight was then recorded as W_1. Each culture bottle contained 150 g of sieved soil and 70 mL of distilled water. Feeder strips of pine wood (35 mm long × 28 mm wide × 3 mm thick) were placed on top of the soil in each culture bottle for fungal inoculation. The bottles with the feeder strips were loosely capped and steam-sterilized at 121 °C for 20 min before being cooled in laminar flow under ultraviolet light at room temperature.

The fungus was placed at the corner of the feeder strip after cooling. Then, the bottles were incubated at a temperature of 27 ± 2 °C and 70% relative humidity for approximately 3 weeks until the mycelium fully covered the feeder strips. Then, the test blocks were placed on top of the feeder strips and incubated for 12 weeks of fungal exposure. The test blocks were taken out from the bottles at the end of the incubation period, and all mycelium on the test block surfaces was removed. The degree of fungal attack was estimated by determining the weight loss of the test block. The test blocks were then oven-dried until they reached a constant weight, W_2. After weighing each block, the percentage of weight loss (WL_{decay}) for the test block was calculated using Equation (1):

$$WL_{decay}\ (\%) = 100\ (W_1 - W_2)/W_1 \quad (1)$$

where W_1 is the initial weight of the test block before exposure to fungus (g) and W_2 is the final weight of the test block after exposure to fungus (g).

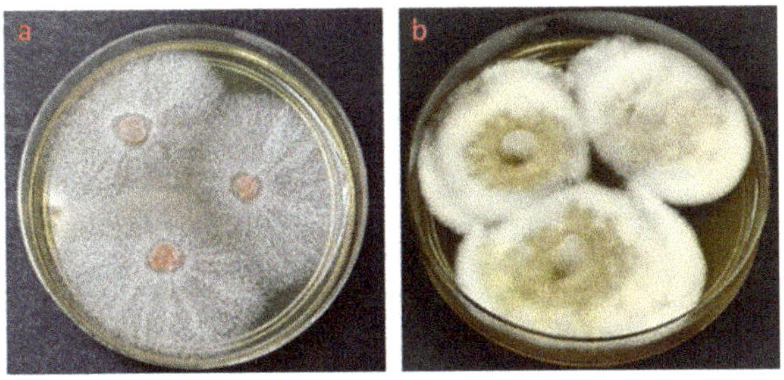

Figure 1. White rot (**a**) (*Pycnoporus sanguineus*) and brown rot (**b**) (*Coniophora puteana*) fungi isolates grown on malt extract agar (MEA) media.

Table 3 lists the assignment of resistance classes based on the mean weight loss of the samples caused by fungal decay.

Table 3. Classification of decay resistance according to weight loss values [c].

Sample Condition	Mean Weight Loss (%)	Resistance Class
Highly resistant	0–10	I
Resistant	11–24	II
Moderately resistant	25–44	III
Slightly resistance or non-resistant	≥45	IV

[c] Adapted from ASTM standard D2017 [29].

2.2.2. Termite Resistance

A no-choice termite resistance test with *Coptotermes curvignathus* Holmgren (subterranean termites) was performed according to the American Wood Protection Association (AWPA) standard E1-09 procedure [31]. The termites were collected from an infested pine plantation at Institute of Biosciences, Universiti Putra Malaysia. The termite species was confirmed both morphologically and genetically with references from previous study [32]. *C. curvignathus*, as shown in Figure 2, is one of the most common species of termite [33], and causes severe damage to forest trees, wooden structures, and buildings [30,34,35]. They are also the only species of termite capable of establishing secondary nests that allow them to attack high-level buildings [36]. Living trees can also be attacked by this species in their inner and outermost parts. A total of 20 replicates (20 mm × 20 mm × 7mm) for each sample of severity class were tested. A sample was placed in a glass container containing 200 g of sterilized sand and 30 mL of distilled water with 1 ± 0.05 g of *C. curvignathus*. *C. curvignathus*, consisting of 90% workers and 10% soldiers, were weighed and placed in each bottle.

In this study, only vigor and active termites were used. A sample from each severity class was randomly selected and conditioned at 20 °C until it reached a constant weight. The initial weight of the blocks, W_i, was measured before they were placed in sterilized test bottles filled with sand. The cultured bottles were incubated and maintained at a temperature of 27 ± 2 °C and 70% relative humidity for approximately 4 weeks. After 4 weeks, the test block samples were taken out of the cultured bottles and brushed off to eliminate sand particles. All samples were oven-dried until a constant weight, W_f, was achieved. The efficacy of severity class infection was evaluated based on the percentage weight loss of the materials due to termite attack. Using Equation (2), each test block was examined based on the percentage of weight loss ($WL_{termites}$). Classification of the durability classes was carried out according to the Indonesian National Standard SNI 01.7207 [37], based on the weight loss criteria developed (Table 4). Each sample was also rated based on the visual evaluation system provided by the AWPA standard E1-09 [31] in Table 5.

Figure 2. Subterranean termites (*Coptotermes curvignathus*) collected from infested pine plantation at Institute of Bio-science, UPM.

$$WL_{termites}\ (\%) = 100\ (W_i - W_f)/W_i \quad (2)$$

where W_i is the initial weight of the test block before exposure to termites (g) and W_f is the weight of the test block after exposure to termites (g).

Table 4. Classification of termite resistance according to weight loss values [d].

Sample Condition	Mean Weight Loss (%)	Resistance Class
Very resistant	<3.52	I
Resistant	3.52–7.50	II
Moderate	>7.50–10.96	III
Poor	>10.96–18.94	IV
Very poor	>18.94	V

[d] Adapted from standard SNI 01.7207 [37] and Terzi et al. [38].

Table 5. Rating system for visual evaluations of termite resistance [e].

Visual Rating Classification	Rating
Sound	10
Trace, surface nibbles permitted	9.5
Slight attack, up to 3% of cross-sectional area affected	9
Moderate attack, 3%–10% of cross-sectional area affected	8
Moderate/severe attack, penetration, 10%–30% of cross-sectional area affected	7
Severe attack, 30%–50% of cross-sectional area affected	6
Very severe attack, 50%–75% of cross-sectional area affected	4
Failure	0

[e] Adapted from AWPA standard E1-09 [31].

2.3. Statistical Analysis

To determine how the different infection classes affected the chemical changes and durability of the *E. urograndis* wood, a one-way analysis of variance (ANOVA) was used to analyze and interpret the data. The discrepancies between the mean values of each severity class were investigated further using Duncan's multiple range (DMR) test at $p \leq 0.05$. Pearson's correlation coefficients were used to assess the relationships between infection classes and weight loss due to fungal decay (WL_{decay}) and termite attack ($WL_{termites}$). All statistical analyses were performed using SPSS version 22.0. (IBM, Armonk, NY, USA).

3. Results and Discussion

3.1. Chemical Properties of Healthy and Infected Wood

Table 6 shows the chemical composition of 11-year-old healthy and infected *E. urograndis* wood. *E. urograndis* has a cellulose content of more than 45% on average. A maximum cellulose content of 53.2% was determined in the healthy (class 1) *E. urograndis* compared to the infected (class 2, 3 and 4) trees, which had cellulose contents of 50.9%, 49.8%, and 45.4%, respectively. A downward trend was observed as the cellulose content decreased, as the extent of the trees' infections were becoming severer. Similar to cellulose, hemicellulose content was also reduced as the severity of the infection increased. Meanwhile, lignin and extractive content increased. However, the changes were interdependent, as a decrease in hemicellulose content was accompanied by an increase in lignin, and vice versa.

Table 6. Mean values of chemical composition of *Eucalyptus urograndis* wood.

Severity Classes	Chemical Composition (%)			
	Cellulose	Hemicellulose	Lignin	Other Component
1 (Healthy)	53.2	14.1	18.1	14.6
2 (Moderate)	50.9	11.2	19.9	18.0
3 (Severe)	49.8	11.6	19.0	19.6
4 (Very severe)	45.4	13.9	20.5	20.2

Aromatic rings with multiple potential branches characterize amorphous lignin. It acts as a binding agent for individual cells, as well as for the fibrils that make up the cell wall. Lignin first forms between adjacent cells in the middle lamella, tightly bonding them to form a tissue. It then spreads through the cell wall, penetrating hemicellulose and bonding cellulose fibrils. Lignin provides compressive strength and stiffness to the cell wall of tree tissue and individual fibers, protecting the carbohydrate from physical and chemical damage. The amount of lignin effects the structure, properties, morphology, and flexibility of wood [39]. According to Ferrari et al. [40] and Savory and Pinion [41], the ascomycete fungi are able to break down lignin, albeit inefficiently. Contrary to that, Andlar et al. [42], Janusz et al. [43], and Kirk and Farrell [44] stated that ascomycete fungi are unable to degrade lignin and only consume the easy to-access hemicellulose and cellulose. In this study, it can be seen in Table 6 that the lignin content of infected *E. urograndis* is higher (19.9%, 19.0%, and 20.5%) than in the healthy (18.1%) sample. This fact is in agreement with previous studies by Mafia et al. [45] and Foelkel et al. [46]. The authors observed the increase in the proportions of lignin and extractives when they analyzed the wood from Eucalyptus trees with canker caused by *Cryphonectria cubensis* (Bruner) Hodges.

Wood extractives are natural products that exist independently of a lignocellulose cell wall. They are present within the cell wall, but are not chemically connected to it. Aromatic phenolic compounds, aliphatic compounds (fats and waxes), terpenes, and terpenoids, as well as a variety of other minor organic compounds, are among these compounds [47,48]. In their study, Shebani et al. [49] discovered that *Eucalyptus* spp. wood contains more polar extractives, which include tannins, gums, sugars, starches, and colored matter. The phenolic compounds affect fungal physiology directly. As previously stated, aliphatic compounds can act as surfactants, limiting fungal adhesion to wood surfaces. The phenolic compounds are classified into four groups: lignans, stilbenes, flavonoids, and tannins [50,51]. These extractives are molecules produced by trees to protect themselves from biotic and abiotic stressors. The proportion of other components in infected samples classes 2, 3, and 4 was higher than in healthy samples (class 1). As the infection progressed, the extractive content increased steadily from 18.6% to 20.2%, compared to 14.6% in the healthy samples. A similar result was found in a study on a *Eucalyptus grandis* hybrid by Mafia et al. [45]. According to the author, after being infected with *Ceratocystis fimbriata* wilt, the proportion of extractives and lignin increased.

Based on this result, it can be said that *C. deuterocubensis* causes a chemical change to *E. urograndis* wood, as the infected wood contained less cellulose and hemicellulose than the healthy wood. This fungus enters the tree via injuries to the bark. It then spreads to the underlying vascular cambium, destroying these tissues as it advances, degrading some or all major cell wall components, and absorbing breakdown products of cellulose or hemicellulose. This finding was supported by the previous study conducted by Dahali et al. [24] on the physical properties of *E. urograndis,* where it was found that the fiber saturation point, equilibrium moisture content, density, and volumetric shrinkage of infected wood were lesser than in healthy trees. A work conducted by Gunduz et al. [52], Fernandez et al. [53], and Mafia et al. [54] also obtained similar results. These authors reported that the holocellulose content of infected wood decreased compared to that of the healthy trees. It is presumed that the decrease in cellulose and hemicellulose in cell material was caused by a fungal disease utilizing cellulose and hemicellulose as energy and carbon sources for its colonization progresses.

In contrast, infected samples contained higher levels of extractives and lignin [53]. Other studies, including Foelkel et al. [46] and Souza et al. [55], have demonstrated the influence of the *Chrysoporthe cubensis* disease on the quality of Kraft pulp. It was found that the infected wood of *E. grandis* and *E. saligna* presented higher levels of extractives and lignin content compared with healthy wood, causing a 1% loss in the pulp yield and a 10% loss in volume. The increase in these components could be due to the activation of the tree's defense mechanisms, which also results in the production of chemicals, gums, and tyloses [56]. In response to infection, the plant may trigger reactions that prevent fungus penetration or limit fungus colonization in host tissues, such as extractive, lignin, and other phenolic compound accumulation [57,58]. According to Kuc [59], lignification in plants can increase the pathogen's inability to penetrate the cell walls. Trees' defense mechanisms can prevent the invasion of pathogens or wood-rotting fungi, and response zones containing lignin act as barriers to their invasion and colonization [60]. The reaction zones, however, are not always effective against microbial penetration. Changes in the cell wall and occlusion of xylem elements are examples of activated defense mechanisms. These plant responses may limit pathogen development in wood [61]. Histochemical and biochemical studies revealed that phenolic compounds associated with phytoalexin production accumulated in *Platanus acerifolia* plants inoculated with *Ceratocystis platani* [62].

3.2. Fourier-Transform Infrared Spectroscopy (FTIR) Analysis

The FTIR was run to qualitatively determine the alteration of chemical properties and functional groups present in healthy and infected *E. urograndis*. Figure 3 illustrates the FTIR spectra that display the common components or functional groups, such as cellulose, hemicellulose, and lignin, at a wavenumber ranging from 4000 to 525 cm^{-1}.

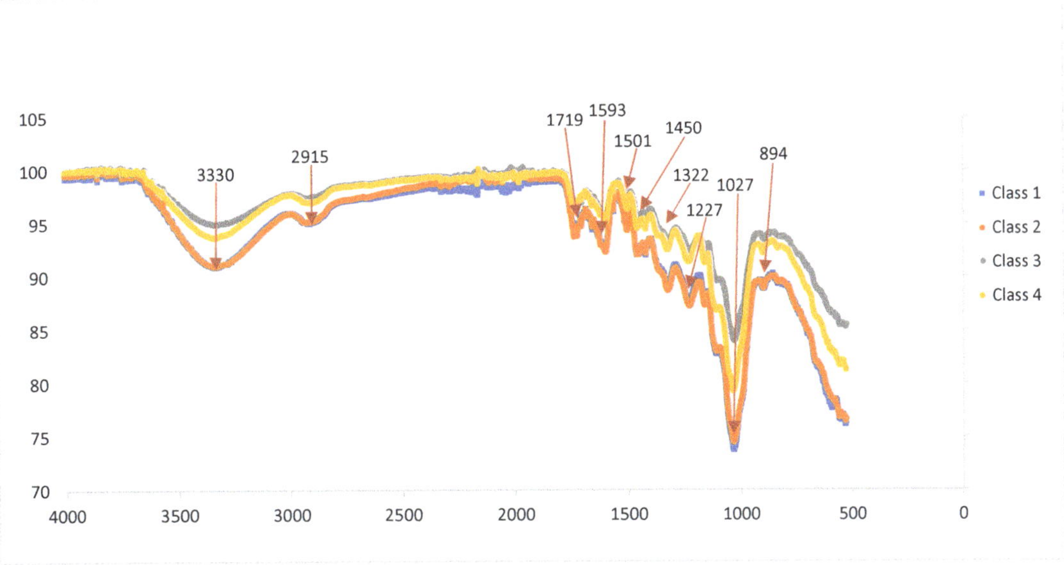

Figure 3. FTIR spectra (4000 cm^{-1}–525 cm^{-1}) of healthy (class 1) and infected samples *Eucalyptus urograndis* (class 2, 3 and 4).

Table 7 depicts the important bands that can be found in an infrared spectrum of wood and includes their functional assignment. The result demonstrated that numerous peaks were detected, and that it is difficult to interpret the differences between the spectra of healthy and infected wood since there are several reactions occurring at the same time. The first single bond region showed that a broadening absorption of the band (3570–3200 cm^{-1}) is assigned to stretching vibrations of intra-molecular hydrogen bonds in the crystalline regions of cellulose, which shift from 3340–3330 cm^{-1} in the spectrum [63]. This band also confirms the existence of hydrate (H_2O), hydroxyl group (-OH), ammonium, or amino. This broadening might be due to the reduction in O-H stretch, which represents decrement of polysaccharides as a result of infection by *C. deuterocubensis*. Like humans, this fungus needs carbohydrates (celluloses and hemicelluloses) for carbon skeletons and energy sources. They also require amino acids to combine with carbon skeletons to synthesize proteins for their growth and enzymes. As a result, infected classes 2, 3, and 4 lost density and moisture content because the components of the cell walls were degraded, resulting in fewer free OH- groups and fewer water molecules binding to the cells. This indicated that the samples had lower hydrophilic properties [24,64]. Simultaneously, O-H from phenolic groups in lignin increases, as it is well known that the lignin percentage of the ratio increases due to carbohydrate degradation [65].

Table 7. Functional group in FT-IR spectra of the studied sample. Data were adopted from references [28,64,66,67].

Wavenumber (cm^{-1})	Functional Group/Band Assignment
3570–3200 (broad)	-OH stretching hydrogen bonds in cellulose
2935–2915	Asymmetric -CH stretching of methylene (CH_2) in lignin
1750–1700	C=O stretching of non-conjugated carbonyl compound in xylan
1615–1580	C=O aromatic ring stretching

Table 7. *Cont.*

Wavenumber (cm^{-1})	Functional Group/Band Assignment
1515–1500	Aromatic ring (benzene) stretching vibrations
1460	C-H deformations in xylan
1420	Aromatic ring of lignin and C-H bending in cellulose
1324–1322	CN stretching in ether and oxy compound, and stretching P=O in simple hetero-oxy compound
1225–950 (several)	Aromatic C-H in-plane bend
1190–1130	Secondary amine, CN stretch
1047–1004	C-O stretching in cellulose I and cellulose II
896	Asymmetric C-H out-of-plane bending deformation in cellulose and hemicellulose

The band at 2920 to 2915 cm^{-1} corresponds to asymmetric -CH stretching vibration in the aromatic methyl and methylene groups of side chains generally found around 2935–2915 cm^{-1}. According to Moharram and Mahmoud [68] and Spiridon et al. [69], the apparent shift in frequency of the -CH band is due to structural and relative composition changes, specifically changes in the crystallinity level of the cellulose. The double bond region (1750 and 1700 cm^{-1}) represents the C=O stretching of non-conjugated carbonyl compounds that occurs in xylan, such as carboxylic acid (1725–1700 cm^{-1}), ketones (1725–1705 cm^{-1}), aldehydes (1740–1725 cm^{-1}), esters (1750–1725 cm^{-1}), and 6 membered rings of lactone (1735 cm^{-1}) [28]. In comparison to healthy wood, the intensities of these bands increased as the infection class became more severe. The increase in lignin content that derived from the carbohydrate loss could be observed as the increase in these bands, which is due to the formation of carbonyl groups in lignin [70].

According to Nandiyanto et al. [28] and Esteves et al. [67], the absorption peak which occurred around 1615–1580 cm^{-1} was due to vibrations in the aromatic ring of lignin and C=O stretching. The band at 1593 cm^{-1} shifted to about 1596 cm^{-1} for healthy and infected wood, respectively. This band increases due to an increase in the percentage of lignin in the infected wood [71]. The band at the fingerprint region of 1460 cm^{-1} corresponds to the asymmetric deformation of the C-H bond of xylan, while the band at 1420 cm^{-1} corresponds to the vibration of the lignin's aromatic ring, but also to the C-H bending in the cellulose [72].

Aromatic rings exhibit, most of the time, a characteristic band at approximately 1224 to 1228 cm^{-1}. The presence of an absorption in the range of 1156–1158 cm^{-1} was noted in the FTIR spectra, and this belongs to the stretching vibrations of the symmetric cellulose C-O-C (1155–1159 cm^{-1}), which corresponds to the content in crystallized and amorphous cellulose.. In their study, Gelbrich et al. [73] reported that the band at 1031 cm^{-1} (class 1) was assigned to bonds of holocellulose (1047–1004 cm^{-1}) and that the intensity of the band decreased as the infection became more severe, indicating the lessened polysaccharide content. The attack of a fungal disease infection on polysaccharides (sugar ring tension) is clear by the decreased band intensities at 896–893 cm^{-1}, corresponding to the pyranose ring opening. Similar results were obtained by Kotilainen et al. [71] and Pena et al. [74]. Meanwhile, the fungi's enzymatic activity is primarily responsible for the decrease in intensities of asymmetric and C-H out-of-plane stretching of cellulose and hemicellulose.

3.3. *Durability*

3.3.1. Fungal Decay

Figure 4 shows that mycelium covered the surfaces of the test blocks, particularly the healthy one (class 1). It can be observed that the healthy samples were heavily colonized by fungi when compared to the infected test blocks (class 2, 3, and 4).

Figure 4. Visual appearance of healthy and infected *E. urograndis* against *P. sanguineus* (upper): **1a** = class 1, **1b** = class 2, **1c** = class 3, and **1d** = class 4; and *C. puteana* (lower): **2a** = class 1, **2b** = class 2, **2c** = class 3, **2d** = class 4.

Table 8 shows the durability *E. urograndis* of classes 1, 2, 3, and 4 after being exposed to *P. sanguineus* and *C. puteana* for about 12 weeks. As presented in Table 8, the results revealed that the WL_{decay} decreased progressively. *P. sanguineus* caused more WL_{decay} than *C. puteana*, with values ranging from 14.4% to 9.5% and 11.2% to 6.1% for healthy and infected samples, respectively. It was observed that the WL_{decay} of the infected samples caused by both *P. sanguineus* and *C. puteana* was significantly lower than that of the healthy samples. It was noted that the white rot fungi caused severer damage on the *Eucalyptus* trees compared to brown rot fungi. The results show that a significant improvement was recorded for the resistance of infected samples against both *P. sanguineus* and *C. puteana*, as indicated by the lower WL_{decay} found in the infected samples (class 2, 3, and 4). Thus, the durability classes of infected samples were also improved, from resistant (class 1) to highly resistant (class 4), against *P. sanguineus* and *C. puteana*.

Table 8. Mean weight loss and durability classes of infected and healthy *Eucalyptus urograndis* against white rot and brown rot fungal decay.

Infection Classes	Value	WL_{decay} (%)		Resistance Classes	
		P. sanguineus	*C. puteana*	*P. sanguineus*	*C. puteana*
1 (Healthy)	Mean SD	14.4 ^{a} 2.7	11.2 ^{a} 2.8	Resistant (II)	Resistant (II)
2 (Moderate)	Mean SD	12.3 ^{b} 3.0	8.0 ^{b} 2.3	Resistant (II)	Highly resistant (I)
3 (Severe)	Mean SD	10.3 ^{c} 2.5	7.8 ^{c} 3.1	Highly resistant (I)	Highly resistant (I)
4 (Very severe)	Mean SD	9.5 ^{c} 1.9	6.1 ^{c} 2.6	Highly resistant (I)	Highly resistant (I)
p-value		0.000 ***	0.000 ***		

Notes: Classification of decay resistance according to ASTM standard D2017 [29]. Means followed by the same letter (a, b, c) in the same column were not significantly different at $p < 0.05$ according to Duncan's multiple range test. ***, high significance ($p < 0.01$); ***, very high significance ($p < 0.001$).

This observation was made due to the increase in lignin and extractive content of *E. urograndis* after *C. deuterocubensis* tree infection, which helped to deter the wood from decay. Samples of classes 3 and 4 showed high resistance against white rot fungi, while classes 2, 3 and 4 showed high resistance against brown rot fungi. Owing to the lower availability of polysaccharides to consume by the *P. sanguineus* and *C. puteana*, the increase in lignin instead of cellulose and hemicellulose is not favorable for the growth of *C. puteana* (brown rot) fungi. Brown rot fungi mainly degrade cellulose and hemicelluloses and leave most of the lignin undegraded [75,76]. Another plausible reason for the decrement of WL_{decay} was the lower density and equilibrium moisture content (EMC) of infected than healthy wood. In our previous study [24] we found that infected samples of classes 2, 3, and 4 had low density (618.9, 706.8, and 542.3 kg/m^3) and moisture content (10.1, 10.2, and 9.7%) values, respectively. This phenomenon is thought to be stressful for fungal growth. Moisture is required for all processes in wood infected by decay fungi, including spore arrival and germination, mycelial growth, and wood metabolism. As a result of the lower hydrophilic properties of infected wood (decrease in MC of wood), the growth and colonization of wood-decaying fungi were minimized, making decaying fungi less able to grow and degrade wood [77].

A study by Gunduz et al. [52] on chestnut wood infected by *Cryphonectria parasitica* (Murrill) found that vessel diameters and vessel frequency in the transverse section became smaller, and the lengths of the vessel elements were shorter. The formation of granular layers was observed via scanning electron microscopy (SEM) in the inner walls of the vessels. Moreover, Mafia et al. [55] found there was deposition of residual stromatic hyphae tissue in vessel of *E. urograndis,* caused by intense growth of *Hypoxylon* spp. It is likely that these structures resulted from the colonization activities of the *Hypoxylon* spp. disease. Thus, in this study, the penetration of both types of fungal decay (*P. sanguineus* and *C. puteana*) into wood vessels for colony growth may be limited due to vessel obstruction from a previous infection by *C. deuterocubensis*. A similar finding of the study by Clerivet and El Modafar [78] reported alteration in xylem vessels, including thickening of vessel walls, occlusion of pores, deposition of gels, gum, and formation of tyloses after inoculation of *Platanus acerifolia* with *Ceratocystis fimbriata* f. sp. platani. A study by Mafia et al. [45], by using SEM, found the presence of the pathogen structures and tyloses in *E. grandis* hybrid infected by *Ceratocystis fimbriata*. Tyloses are considered an active defense mechanism which involves partial or complete occlusion of xylem vessels through the deposition of gels and gummosis (kino) to limit microbial growth [79,80]. According to MTIB [81], the vessel of *Eucalyptus* spp. contained a moderate amount of tyloses and gums. *Eucalyptus* spp. is known as the "gum tree", as it presents a gum canal (kino veins) and secretes a gooey substance. Timor white gum is known to come from *E. urophylla* [82], while flooded gum is known to come from *E. grandis* [83]. When the trunk surface of a gum tree is damaged, it oozes visible amounts of thick, gummy (resinous liquid) sap.

3.3.2. Termite Attack

The mean weight loss ($WL_{termites}$), increment of resistance, block visual rating, and class of resistance of healthy (class 1) and infected wood samples (class 2, 3, and 4) after 4 weeks of exposure to *C. curvignathus* are presented in Table 9. The results show that the $WL_{termites}$ values decreased consistently from healthy to infected (class 4) samples.

From the experimental results, all sample blocks tested were attacked by termites, as reflected in Figure 5. The healthy sample (class 1), which had the highest $WL_{termites}$ value of 20.1%, was attacked the most, followed by the infected samples in classes 2 (16.4%), 3 (12.5%), and 4 (9.9%). Significant resistance improvement was observed for the infected class 2 to class 4 samples, from very poor (V) to moderately resistant (III).

Based on the visual rating classification of termite attacks, the samples from classes 1, 2, 3, and 4 had rating values of 7, 7, 7, and 8, respectively. The results demonstrated a slight improvement for infected *E. urograndis*. This shows that as the lignin and extractive proportion and concentration increased, so did the resistance to the *C. curvignathus* attack [46]. The toxicity of lignin and extractives may have increased resistance and provided protection to infected *E. urograndis* against *C. curvignathus* termite feeding in our study. Another possibility is that the tree's defense mechanisms were activated while or after the *E. urograndis* tree was infected with *C. deuterocubensis*, resulting in the formation of chemical substances, such as gums, tyloses, and other phenolic compounds [57], which act as slow poisons to termites.

Table 9. Mean weight loss and durability classes of infected and healthy *Eucalyptus urograndis* against termite attacks.

Infection Classes	Value	$WL_{termite}$ (%)	Visual Rating	Resistant Classes
1 (Healthy)	Mean SD	20.12 [a] 4.5	7	Very poor (V)
2 (Moderate)	Mean SD	16.44 [ab] 8.8	7	Poor (IV)
3 (Severe)	Mean SD	12.49 [bc] 5.5	7	Poor (IV)
4 (Very severe)	Mean SD	9.86 [c] 6.0	8	Moderately resistant (III)
p-value		0.000 ***		

Notes: Classification of termite resistant according to SNI 01.7207 [37] and rating system according to AWPA stand ard E1-09 [31]. Means followed by the same letter (a, b, c) in the same column are not significantly different at $p < 0.05$ according to Duncan's multiple range test. ***, high significance ($p < 0.01$); ***, very high significance ($p < 0.001$).

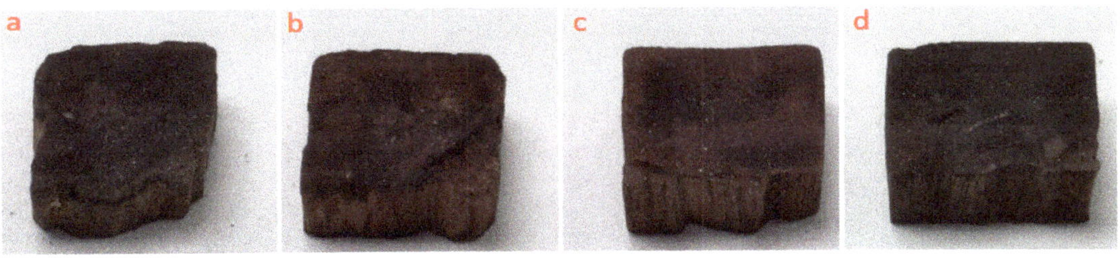

Figure 5. Condition of samples after exposure to termite attack; (**a**) = class 1, (**b**) = class 2, (**c**) = class 3, and (**d**) = class 4.

In addition, Bayle [84] and FAO [85] stated that more than 300 species of Eucalyptus are interspecific hybrid containing volatile oils, which consist of mixtures of hydrocarbons (terpenes and sesquiterpenes) and oxygenated compounds (alcohols, esters, ethers, aldehydes, ketones, lactones, phenols, and phenol ethers) [85–87]. These volatile oils have favorable properties that have insecticidal and antimicrobial properties [84,85]. Furthermore, vessel obstruction caused by *C. deuterocubensis* fungus infection may be another factor that deters termite attack.

3.4. Correlation between Infection Class and Weight Loss against Fungal Decay and Termite Attacks

Pearson's correlation between the infection class and WL_{decay} against *P. sanguineus* and *C. puteana*, as well as that between the infection class, $WL_{termites}$, and *C. curvignathus* in *E. urograndis* wood was conducted and tabulated in Table 10. The correlation was moderate, but highly significant ($p = 0.000$), in all durability tests. Generally, the weight loss from both fungal and termite tests decreased in trees of higher infection classes, as indicated by the weak negative correlations of WL_{decay}, *P. sanguineus* ($r = -0.594$), and *C. puteana* ($r = -0.540$), and of $WL_{termites}$ and *C. curvignathus* ($r = -0.528$).

Table 10. Correlation between infection classes with weight loss against fungal decay and termite attack.

Correlations					
		Class	Decay		Termite
			P. sanguineus	*C. puteana*	*C. curvignathus*
Class	Pearson correlation	1	−0.594 **	−0.540 **	−0.528 **
	Sig. (2-tailed)		0.000	0.000	0.000
	N	80	80	80	80

** Correlation is significant at the 0.01 level (2-tailed). ** Correlation is significant at the 0.05 level (2-tailed).

4. Conclusions

The chemical properties and biological durability of the *E. urograndis* trees infected by canker disease (*C. deuterocubensis*) were investigated in this study. The findings revealed that the cellulose and hemicellulose content decreased with increasing infection severity. On the contrary, lignin and extractive content increased as the severity of the infection increased. Consequently, the resistance of the infected tree against white rot and brown rot fungi was enhanced, probably due to the lower availability of polysaccharides in the infected wood on which the fungi could thrive. The resistant of the *E. urograndis* wood against subterranean termites was also improved. The resistance against termites improved from very poor in healthy samples (class 1) to moderately resistant in the most severely infected samples (class 4). Overall, the results showed that the infected wood had a high extractive content, which made it unsuitable for pulp and paper production. However, it may have some potential in lumber production, as it has more added value and a lower cost of production than pulp and paper products.

The current study expands on previous research into the physical, mechanical, and machining properties of infected *E. urograndis* wood. When compared to healthy trees, infected trees showed only a minor reduction in strength. Because of the minor strength reduction, they are appropriate for non-structural applications such as furniture, interior finishing, window frames, and doors. Furthermore, the infected wood has acceptable machining properties, justifying its use in furniture production. When combined with the current study's findings that the infected wood demonstrated improved biological durability, the case for using infected wood in non-structural applications is strengthened. This may assist the planter in deciding whether to extract the trees for timber production when the first signs of infection appear. Future research should concentrate on the anatomical properties and cell structure of wood in order to clearly observe the level of infection by *C. deuterocubensis* for better infection severity classification.

Author Contributions: Conceptualization, R.D., P.M.T. and S.S.; methodology, P.M.T., R.D., A.C.I., M.S.H. and P.S.K.; formal analysis, R.D. and S.H.L.; investigation, R.D., M.S.H. and P.S.K.; data curation, R.D., P.M.T., S.H.L. and P.S.K.; writing—original draft preparation, R.D., S.H.L., S.S. and P.A.; writing—review and editing, R.D., S.H.L., T.K. and P.A.; supervision, S.H.L., P.M.T. and S.S.; project administration, P.M.T.; funding acquisition, P.M.T., S.H.L., T.K. and S.S. All authors have read and agreed to the published version of the manuscript.

Funding: This research was funded by the Trans-disciplinary Research Grant Scheme (TRGS 2018-1), Reference code: TRGS/1/2018/UPM/01/2/3 (vote number: 5535800) by Ministry of Higher Education (MOHE), Malaysia.

Institutional Review Board Statement: Not applicable.

Informed Consent Statement: Not applicable.

Data Availability Statement: Not applicable.

Acknowledgments: The authors would like to express their gratitude to Sabah Softwoods Berhad (SSB) staff and management for providing the wood materials, and to the Institute of Tropical Forestry and Forest Products and Faculty of Forestry and Environment, Universiti Putra Malaysia for the facilities and assistance provided.

Conflicts of Interest: The authors declare no conflict of interest.

References

1. Luo, J.; Arnold, R.; Ren, S.; Jiang, Y.; Lu, W.; Peng, Y.; Xie, Y. Veneer grades, recoveries, and values from 5-year-old eucalypt clones. *Ann. For. Sci.* **2013**, *70*, 417–428. [CrossRef]
2. Ahmad, M.F.; Hishamuddin, M.S. Trees Diseases and Disorders in Urban Forests of Peninsular Malaysia. In *Urban Forestry and Arboriculture in Malaysia "An Interdisciplinary Research Perspective"*; Maruthaveeran, S., Chen, W.Y., Morgenroth, J., Eds.; Springer: Berlin/Heidelberg, Germany, 2022; Volume 1, pp. 83–104.
3. Alves, I.C.N.; Gomide, J.L.; Colodette, J.L.; Silva, H.D. Technological characterization of *Eucalyptus benthamii* wood for kraftpulp production. *Cienc. Florestal.* **2011**, *21*, 167–174. [CrossRef]
4. Bassa, A.G.M.C.; Francides, G.D.S.J.; Sacon, V.M. Mixtures of *Eucalyptus grandis* x *Eucalyptus urophylla* and *Pinus taeda* woodchips for kraft pulp production by Lo-Solids process. *Scientia Forestalis.* **2007**, *75*, 19–30.
5. Nakabonge, G.; Roux, J.; Gryzenhout, M.; Wingfield, M.J. Distribution of *Chrysoporthe* canker pathogens on *Eucalyptus* and *Syzygium* spp. in eastern and southern Africa. *Plant Dis.* **2006**, *90*, 734–740. [CrossRef]
6. Leonardi, G.D.A.; Carlos, N.A.; Mazzafera, P.; Balbuena, T.S. *Eucalyptus urograndis* stem proteome is responsive to short-term cold stress. *Genet. Mol. Biol.* **2015**, *38*, 191–198. [CrossRef]
7. Lee, S.H.; Lum, W.C.; Antov, P.; Kristak, L.; Paridah, M.T. Engineering Wood Products from *Eucalyptus* spp. *Adv. Mater. Sci. Eng.* **2022**, *2022*, 8000780.
8. Carvalho, A.M. The valuation of *Eucalyptus grandis* x *Eucalyptus urophylla* hybrid wood through the production of small dimension sawn wood, pulpwood and fuelwood. *Sci. Forestalis.* **2000**, *59*, 61–76.
9. Jiang, X.M.; Ye, K.L.; Lu, J.X.; Zhao, Y.K.; Yin, Y.F. *Guide on Utilisation of Eucalyptus and Acacia Plantations in China for Solid Wood Products. The Final Technical Report for: Improved and Diversified Use of Tropical Plantation Timbers in China to Supplement Diminishing Supplies from Natural Forests "ITTO ProjectPD69/01 REV.2"*; Science Press: Beijing, China, 2007; p. 181.
10. Labate, C.A.; de Assis, T.F.; Oda, S.; Mello, E.J.; Mori, E.S.; de Moraes, M.L.T.; Barrueto, L.P.; Gonzalez, E.R.; Alfenas, A.C.; Edival, A.; et al. Eucalyptus. In *Compendium of Transgenic Crop Plants: Transgenic Forest Tree Species*; Cole, C., Hall, T.C., Eds.; Wiley: New York, NY, USA, 2008; Volume 9, pp. 35–99. ISBN 978-1-405-16924-0.
11. Brigatti, R.A.; Ferreira, M.; Silva, A.P.; Freitas, M. Comparative study of the behavior of some *Eucalyptus* spp. hybrids. *Silvicultura* **1983**, *32*, 761–764.
12. Ikemori, Y.K.; Campinhos, E. *Eucalyptus urophylla* x *Eucalyptus grandis* seed production by open pollination—Preliminaries results. *Silvicultura* **1983**, *8*, 306–308.
13. Bertolucci, F.; Rezende, G.; Penchel, R. Production and use of eucalyptus hybrids. *Silvicultura* **1995**, *51*, 12–16.
14. Sembiring, N.; Napitupulu, H.L.; Sembiring, M.T.; Ishak, A.; Gunawan, H.A. Fulfilling Eucalyptus raw materials for pulp and paper production plants. In *IOP Conference Series: Earth and Environmental Science*; IOP Publishing: Bristol, UK, 2021; Volume 912, p. 012008. [CrossRef]
15. Scanavaca, L., Jr.; Garcia, J.N. Yield in sawed wood of *Eucalyptus urophylla*. *Sci. For.* **2003**, *63*, 32–43.
16. Ahmad, Z.Y. Planting of Eucalyptus in Malaysia. *Acta Sci. Agric.* **2020**, *4*, 139–140.
17. Ahmad, Z.Y.; Hassan, N.H.; Loon, N.T.; Heng, L.H.; Zorkarnain, F.A. Comparing the early growth performance of plantation–grown Eucalyptus hybrid and *Eucalyptus pellita*, South Johore, Peninsular Malaysia. *WJARR* **2020**, *6*, 234–238.
18. Muhammad, S.A.R. Isolation and Identification of Causal Disease of *Eucalyptus pellita*. Bachelor's Thesis, Universiti of Malaysia, Sarawak, Kota Samarahan, Malaysia, 2012.
19. Suzuki, H.; Marincowitz, S.; Wingfield, B.D.; Wingfield, M.J. Genetic diversity and population structure of *Chrysoporthe deuterocubensis* isolates from Melastoma and Eucalyptus in Malaysia and Indonesia. *For. Pathol.* **2022**, *52*, 12762. [CrossRef]
20. Rauf, M.R.B.A.; McTaggart, A.R.; Marincowitz, S.; Barnes, I.; Japarudin, Y.; Wingfield, M.J. Pathogenicity of *Chrysoporthe deuterocubensis* and *Myrtoporthe bodenii* gen. et sp. nov. on Eucalyptus in Sabah, Malaysia. *Australas. Plant Pathol.* **2019**, *49*, 53–64. [CrossRef]
21. Gezahgne, A. Main Diseases of Eucalyptus Species in Ethiopia. In *Eucalyptus Species Management, History, Status and Trends in Ethiopia*; Wubalem, G., Lopez, T., Eds.; Technical University of Madrid: Madrid, Spain, 2010; pp. 351–369.
22. Lee, S.S. Observations on the successes and failures of acacia plantations in Sabah and Sarawak and the way forward. *J. Trop. For. Sci.* **2018**, *30*, 468–475. [CrossRef]
23. Forestry Economics and Policy Division. *Global Review of Forest Pests and Diseases*; Food and Agriculture Organization (FAO): Rome, Italy, 2009; p. 235. ISBN 978-92-5-106208-1.
24. Dahali, R.; Md Tahir, P.; Roseley, A.S.M.; Hua, L.S.; Bakar, E.S.; Ashaari, Z.; Abdul Rauf, M.R.; Zainuddin, N.A.; Mansoor, N.S. Influence of *Chrysoporthe deuterocubensis* Canker Disease on the Physical and Mechanical Properties of *Eucalyptus urograndis*. *Forests* **2021**, *12*, 639. [CrossRef]

25. Dahali, R.; Lee, S.H.; Md Tahir, P.; Bakar, E.S.; Muhammad Roseley, A.S.; Ibrahim, S.A.; Mohd Yusof, N.; Mohammad Suffian James, R. Tahir, P.; Bakar, E.S.; Muhammad Roseley, A.S.; Ibrahim, S.A.; Mohd Yusof, N.; Mohammad Suffian James, R. Influence of *Chrysoporthe deuterocubensis* Canker Disease on the Machining Properties of *Eucalyptus urograndis*. *Forests* **2022**, *13*, 1366. [CrossRef]
26. Alvares, C.A.; Stape, J.L.; Sentelhas, P.C.; Goncalves, J.L.D.M.; Sparovek, G. Koppen's climate classification map for Brazil. *Meteorol. Z.* **2014**, *22*, 711–728. [CrossRef]
27. *TAPPI Standard, T 257 cm-02*; Sampling and Preparing Wood for Chemical Analysis. TAPPI: Atlanta, GA, USA, 2002.
28. Nandiyanto, A.B.D.; Oktiani, R.; Ragadhita, R. How to Read and Interpret FTIR Spectroscope of Organic Material. *Indones. J. Sci. Technol.* **2019**, *4*, 97–118. [CrossRef]
29. *ASTM Standard D2017*; Standard Test Method of Accelerated Laboratory Test of Natural Decay Resistance of Woods. ASTM International: West Conshohocken, PA, USA, 2012.
30. Bakar, E.S.; Hao, J.; Ashaari, Z. Durability of phenolic-resin-treated oil palm wood against subterranean termites a white-rot fungus. *Int. Biodeterior. Biodegrad.* **2013**, *85*, 126–130. [CrossRef]
31. *AWPA Standard E1-09*; Standard Method for Laboratory Evaluation to Determine Resistance to Subterranean Termites. American Wood Protection Association: Birmingham, AL, USA, 2012.
32. Arinana, A.; Philippines, I.; Koesmaryono, Y.; Sulaeha, S.; Maharani, Y.; Indarwatmi, M. The daytime indoor and outdoor temperatures of the subterranean termite *Coptotermes curvignathus* Holmgren (*Isoptera: Rhinotermitidae*) tunnel. In *IOP Conference Series: Earth and Environmental Science*; IOP Publishing: Bristol, UK, 2021; Volume 807, p. 022027. [CrossRef]
33. Kuswanto, E.; Ahmad, I.; Dungani, R. Threat of subterranean termites attack in the Asian countries and their control: A review. *Asian J. Appl. Sci.* **2015**, *8*, 227–239. [CrossRef]
34. Dahali, R.; Hua, L.S.; Ashaari, Z.; Bakar, E.S.; Ariffin, H.; San, K.P.; Bawon, P.; Salleh, Q.N. Durability of superheated steam-treated light red meranti (*Shorea* spp.) and kedondong (*Canarium* spp.) wood against white rot fungus and subterranean termite. *Sustainability* **2020**, *12*, 4431. [CrossRef]
35. Anantharaju, T.; Kaur, G.; Gajalakshmi, S.; Abbasi, S.A. Sampling and identification of termites in Northeastern. *Puducherry J. Entomol. Zool. Stud.* **2014**, *2*, 225–230.
36. Rilatupa, J. 2006 Kondisi komponen konstruksi bangunan tinggi dan hubungannya dengan karakteristik serangan rayap. *J. Sains Teknol. EMAS* **2006**, *16*, 71–86.
37. *SNI Standard 01.7207*; Standard Method for Test of Resistance Wood and Wood Products against Wood Destroying Organisms "Ujian ketahanan kayu dan produk kayu terhadap organisme perusak kayu". Indonesian National Standard (Standar Nasional Indonesia): Jakarta, Indonesia, 2006.
38. Terzi, E.; Kartal, S.N.; Muin, M.; Hassanin, A.H.; Hamouda, T.; Kılic, A.; Candan, Z. Biological Performance of Novel Hybrid Green Composites Produced from Glass Fibers and Jute Fabric Skin by the VARTM Process. *BioResources* **2018**, *13*, 662–677. [CrossRef]
39. Reddy, N.; Yang, Y. Biofibers from agricultural byproducts for industrial applications. *Trends Biotechnol.* **2005**, *23*, 22–27. [CrossRef]
40. Ferrari, R.; Gautier, V.; Silar, P. Lignin degradation by ascomycetes Wood Degradation and Ligninolytic Fungi. *Adv. Bot. Res.* **2021**, *9*, 77–113.
41. Savory, J.G.; Pinion, L.C. Chemical aspects of decay of beech wood by *Chaetomium globosum*. *Int. J. Biol. Chem. Phys. Technol. Wood* **1958**, *12*, 99–103.
42. Andlar, M.; Rezic, T.; Mardetko, N.; Kracher, D.; Ludwig, R.; Santek, B. Lignocellulose degradation: An overview of fungi and fungal enzymes involved in lignocellulose degradation. *Eng. Life Sci.* **2018**, *18*, 768–778. [CrossRef]
43. Janusz, G.; Pawlik, A.; Sulej, J.; Swiderska-Burek, U.; Jarosz-Wilkolazka, J.; Paszczyoski, A. Lignin degradation: Microorganisms, enzymes involved, genomes analysis and evolution. *FEMS Microbiol. Rev.* **2017**, *41*, 941–962. [CrossRef]
44. Kirk, T.K.; Farrell, R.L. Enzymatic "combustion": The microbial degradation of lignin. *Annu. Rev. Microbiol.* **1987**, *41*, 465–505. [CrossRef]
45. Mafia, R.G.; Ferreira, M.A.; Zauza, E.A.V.; Silva, J.F.; Colodette, J.L.; Alfenas, A.C. Impact of *Ceratocystis* wilt on Eucalyptus tree growth and cellulose pulp yield. *For. Pathol.* **2013**, *43*, 379–385. [CrossRef]
46. Foelkel, C.E.B.; Zvinakevicius, C.; Andrade, J.M. A qualidade do eucalipto. *Silvicultura* **1978**, *2*, 53–62.
47. Pereira, B.L.; Carneiro, A.C.; Carvalho, A.M.; Colodette, J.L.; Oliveira, A.C.; Fontes, M.P. Influence of Chemical Composition of Eucalyptus Wood on Gravimetric Yield and Charcoal Properties. *Bioresources* **2013**, *8*, 4574–4592. [CrossRef]
48. Rowell, R.M.; Pettersen, R.; Han, J.S.; Rowell, J.S.; Tshabalala, M.A. *Cell wall chemistry, In Handbook of Wood Chemistry and Wood Composites*; Rowell, R.M., Ed.; CRC Press: Boca Raton, FL, USA, 2005.
49. Shebani, A.N.; Van Reenen, A.J.; Meincken, M. The effect of wood extractives on the thermal stability of different wood species. *Thermochim. Acta* **2008**, *471*, 43–50. [CrossRef]
50. Fengel, D.; Wegener, G. *Wood, Chemistry, Ultrastructure, Reactions*; Walter de Gruyter: New York, NY, USA, 1984.
51. Stenius, P. *Paper Making Science and Technology: Forest Products Chemistry (Book 3)*; Fapet: Helsinki, Finland, 2000.
52. Gunduz, G.; Oral, M.A.; Akyuz, M.; Aydemir, D.; Yaman, B.; Asik, N.; Bulbul, A.S.; Allahverdiyev, S. Physical, morphological properties and raman spectroscopy of chestnut blight diseased *Castanea sativa* Mill. wood. *CERNE* **2016**, *22*, 43–58. [CrossRef]
53. Fernandes, B.V.; Zanuncio, A.J.V.; Furtado, E.L.; Andrade, H.B. Damage and Loss Due to Ceratocystis fimbriata in Eucalyptus Wood for Charcoal Production, "Eucalyptus fungal loss". *BioResources* **2014**, *9*, 5473–5479. [CrossRef]

54. Mafia, R.G.; Santos, P.C.; Demuner, B.J.; Massoquete, A.; Sarto rio, R.C. Eucalyptus wood decay: Effects on productivity and quality of cellulose. *For. Pathol.* **2012**, *42*, 321–329. [CrossRef]
55. Souza, S.E.; Sansigolo, C.A.; Furtado, E.L.; de Jesus, W.C.; Oliveira, R.R. Influencia do cancro basal em *Eucalyptus grandis* nas propriedades da madeira e polpacao Kraft. *Sci. Florest.* **2010**, *38*, 447–457.
56. Leśniewska, J.; Öhman, D.; Krzesłowska, M.; Kushwah, S.; Barciszewska-Pacak, M.; Kleczkowski, L.A.; Sundberg, B.; Moritz, T.; Mellerowicz, E.J. Defense responses in aspen with altered pectin methylesterase activity reveal the hormonal inducers of tyloses. *Plant Physiol.* **2017**, *173*, 1409–1419. [CrossRef] [PubMed]
57. Nicholson, R.L.; Hammerschmid, T.R. Phenolic compounds and their role in disease resistance. *Ann. Rev. Phytopathol.* **1992**, *30*, 369–389. [CrossRef]
58. Vance, C.P.; Kirk, T.K.; Sherwood, R.T. Lignification as a mechanism of disease resistance. *Ann. Rev. Phytopathol.* **1980**, *18*, 259–288. [CrossRef]
59. Kuc, J. Concepts and direction of induced systemic resistance in plants and its application. *Eur. J. Plant Pathol.* **2001**, *107*, 7–12. [CrossRef]
60. Shigo, A.I.; Marx, H.G. 1977: Compartimentalization of decay in trees. *USDA Agric. Inform. Bull.* **1977**, *405*, 1–73.
61. Elgersma, D.M. Tylose formation in elms after inoculation with *Ceratocystis* ulmi, a possible resistance mechanism. *Eur. J. Forest Pathol.* **1973**, *79*, 218–220. [CrossRef]
62. El Modafar, C.; Clerivet, A.; Macheix, J.J. Flavan accumulation in stems of *Platanus acerifolia* seedlings inoculated with *Ceratocystis fimbriata* f. sp. platani, the canker stain disease agent. *Can. J. Bot.* **1996**, *74*, 1982–1987. [CrossRef]
63. Broda, M.; Popescu, C.M.; Curling, S.F.; Timpu, D.I.; Ormondroyd, G.A. Effects of Biological and Chemical Degradation on the Properties of Scots Pine Wood-Part I: Chemical Composition and Microstructure of the Cell Wall. *Materials* **2022**, *15*, 2348. [CrossRef]
64. Hua, L.S.; Ashaari, Z.; Ang, A.F.; Halip, J.A.; Lum, W.C.; Dahali, R.; Halis, R. Effects of two-step post heat-treatment in palm oil on the properties of oil palm trunk particleboard. *Ind. Crop. Prod.* **2018**, *116*, 249–258.
65. Esteves, B.; Videira, R.; Pereira, H. Chemistry and ecotoxicity of heat-treated pine wood extractives. *Wood Sci Technol.* **2011**, *45*, 661–676. [CrossRef]
66. Coates, J. Interpretation of infrared spectra, a practical approach. *Encycl. Anal. Chem.* **2000**, *12*, 10815–10837.
67. Esteves, B.; Marques, A.V.; Domingos, I.; Pereira, H. Chemical changes of heat-treated pine and eucalypt wood monitored by FTIR. *Maderas. Cienc. Y Tecnol.* **2013**, *15*, 245–258. [CrossRef]
68. Moharram, M.; Mahmoud, O. FTIR Spectroscopic Study of the Effect of Microwave Heating on the Transformation of Cellulose I into Cellulose II during Mercerization. *J. Appl. Pol. Sci.* **2008**, *107*, 30–36. [CrossRef]
69. Spiridon, I.; Teaca, C.; Bodirlau, R. Structural changes evidenced by FTIR spectroscopy in cellulosic materials after pre-treatment with ionic liquid and enzymatic hydrolysis. *Bioresources* **2011**, *6*, 400–413. [CrossRef]
70. Li, J.; Li, B.; Zhang, X. Comparative studies of heat degradation between larch lignin and Manchurian ash lignin. *Polym. Degrad. Stab.* **2002**, *78*, 279–285. [CrossRef]
71. Kotilainen, R.; Toivannen, T.; Alen, R. FTIR monitoring of chemical changes in softwood during heating. *J. Wood Chem. Technol.* **2002**, *20*, 307–320. [CrossRef]
72. Huang, Y.; Wang, L.; Chao, Y.; Nawawi, D.S.; Akiyama, T.; Yokoyama, T.; Matsumoto, Y. Analysis of lignin aromatic structure in wood based on the IR spectrum. *J. Wood Chem. Technol.* **2012**, *32*, 294–303. [CrossRef]
73. Gelbrich, J.; Mai, C.; Militz, H. Chemical changes in wood degraded by bacteria. *Int. Biodeterior. Biodegrad.* **2008**, *61*, 24–32. [CrossRef]
74. Pena, M.M.G.; Curling, S.F.; Hale, M.D.C. On the effect of heat on the chemical composition and dimensions of thermally-modified wood. *Polym. Degrad. Stab.* **2009**, *94*, 2184–2193. [CrossRef]
75. Riley, R.; Salamov, A.A.; Brown, D.W.; Nagy, L.G.; Floudas, D.; Held, B.W.; Levasseur, A.; Lombard, V.; Morin, E.; Otillar, R.; et al. Extensive sampling of basidiomycete genomes demonstrates inadequacy of the white-rot/brown-rot paradigm for wood decay fungi. *Proc. Natl. Acad. Sci. USA* **2014**, *111*, 9923–9928. [CrossRef]
76. Schilling, J.S.; Kaffenberger, J.T.; Held, B.W.; Ortiz, R.; Blanchette, R.A. Using wood rot phenotypes to illuminate the "gray" among decomposer fungi. *Front. Microbiol.* **2020**, *11*, 1288. [CrossRef] [PubMed]
77. Rouhier, M.M. Wood as a hostile habitat for ligninolytic fungi "Wood Degradation and Ligninolytic Fungi". *Adv. Bot. Res.* **2021**, *99*, 115–149.
78. Clerivet, A.; El Modafar, C. Vascular modifications in *Platanus acerifolia* seedlings inoculated with *Ceratocystis fimbriata* f. sp. platani. *Eur. J. Forest Pathol.* **1994**, *24*, 1–10. [CrossRef]
79. Rioux, D.; Nicole, M.; Simard, M.; Ouellette, G.B. Immunocytochemical evidence that secretion of pectin occurs during gel (gum) and tylosis formation in trees. *Phytopathology* **1998**, *88*, 494–505. [CrossRef] [PubMed]
80. Clerivet, A.; Deon, V.; Alami, I.; Lopez, F.; Geiger, J.P.; Nicole, M. Tyloses and gels associated with cellulose accumulation in vessels are responses of plane tree seedlings (*Platanus acerifolia*) to the vascular fungus *Ceratocystis fimbriata* f. sp. platani. *Trees* **2000**, *15*, 25–31. [CrossRef]
81. Malaysian Timber Industrial Board (MTIB). Medium Hardwood Eucalyptus. In *100 Malaysian Timbers*, 2010th ed.; Malaysian Timber Industry Board: Kuala Lumpur, Malaysia, 2010; pp. 132–133.
82. Eucalyptus Urophylla. Available online: https://en.wikipedia.org/wiki/Eucalyptus_urophylla (accessed on 20 December 2022).

83. Eucalyptus grandis. Available online: https://en.wikipedia.org/wiki/Eucalyptus_grandis (accessed on 20 December 2022).
84. Bayle, G.K. Ecological and social impacts of Eucalyptus tree plantation on the environment. *Biodivers. Conserv. Bioresour. Manag.* **2019**, *5*, 93–104. [CrossRef]
85. Food and Agriculture Organization (FAO). *Flavours and Fragrances of Plant Origin*; Food and Agriculture Organization: Rome, Italy, 1995.
86. Toloza, A.C.; Lucia, A.; Zerba, Z.; Masuh, H.; Picollo, M.I. Interspecific hybridization of Eucalyptus as a potential tool to improve the bioactivity of essential oils against permethrin-resistant head lice from Argentina. *Bioresour. Technol.* **2008**, *99*, 7341–7347. [CrossRef]
87. Guenther, E. *The Essential Oils*; Krieger Publishing Company: Malabar, FL, USA, 1972.

MDPI
St. Alban-Anlage 66
4052 Basel
Switzerland
Tel. +41 61 683 77 34
Fax +41 61 302 89 18
www.mdpi.com

Forests Editorial Office
E-mail: forests@mdpi.com
www.mdpi.com/journal/forests

www.ingramcontent.com/pod-product-compliance
Lightning Source LLC
Chambersburg PA
CBHW061458210926
51850CB00008B/1748